AF752812

Supernova Cosmology for the 21st Century

Konstantin Karchev

Supernova Cosmology for the 21st Century

How I Learnt to Stop Worrying About Likelihoods and Train a Neural Network

Foreword by Roberto Trotta

Springer

Konstantin Karchev
Institute of Cosmos Sciences (ICC)
University of Barcelona
Barcelona, Spain

ISBN 978-3-032-15071-4 ISBN 978-3-032-15072-1 (eBook)
https://doi.org/10.1007/978-3-032-15072-1

This work originally was accepted for the Springer Theses series but due to technical reasons is published as a standalone/ in non-series.

© The Editor(s) (if applicable) and The Author(s), under exclusive license to Springer Nature Switzerland AG 2026

This work is subject to copyright. All rights are solely and exclusively licensed by the Publisher, whether the whole or part of the material is concerned, specifically the rights of translation, reprinting, reuse of illustrations, recitation, broadcasting, reproduction on microfilms or in any other physical way, and transmission or information storage and retrieval, electronic adaptation, computer software, or by similar or dissimilar methodology now known or hereafter developed.
The use of general descriptive names, registered names, trademarks, service marks, etc. in this publication does not imply, even in the absence of a specific statement, that such names are exempt from the relevant protective laws and regulations and therefore free for general use.
The publisher, the authors and the editors are safe to assume that the advice and information in this book are believed to be true and accurate at the date of publication. Neither the publisher nor the authors or the editors give a warranty, expressed or implied, with respect to the material contained herein or for any errors or omissions that may have been made. The publisher remains neutral with regard to jurisdictional claims in published maps and institutional affiliations.

This Springer imprint is published by the registered company Springer Nature Switzerland AG
The registered company address is: Gewerbestrasse 11, 6330 Cham, Switzerland

If disposing of this product, please recycle the paper.

Foreword

Progress in science requires the simultaneous advancement of theory, observation and the all-important bridge between them — statistical inference. On the cusp of the vertiginous increase of the quantity and complexity of astronomical data promised by upcoming space- and ground-based observatories, the doctoral work of Dr Konstantin Karchev represents a landmark contribution to modern cosmology and data science.

This thesis pioneers the use of simulation-based inference and machine learning to transform how cosmological information is extracted from type Ia supernovae (SNæ Ia). It constitutes a profound methodological and conceptual leap forward that will define the inferential foundations of the next generation of cosmological surveys, thus enabling discoveries otherwise unattainable.

Dr Karchev's research delivers the first practical, fully scalable SBI framework capable of handling the vast data volumes expected from upcoming facilities such as the Vera Rubin Observatory's Legacy Survey of Space and Time. No other research group has achieved to date this level of scalability or demonstrated this degree of reliability and interpretability in cosmological inference at such scale.

Dr Karchev's work demonstrates that SBI methods, and in particular truncated neural ratio estimation, can outperform traditional likelihood-based techniques by orders of magnitude in speed and efficiency, while retaining full Bayesian rigour. This innovation allows complex astrophysical effects (such as selection functions, volumetric occurrence rates and photometric uncertainties) to be modeled without simplifications that historically introduced bias. From a methodological perspective, this novel approach elegantly unifies realism, scalability, and statistical rigour, ensuring both accuracy and computational feasibility for the era of massive astronomical datasets.

The results of this thesis are far-reaching. Dr Karchev's pioneering papers collectively introduce a suite of new algorithms that redefine the standard of precision cosmology. His methods enable principled Bayesian model comparison in high-dimensional parameter spaces. The STAR NRE framework, in particular, stands as a decisive step towards unbiased inference from purely photometric supernova samples – a necessity for next-generation sky surveys. The originality, depth, and reach of his contributions have already begun to reshape the field, providing the cosmological community with the tools it needs to fully exploit the data revolution now underway.

This thesis is written and presented to an exemplary standard, combining lucid exposition, impeccable figures, and a deep understanding of machine learning, statistics, and astrophysics. I believe it will become a reference work for researchers seeking to master the intersection of cosmology and data science.

In sum, this doctoral thesis embodies what modern astrophysics should aspire to: methodological innovation underpinned by physical understanding, executed with precision and intellectual elegance.

It stands as an exemplar of excellence in scientific creativity, rigour, and impact.

Prof Roberto Trotta
Trieste, October 2025

Abstract

Type Ia supernovæ (SNæ Ia) are extremely powerful stellar explosions used for measuring cosmographical distances and constraining parameters of the cosmological model via standardisation: the process of inferring their intrinsic brightness from properties of the observed light curves. Inference from current data sets (≈2000 objects) is already dominated not by statistical noise but by systematic uncertainties and modelling choices. The very near future promises vast amounts of new data (∼100 000 SNæ Ia), accompanied by new modelling challenges like the unavailability of spectroscopic classification and precise redshift measurements.

Embedded within the framework of neural simulation-based Bayesian inference (SBI), this thesis presents solutions for no-compromise analyses of future large SN surveys on three fronts: model realism, scalability, and probabilistic rigour. We develop a modern GPU-accelerated simulator for SN light curves that incorporates realistic uncertainties in the SN Ia flux template and physically motivated dust extinction in the host and Milky Way. We then use it to analyse a low-redshift SN Ia sample, inferring simultaneously all global and object-specific parameters with truncated marginal neural ratio estimation in excellent agreement with conventional methods. Moreover, we describe a procedure to construct calibrated regions with exact frequentist confidence from the approximate Bayesian results. With minimal extra training and re-using simulations, we also perform fully Bayesian model comparison of host mass-dependent standardisation and dust models, deemed extremely challenging computationally for high-dimensional problems. We furthermore demonstrate scalable set-based neural inference from up to 100 000 mock SNæ Ia, elucidating the biases introduced by model simplifications used for handling photometric redshift uncertainties and selection effects. Finally, we combine SN and host-galaxy modelling in a one-stop SBI framework for SN cosmology.

Publications

SICRET [269]	Supernova Ia cosmology with truncated marginal neural Ratio EStimation
AADcosmo [264]	Analytic auto-differentiable ΛCDM cosmography
SIDE-real [270]	Supernova Ia Dust Extinction with truncated marginal neural ratio estimation applied to real data
SimSIMS [268]	Simulation-based supernova Ia Model selection with thousands of latent variables
RESSET [265]	Solving supernova selection effects with set-based Truncated Auto-Regressive Neural Ratio Estimation

Contents

II Supernova cosmology

III Simulators for supernova cosmology

IV Science

Appendices

Acknowledgements

I would like to extend my heartfelt gratitude

to Roberto Trotta: for all the support (from day −14) and opportunities to travel and work beyond the PhD project, for encouraging me to pursue independent lines of research, for tolerating my idiosyncrasies and the many collaboration platforms I use, for keeping me on the rails when absolutely necessary, and for allocating me the office with the best view in the world (after the Director's); may our common research output continue expanding like the group that you built in SISSA!

to Christoph Weniger: for continuing to believe in me and for the many intense Slack chats that we've had through the years;

to Kaisey Mandel[*] and Matt and Ben: for their GIGANTIC domain expertise, the healthy competition and fruitful collaboration;

to Raul Jimenez: for a provocative question and hosting me in Barcelona;

to all the professors who bought me cocktails around the world;

to the city lights of Paris, Cambridge, Amsterdam, New Amsterdam, and Heidelberg and to the food in Sicily and Toscana (and of course, to the dense scientific programmes during the days);

to my colleagues—and friends—from day 0: for a pair of scissors and a stapler, and for occasionally switching off their me-filter (and the lights);

to Max Autenrieth, for fruitful discussions on linear regression and losing a game of chess (you can keep the ten cents);

to ~~office~~ forest 657: for a chair, a lot of square papers, and barely a single prank on my desk (two if we count the cup);

to the inhabitants of the Carso and of that haunted city apartment: for the legendary times dopo, prima e durante lavoro;

to everyone who contributed to my Zoom background, left me thank-you notes, complimented my photos, called me crazy (but in a good way), and cleared my vision;

to `egreg`, `David Carlisle`, *et al.*: for the strife to perfection;

to Muggia, Duino, and Oggi (in that order): for gelato;

to the streets of Trieste, the sea, and the Bora at night (and early dawn);

and to my family: for only rarely inquiring as to the purpose of it all.

[*] and for footnote 8 in Mandel et al. [346]

to the sunsets
—and everyone I've shared them with—
for their incessancy

In the asymptotic limit…

Preface: The story so far

In the beginning, the Universe was created. *the Universe*

This made cosmologists very happy—because they suddenly had something to study the evolution of—but also rather sad—because they only had one universe to study.

This Universe has—or rather had—a problem, which was this: over time, the expansion set off by the Big Bang™ was slowing down under the gravitational influence of matter, threatening to un-create the Universe in a MTgnaB giB. Many solutions were sought for this problem, but most of these were largely concerned with the movements of stars within galaxies or galaxies within clusters, which is odd because on the whole it wasn't the stars and galaxies that were slowing down.

And so the problem remained; matter was clumping together, and even some stars began collapsing on one another![citation needed]

Many of them were of the opinion that they'd all made a big mistake in being gravitationally bound in the first place. And some said that even forming atoms had been a bad move, and that they should go off in grand cosmic explosions called supenovæ.

And then, one Thursday, 80 years after one stone had tried to stop the Universe from collapsing by adding a Greek letter to his equation, a group of astronomers enjoying this remarkably homogeneous firework display discovered that there had been no problem all along, and the Universe had been accelerating for the past four billion years.[1] This time it was right, it would work, and we would all eventually fade out of causality in a Big Rip™.

Sadly, however, before they could figure out the exact mechanism behind the supernova eruptions, or why a certain subset of them seemed to have similar intrinsic brightnesses, many new ones started being discovered, and the prospect of doing proper[clarification needed] data analysis on them was lost forever.

This is not their story.

But it is the story of that ever intensifying fireworks display and some of its ramifications.

---

[1] This made many theorists very happy because they suddenly had something to theorise about; and others very perplexed because what they had already theorised was wrong by about 120 orders of magnitude.

It is also the story of a framework, a framework for Bayesian inference called simulation-based inference: not a traditional framework and before this thesis rarely heard of or used by supernova cosmologists.

Nevertheless, a remarkably powerful framework. Indeed, in many of the more relaxed areas of science, it has already supplanted the great χ^2 fit as the standard inference methodology, for though it is approximate and uses black magic, or at least black-box estimators, it scores over the older, more pedestrian work in two important respects. First, it is immensely more flexible and faster; and secondly, it has the buzzwords "neural networks" in its description.

But the story of this framework and all the issues in supernova cosmology that it can resolve begins simply. It begins with Bayes' theorem.

Part I

Simulation-based inference

Chapter 1

Bayesian inference

*The Bayes Drip**
by Stable Diffusion
"Thomas Bayes in a white puffer jacket on his way to drop the sickest theorem in the history of statistics"

Bayesianism is a philosophical approach to science founded on Bayes's reasoning about the information contained in causes about effects *and* in effects about causes [36]. At its core, it puts both—causes and effects—on an equal footing, basing all considerations on their *joint* probability,[2] which thus embodies the entirety of a given *model* for the studied phenomenon. Bayes' famous theorem is nothing but a consequence of the two ways to decompose the joint probability of two random variables[3] (call them $\boldsymbol{\theta}$ and $\mathbf{d}$):

joint
model

$$\mathrm{p}(\boldsymbol{\theta}, \mathbf{d}) = \mathrm{p}(\mathbf{d} \mid \boldsymbol{\theta})\, \mathrm{p}(\boldsymbol{\theta}) = \mathrm{p}(\boldsymbol{\theta} \mid \mathbf{d})\, \mathrm{p}(\mathbf{d}). \tag{1.1}$$

The symmetry between $\boldsymbol{\theta}$ and $\mathbf{d}$ can be broken—and an interpretation of their roles made—only by an empirical observation that identifies, arbitrarily, $\mathbf{d}$ as *data*, to which a concrete value $\mathbf{d}_\mathrm{o}$ can be assigned. *Inference* is then concerned with the probability of $\boldsymbol{\theta}$ (now identified as *unknown parameters*) conditioned on that value, i.e. the *posterior* $\mathrm{p}(\boldsymbol{\theta} \mid \mathbf{d} = \mathbf{d}_\mathrm{o})$ and quantities derived from it (e.g. mean, standard deviation, etc.).

data
inference
parameters
posterior

Along this line of interpretation, $\mathrm{p}(\mathbf{d} \mid \boldsymbol{\theta})$, called the *sampling distribution* of the data, represents the stochastic process—in the sense of the physical procedure[4]—from which Bayesians philosophise $\mathbf{d}_\mathrm{o}$ was realised.[5] As such, it is the "objective" part of the model, which recounts the ontology of an analysis.

sampling distribution

[2] or, in the common case of continuous random variables, *probability density*

[3] For all the pomp and attempted cleverness throughout this thesis, I will, for the most part, allow myself not to make a graphical distinction between random variables and their values. Furthermore, usually I will consider them to be multi-valued, i.e. bolded "vectors", combining many quantities from the model.

[4] a *statistical* process, in contrast, describes the distribution/probability of a function, i.e. an infinite-dimensional random variable

[5] This view has an important consequence: it admits the possibility of *other* realisations $\mathbf{d}_\mathrm{o}$ of the same random variable. Such—*counterfactual* after the fact of observing $\mathbf{d}_\mathrm{o}$—data produced by an artificial recreation of the data-generating process (a forward simulator) lies at the heart of simulation-based inference; but more on that later.

© The Author(s), under exclusive license to Springer Nature Switzerland AG 2026
K. Karchev, *Supernova Cosmology for the 21st Century*,
https://doi.org/10.1007/978-3-032-15072-1_1

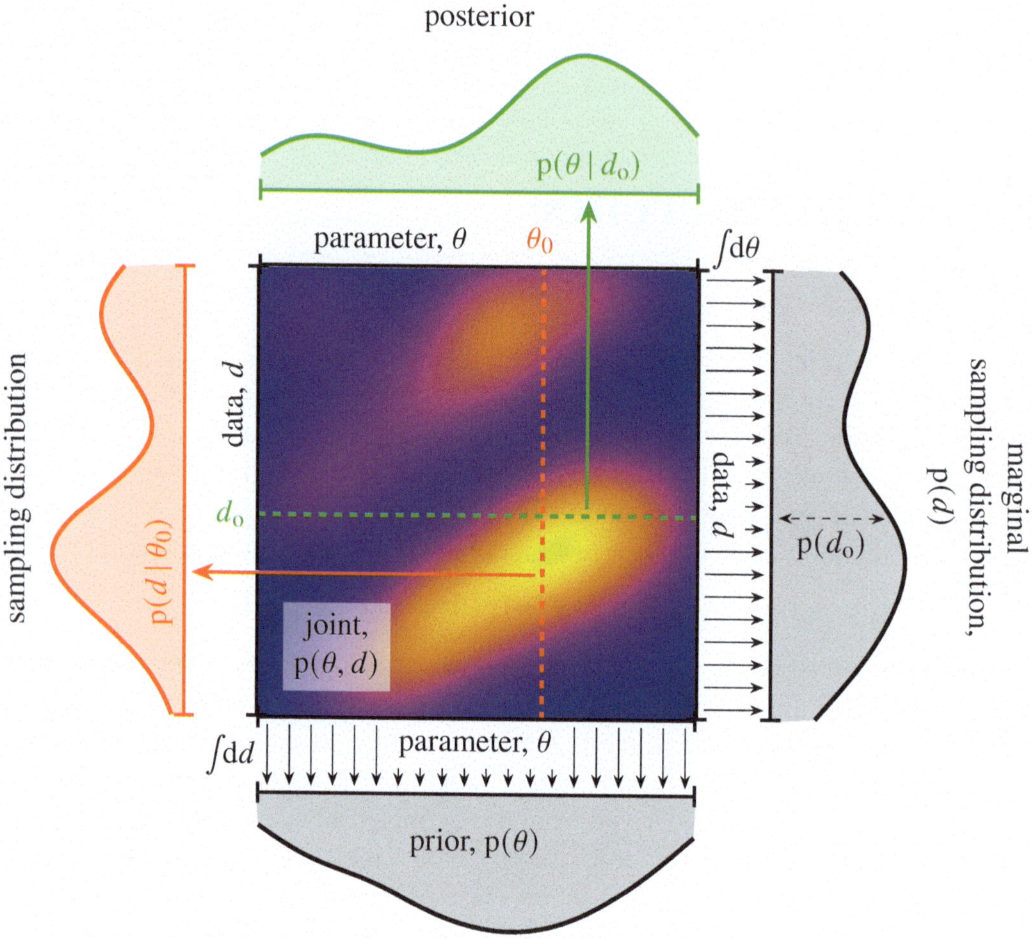

Figure 1.1: Elements of Bayesian reasoning: the joint probability (density depicted as colour in the central panel for a model with a single parameter and single datum) and four perspectives onto it: prior and evidence, resulting from marginalisation of d or θ, respectively, and posterior and sampling distribution, resulting from conditioning on d_o or θ_o (dashed lines). Note that the *likelihood* ($p(d_o \mid \theta)$), a function of parameters for fixed data, is not present.

However, $\mathrm{p}(\mathbf{d}\,|\,\boldsymbol{\theta})$ is not (typically) used by Bayesians in its capacity of a distribution[6] but rather as a measure of probability, always conditioned on the observed data: $\mathrm{p}(\mathbf{d}=\mathbf{d}_\mathrm{o}\,|\,\boldsymbol{\theta})$. This is called the *likelihood function* and often labelled $L_{\mathbf{d}_\mathrm{o}}(\boldsymbol{\theta})$; its absence is a prominent characteristic of this thesis.

likelihood function

This leaves us with the two terms $\mathrm{p}(\boldsymbol{\theta})$ and $\mathrm{p}(\mathbf{d})$, referred to as *marginals* because they represent the probabilities of one variable irrespective of the value of the other. The latter is the *marginal* (average) *likelihood*:

marginals

marginal likelihood

$$\mathrm{p}(\mathbf{d}=\mathbf{d}_\mathrm{o}) \equiv \int \mathrm{p}(\boldsymbol{\theta},\mathbf{d}=\mathbf{d}_\mathrm{o})\,\mathrm{d}\boldsymbol{\theta} = \int \mathrm{p}(\mathbf{d}=\mathbf{d}_\mathrm{o}\,|\,\boldsymbol{\theta})\,\mathrm{p}(\boldsymbol{\theta})\,\mathrm{d}\boldsymbol{\theta}\,. \tag{1.2}$$

However, a more common name for it is *evidence*, alluding to its use in Bayesian model comparison (see section 3.1). Notice that eq. (1.2) represents a constant (that ensures proper normalisation of the posterior) entirely determined by the observed $\mathbf{d}_\mathrm{o}$ (*under a given model* $\mathrm{p}(\boldsymbol{\theta},\mathbf{d})$). As such, it is often ignored in Bayesian parameter inference.

evidence

On the other hand, the average posterior:

$$\mathrm{p}(\boldsymbol{\theta}) \equiv \int \mathrm{p}(\boldsymbol{\theta},\mathbf{d})\,\mathrm{d}\mathbf{d} = \int \mathrm{p}(\boldsymbol{\theta}\,|\,\mathbf{d})\,\mathrm{p}(\mathbf{d})\,\mathrm{d}\mathbf{d}\,, \tag{1.3}$$

is practically exclusively called the *prior* probability since it represents the (non-)existence of *knowledge* about the "unknown" parameters.

prior

Priors

The prior is often viewed as a modelling *choice*, owing to its independence from data. This raises a philosophical question: what is the influence of this—seemingly arbitrary/subjective [244]—probability on the final result of Bayesian inference? That is, how *informative* the prior is of the "unknown" parameters. Model builders have addressed this issue in a variety of ways, motivated either by *progress*, *integrity*, or *convenience*.

Staying true to nomenclature and leaning on the rules of conditional probability (i.e. eq. (1.1)), some studies explicitly incorporate empirical knowledge, in the form of the posterior from a previous analysis of independent data $\mathbf{d}^{(I)}$, as a prior when confronting $\mathbf{d}^{(II)}$:

$$\mathrm{p}\left(\boldsymbol{\theta},\mathbf{d}^{(I)},\mathbf{d}^{(II)}\right) = \mathrm{p}\left(\boldsymbol{\theta},\mathbf{d}^{(II)}\middle|\mathbf{d}^{(I)}\right)\cancel{\mathrm{p}\left(\mathbf{d}^{(I)}\right)} \propto \mathrm{p}\left(\mathbf{d}^{(II)}\middle|\boldsymbol{\theta},\cancel{\mathbf{d}^{(I)}}\right)\underbrace{\mathrm{p}\left(\boldsymbol{\theta}\middle|\mathbf{d}^{(I)}\right)}_{\text{previous posterior}}, \tag{1.4}$$

[6] This is reserved for—and the basis of—*frequentist* statistics and only used in a small area of the Bayesian realm: posterior predictive considerations that derive the sampling distribution of subsequent analyses, conditioned on the outcome of the present one, i.e. $\mathrm{p}(\mathbf{d}\,|\,\mathbf{d}_\mathrm{o})$.

where $p\left(\mathbf{d}^{(I)}\right)$ is ignored as constant as usual, and the observation of a particular value for $\mathbf{d}^{(I)}$ is assumed to not influence $\mathbf{d}^{(II)}$, i.e. they are *conditionally independent*: $\mathbf{d}^{(I)} \perp\!\!\!\perp \mathbf{d}^{(II)} \mid \boldsymbol{\theta}$.

conditional independence

uninformative prior

uniform distribution

re-parametrisation

Jeffreys prior

On the other hand, *uninformative priors* are *designed* to specifically reflect a lack of[7] prior knowledge. Many a rookie Bayesian have been tempted to stipulate "*uniform prior probability* over possible parameter values", but such a statement rarely has a concrete meaning due to the possibility of *re-parametrisation*. In fact, by re-defining the parameter space using a specifically chosen transformation $\boldsymbol{\theta} \to \boldsymbol{\theta}'$, one can turn $\boldsymbol{\theta} \sim \mathcal{U}(\boldsymbol{\theta})$ into any $p(\boldsymbol{\theta}')$.[8] However, re-parametrisation can also be exploited to the benefit of objectivity by the constructing the *Jeffreys prior*: a distribution that is invariant under a particular symmetry (e.g. translation or scaling for location- and size-related parameters, respectively) expected of the parameter space [see e.g. 244]. A final technique for objectivising the prior, again due to Jaynes [244], is entropy maximisation, which, according to Shannon [476], minimises its information content. This method can even incorporate conditions like a bounded domain or previous constraints that do not take the form of a full posterior distribution (e.g. point estimates).

conjugate prior

Finally, a particular prior probability can be *chosen* in the interest of tractability, i.e. the ease of deriving the posterior. Since this depends also on the likelihood, the prior needs to be coordinated with it, e.g. via *conjugation*: the adoption, given a likelihood function, of a form for the prior which results in a posterior of the same form [see e.g. 150]. This "conjugate update" procedure also facilitates reuse of the posterior as a new prior for analysis of further data under the same likelihood (cf. eq. (1.4)).

In this thesis, the question of the prior choice will be only of marginal importance; that is, we will not be designing priors. We will simply be manipulating the ones given by model builders as part of complete joint distributions, and at that we will aim to be as non-intrusive as possible. As a rule, we will only make such changes that do not affect the final posterior because they happen in regions of parameter space (almost) completely disfavoured by the observed data, i.e. for which the likelihood (nearly) vanishes: $p(\mathbf{d} = \mathbf{d}_o \mid \boldsymbol{\theta}) \approx 0$. This procedure in all its forms is presented in subsection 2.2.2.

---

[7] or as little of it as possible, since, as Tak et al. [501] argue, "all prior distributions are informative"

[8] This is indeed how random values are drawn from *any* distribution within a computer program: starting with a sample from $\mathcal{U}(0, 1)$.

1.1 Bayesian hierarchical modelling

Ontology

Complicated ontologies require a statistical representation that transcends the simple statement eq. (1.1) of Bayes' theorem. One example are so-called *nuisance parameters* ($\boldsymbol{\nu}$), which are not of scientific interest[9]—e.g. instrumental calibration parameters or the physical model of SN Ia explosions—but are still *a priori* uncertain. In this case, we can only describe the data-generating process (sampling distribution) as $\mathrm{p}(\mathbf{d} \mid \boldsymbol{\theta}, \boldsymbol{\nu})$ and obtain, through eq. (1.1), the *joint posterior* $\mathrm{p}(\boldsymbol{\theta}, \boldsymbol{\nu} \mid \mathbf{d})$ when in fact we are seeking the *marginal posterior*:

nuisance parameters

joint vs. marginal posterior

$$\mathrm{p}(\boldsymbol{\theta} \mid \mathbf{d}) = \int \mathrm{p}(\boldsymbol{\theta}, \boldsymbol{\nu} \mid \mathbf{d}) \, \mathrm{d}\boldsymbol{\nu} \propto \int \mathrm{p}(\mathbf{d} \mid \boldsymbol{\theta}, \boldsymbol{\nu}) \, \mathrm{p}(\boldsymbol{\theta}, \boldsymbol{\nu}) \, \mathrm{d}\boldsymbol{\nu} \tag{1.5}$$

(ignoring the evidence). This marginalisation cannot usually be performed analytically, although in the absence—or in conscious denial—of prior knowledge about $\boldsymbol{\nu}$, one may *assert* a specific prior $\mathrm{p}(\boldsymbol{\theta}, \boldsymbol{\nu})$ that simplifies calculations at the expense of objectivity.[10]

Another example are systems with a hierarchical structure in which stochasticity arises on multiple levels: e.g. a population with uncertain *global parameters* $\boldsymbol{\gamma}$ and *local parameters* $\boldsymbol{\lambda}^i$ that determine the properties of each object i (out of N_{obj}) in the population. If noisy measurements $\{\mathbf{d}_\mathrm{o}^i\}$ of the individual objects are made, inference can be performed using the joint probability

global and local parameters

$$\mathrm{p}(\boldsymbol{\gamma}, \{\boldsymbol{\lambda}^i\}, \{\mathbf{d}^i\}) = \mathrm{p}(\{\mathbf{d}^i\} \mid \{\boldsymbol{\lambda}^i\}, \boldsymbol{\gamma}) \, \mathrm{p}(\{\boldsymbol{\lambda}^i\} \mid \boldsymbol{\gamma}) \, \mathrm{p}(\boldsymbol{\gamma}). \tag{1.6}$$

This is called a *Bayesian hierarchical model (BHM)*, and its key characteristic is the assumption[11] that objects are *independent and identically distributed (i.i.d.)*, resulting in conditional independence for their data *and* associated local parameters:

BHM
i.i.d.

$$\mathrm{p}(\{\mathbf{d}^i\} \mid \{\boldsymbol{\lambda}^i\}, \boldsymbol{\gamma}) \, \mathrm{p}(\{\boldsymbol{\lambda}^i\} \mid \boldsymbol{\gamma}) \to \prod_i^{N_{\mathrm{obj}}} \mathrm{p}(\boldsymbol{\lambda}^i, \mathbf{d}^i \mid \boldsymbol{\gamma}) = \prod_i^{N_{\mathrm{obj}}} \mathrm{p}(\mathbf{d}^i \mid \boldsymbol{\lambda}^i, \boldsymbol{\gamma}) \, \mathrm{p}(\boldsymbol{\lambda}^i \mid \boldsymbol{\gamma}). \tag{1.7}$$

BHMs are powerful tools for inference of the global/population parameters since they allow information from all objects to be *combined* while properly accounting for *two kinds of variance*: population *scatter* of the *latent*[12] *values* $\boldsymbol{\lambda}^i \sim \mathrm{p}(\boldsymbol{\lambda}^i \mid \boldsymbol{\gamma})$ and observational *noise*

pooling of constraints
scatter vs. noise
latent values

[9] to a particular researcher or their audience, that is

[10] One (a mathematician, for example) might argue against this point, for an inspired choice may be considered correct and justified for the same reason that theoretical physicists advocate for particular Lagrangians on the basis of elegance or simplicity.

[11] Otherwise, a distinction between "global" and "local"/"object-specific" cannot really be made, and the model reverts to eq. (1.5).

in $\mathbf{d}^i \sim \mathrm{p}(\mathbf{d}^i \,|\, \boldsymbol{\lambda}^i, \boldsymbol{\gamma})$. Often, therefore, all local parameters are treated as nuisances, whose number—and consequently, that of the possible *a posteriori* correlations that need to be taken into account—scales with the number of objects considered. In this case, analytic marginalisation can drastically reduce the problem complexity and, importantly, decouple its dimensionality from the data set size:

$$\int \prod_i^{N_{\mathrm{obj}}} \mathrm{p}(\mathbf{d}^i \,|\, \boldsymbol{\lambda}^i, \boldsymbol{\gamma})\, \mathrm{p}(\boldsymbol{\lambda}^i \,|\, \boldsymbol{\gamma})\, \mathrm{d}\boldsymbol{\lambda}^i = \mathrm{p}(\{\mathbf{d}^i\} \,|\, \boldsymbol{\gamma}).^{13} \tag{1.8}$$

The convenience this affords the model builder, especially with regards to the availability of established computational methods (section 1.2) for Bayesian inference, might overpower their desire for adherence to nature and lead them to stipulate particular forms for $\mathrm{p}(\mathbf{d}^i \,|\, \boldsymbol{\lambda}^i, \boldsymbol{\gamma})$ and/or $\mathrm{p}(\boldsymbol{\lambda}^i \,|\, \boldsymbol{\gamma})$ that simplify the computation in eq. (1.8).

Graphical representation

As a BHM grows in complexity to take into account more details of the data-generating procedure,[14] it becomes increasingly cumbersome to describe and communicate through an explicit factorisation of the joint probability. Moreover, some model components may be *deterministic* functions of other (stochastic or not variables), while others are treated as *fixed settings* (global $\mathbf{s}$ and local $\{\mathbf{a}^i\}$), under which the joint model is evaluated; the latter can equivalently be conceptualised as stochastic variables with delta-distribution priors

deterministic and fixed quantities

$$\mathrm{p}(\mathbf{s}, \boldsymbol{\gamma}, \{\mathbf{a}^i\}, \{\boldsymbol{\lambda}^i\}, \{\mathbf{d}^i\}) = \mathrm{p}(\boldsymbol{\gamma}, \{\boldsymbol{\lambda}^i\}, \{\mathbf{d}^i\} \,|\, \mathbf{s}, \{\mathbf{a}^i\}) \times \delta(\mathbf{s}) \times \prod_i^{N_{\mathrm{obj}}} \delta(\mathbf{a}^i). \tag{1.9}$$

DAG

In such intricate cases, a graphical representation as a *directed acyclic graph (DAG)*—like the general one we show in fig. 1.2—is usually employed instead. Since any DAG can be topologically sorted, i.e. ordered so that distributions are only conditions on already-encountered quantities, it represents the *forward model* for the data-generating process. Determining the conditional probabilities needed to describe it in the *reverse* direction (data → parameters) is the goal of *inference*.

forward model

reverse model (inference)

latent variable

[12] In the Bayesian literature [e.g. 337, chapter 34], *latent random variables* are any that are not observed, i.e. $\boldsymbol{\psi}$. In this thesis, however, by "latent" we will mean only object-specific parameters which can, in a sense, be interpreted as the "true" values of measured quantities (and will in general refrain from using both terms).

[13] This is again called a "marginal likelihood" like the evidence—a connection we will exploit in chapter 3.

population parameters

[14] For example, the distributions of various object-specific quantities might be controlled by separate *population parameters*, while other global variables (e.g. the cosmological model and instrumental noise) directly influence the data.

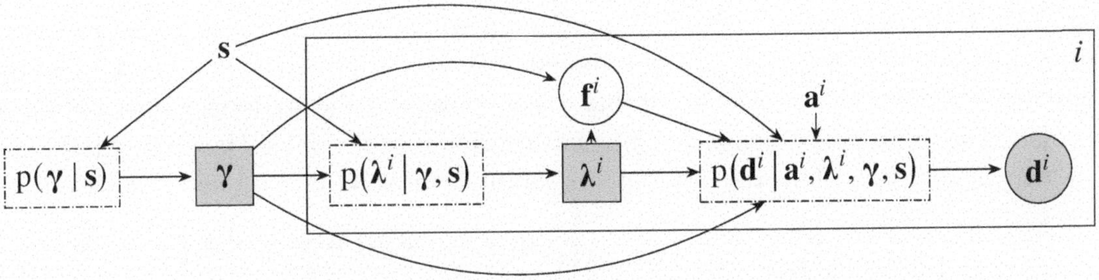

Figure 1.2: Graphical representation of a generic hierarchical model that depicts its global and local parameters ($\boldsymbol{\gamma}$ & $\boldsymbol{\lambda}^i$), the object-specific data ($\mathbf{d}^i$), and the various conditional distributions ($\mathrm{p}(\ldots \mid \ldots)$) relating them. It also includes fixed settings/inputs ($\mathbf{s}$ & $\mathbf{a}^i$) and deterministic quantities ($\mathbf{f}^i$), which are usually omitted from the probabilistic description for brevity, and indicates conditional independence with a "plate" around i.i.d. quantities, specifying the indexing label (i).

1.2 More or less established methods

Episte-mology

For all but the simplest models—hierarchical or not—and without resorting to convenient assertions, assumptions, or approximations, the posterior does not have a simple analytic form, and so Bayesian inference needs to be performed computationally. The "traditional" approach is based on calculating the likelihood function (or its gradient) given the particular observed data and combining it with the prior. The posterior is then represented either through samples from a stochastic chain, or via a trained surrogate approximation that is easy to sample from and/or evaluate. Regardless of the method, in the presence of nuisance parameters, they need to be inferred jointly with those of interest since only the likelihood $\mathrm{p}(\mathbf{d}_\mathrm{o} \mid \boldsymbol{\psi})$ of the full set of parameters $\boldsymbol{\psi} \equiv \boldsymbol{\theta} \cup \boldsymbol{\nu}$ is available for evaluation. The marginalisation in eq. (1.5) can then performed, if required, on the level of samples from $\mathrm{p}(\boldsymbol{\psi} \mid \mathbf{d}_\mathrm{o})$ (by just ignoring their components corresponding to $\boldsymbol{\nu}$) or analytically when a suitable surrogate is used.

likelihood-based methods

Below, we survey the "established" *likelihood-based methods* relevant to SN cosmology, whose performance has been thoroughly studied and optimised in numerous previous works (in SN cosmology and other areas of science and statistics). We will be using them for two purposes. First, to provide *motivation* and *justification* for method development aimed at aspects in which reliance on these techniques is inconvenient (scalability) or restrictive (realism and statistical rigour). And second, to cross-check (only in simplified scenarios!) the results we derive via new methods we propose, thus demonstrating the validity, credibility, and applicability of our approaches.

MCMC

proposal distribution

MCMC Bayesian inference is most commonly performed through *Markov chain Monte Carlo (MCMC)*[15] sampling: a Markov process in which new values ψ_{new} are proposed from a distribution $q(\psi_{\text{new}} \mid \psi_{\text{old}})$ conditioned only on the previous element ψ_{old} in the chain and accepted with a probability that makes $p(\psi \mid \mathbf{d}_o)$ the stationary distribution of the chain. The prevalent choice of acceptance criterion is Metropolis–Hastings (MH) [358, 202], which makes use of the ratio of joint probabilities for new/old parameters and the data:

$$p(\text{accept} \mid \psi_{\text{new}}, \psi_{\text{old}}) = \min\left[1, \frac{q(\psi_{\text{new}} \mid \psi_{\text{old}})\, p(\psi_{\text{old}}, \mathbf{d}_o)}{q(\psi_{\text{old}} \mid \psi_{\text{new}})\, p(\psi_{\text{new}}, \mathbf{d}_o)}\right]. \tag{1.10}$$

exploration vs. exploitation

random walk

ensemble MC

curse of dimensionality

The proposal distribution for MCMC must strike a balance between *independence* and *efficiency*: i.e., it must be able to propose uncorrelated samples in unexplored regions of parameter space (e.g. undiscovered modes) while avoiding unlikely values that will be rejected. A simple strategy based on a *random walk* can achieve this in moderate dimensions by setting the step size (i.e. the spread of a Gaussian proposal centered at ψ_{old}) in proportion to the covariance of the target (and the parameter dimension) [170, 40], which can be estimated from an *ensemble* of chains [e.g. 180]. However, a random walk progresses slowly (with the square root of the number of steps), leading to auto-correlated samples. Moreover, it scales poorly to high dimensions—and is hence extremely inefficient for BHMs—due to the *curse of dimensionality* (the exponential abundance of "distant" space that eventually overwhelms any "locality" of the proposal), and the need to estimate a full covariance matrix.[16]

HMC

mass matrix

HMC *Hamiltonian* (or *hybrid*) *Monte Carlo (HMC)* [133, see also 379] elevates MCMC to higher dimensions and produces largely independent samples via a *global*[17] proposal that follows the trajectory of a "particle" with a random "momentum" vector (introduced as an auxiliary stochastic variable). Its dynamics is defined through a Hamiltonian consisting of a potential field given by the negative logarithm of the target density[18] and a quadratic kinetic term with an arbitrary square *mass matrix* (of dimension equal to the parameter space). The latter corresponds to the (inverse) covariance of the momentum proposal distribution, and as such, it is usually set to an initial estimate of the target covariance (although in very high dimensions, only its diagonal is usually kept [514]).[19]

[15] For a historical overview and extended discussion, refer to Martin et al. [351].

[16] The popular `emcee` package, for example, requests at least twice as many walkers as dimensions.

[17] Beskos et al. [41] show that the number of steps required for an independent HMC sample scale only weakly with the dimension as $d^{1/4}$.

[18] Particles are thus "forced" towards regions of high probability, so HMC explores in linear time, rather than quadratic for a random walk.

[19] In contrast, Girolami & Calderhead [176], adopting the perspective of information geometry, argue for the replacement of the mass matrix with a position-dependent (inverse) metric for the manifold on which Hamil-

HMC's position/parameter proposal relies on repeated evaluations of the model's gradient;[20] $\nabla_\psi \ln p(\psi, \mathbf{d}_o)$ and an energy-preserving integration routine with *tunable settings*[21] like the step size (adjusted based on the acceptance rate[22]) and the trajectory length (for which the no-u-turn sampler (NUTS) [225, 42] scheme is widely adopted). *tuning*

In the past decade, HMC has greatly benefited from the development of frameworks for probabilistic programming (discussed further in chapter 9), which automate the gradient calculation and the tuning procedure and facilitate model re-parametrisations that make the geometry of the problem easier to explore with Hamiltonian dynamics [43]. For all its complexity, HMC is thus still the state-of-the-art exact high-dimensional likelihood-based sampling method.

NS The MH algorithm has two major likelihood-based alternatives. The first, *nested sampling (NS)* [483, 484, see also 20], is a form of MCMC designed to calculate the evidence—the result of a multivariate integral (eq. (1.2)) as per the original intent of Monte Carlo methods—that produces *unequal-weight samples* from the posterior as a by-product. In NS, "live points" (samples) are drawn from the *prior* (usually transformed to the high-dimensional unit cube) subject to a likelihood constraint, i.e. within an iso-likelihood contour. This can be realised through a number of strategies, including rejection sampling (in low dimensions and usually after constructing a simple analytical bounding volume [see 491, and references therein]), slice sampling [378, 199] (in moderate dimensions), or any of the methods discussed above. However, even highly optimised NS implementations scale at least cubically with the number of inferred/integrated parameters, making them unsuitable for high-dimensional inference.[23] Nevertheless, NS remains among the few reliable *NS* *unequal-weight samples*

tonian dynamics is simulated, leading to improved performance, at the expense of a more complicated integration scheme and the need to explicitly specify the metric, for which they choose the expectation (over data) of the joint's Hessian (with respect to parameters): $-\mathbb{E}_{\mathbf{d}\mid\psi}\left[\frac{\partial^2}{\partial\psi^2}\ln p(\psi, \mathbf{d})\right]$.

20 An algorithm essentially consisting of a single gradient step, Langevin Monte Carlo (LMC), also exists; since it needs to be corrected for a non-unary acceptance probability, it is better known as *Metropolis-adjusted Langevin algorithm* (MALA).

21 An algorithm that has no tunable settings (on the surface) is *Gibbs sampling* [171, 169], which iterates over a partition of ψ, sampling each group conditionally on the rest. (In fact, HMC can be viewed as a Gibbs sampler for the joint distribution of parameters and momenta.) In certain cases, these low-dimensional conditionals can be derived analytically, but in others, conditional MCMC sampling (within Gibbs) is still needed. *Gibbs sampling*

22 If energy is truly preserved, acceptance should be perfect, but the MH criterion is still checked in practice to counteract integration errors: see Beskos et al. [41].

23 The difficulty of high-dimensional nested sampling comes from the simplicity (e.g. uniform) of the prior, from which samples are to be drawn, and its discontinuous truncation at the iso-likelihood contour, both of which immobilise the otherwise performant gradient-based methods. In a recent development, Cai et al. [78] work around the hard boundary problem by using *proximal* MCMC [402, 135] to smear the constraint and scale NS to millions of dimensions. However, even their method reverts to a simple random walk in the

methods for likelihood-based evidence estimation for models with up to a few hundred parameters.

VI
proposal distribution
variational parameters
ELBO

VI As a general alternative to sampling methods, *variational Bayesian inference (VI)* [see e.g. 51, 558] can be performed through optimisation of an explicit tractable *proposal distribution* $q_\xi(\psi)$ (within a family with *variational* (or *hyper)parameters* ξ) so that it approximates the posterior (given concrete data), thus avoiding the issues of MCMC. The optimisation (maximisation) objective in VI is the *evidence lower bound (ELBO)* [468, 263][24]:

$$\begin{aligned}\mathrm{ELBO}\left[q_\xi(\psi) \parallel p(\psi, \mathbf{d}_o)\right] &\equiv \mathbb{E}_{\psi\sim q_\xi}\left[\ln p(\psi, \mathbf{d}_o) - \ln q_\xi(\psi)\right] \\ &= \underbrace{\ln p(\mathbf{d}_o)}_{\text{constant}} - \underbrace{\mathbb{E}_{\psi\sim q_\xi}\left[\ln q_\xi(\psi) - \ln p(\psi \mid \mathbf{d}_o)\right]}_{\equiv \mathrm{KL}\left[q_\xi(\psi) \parallel p(\psi \mid \mathbf{d}_o)\right] \geq 0},\end{aligned} \tag{1.11}$$

KL divergence

closely related to the popular measure of similarity between distributions, the *Kullback–Leibler (KL) divergence* [299], which is minimised if and only if $q_\xi(\psi)$ identically matches the target, $p(\psi \mid \mathbf{d}_o)$. In any case, the KL divergence is never negative, which elucidates the name of the ELBO and makes its maximal value useful for approximate evidence-based model optimisation[25] albeit with severe caveats related to the choice of $q_\xi(\psi)$ [267].

Once again, the expectation in eq. (1.11) cannot be evaluated analytically in general, or at least requires lengthy model-specific calculations. Nevertheless, an unbiased estimate of its gradient with respect to the variational parameters:

$$\nabla_\xi \,\mathrm{ELBO} = \mathbb{E}_{\psi\sim q_\xi}\left[\Big(\nabla_\xi \ln q_\xi(\psi)\Big)\Big(\ln p(\psi, \mathbf{d}_o) - \ln q_\xi(\psi)\Big)\right], \tag{1.12}$$

stochastic gradient-based optimisation
re-parametrisation trick

can be easily obtained by sampling from $q_\xi(\psi)$ (tractable by design), which enables optimisation via *stochastic gradient ascent* [452]. Notice that eq. (1.12) does *not* involve the model gradient and can be very noisy when estimated from a finite number of samples. Ranganath et al. [439] proposed several methods to reduce its variance; instead, Kingma & Welling [290] used the *re-parametrisation trick*: $\psi \to f_\xi(\boldsymbol{\varepsilon})$ for a suitably defined differentiable deterministic function f_ξ and $\boldsymbol{\varepsilon}$ sampled from a suitably defined *constant* distribution q.[26] This abstracts the stochasticity away from the hyperparameters, allowing

common case of a prior transformed to a uniform distribution on the unit cube.

[24] more recently re-popularised by Kingma & Welling [290] in a slightly different, but relevant, context

[25] in other words, maximum-likelihood estimation of higher-level hierarchical parameters

[26] Re-parametrisation like this is possible for any continuous distribution, e.g. through its cumulative distribution function and a uniform distribution on the unit interval.

the expectation and gradient to commute:

$$\nabla_{\boldsymbol{\xi}}\,\text{ELBO} \to \mathbb{E}_{\boldsymbol{\varepsilon}\sim q}\left[\nabla_{\boldsymbol{\xi}}\left(\ln p\left(f_{\boldsymbol{\xi}}(\boldsymbol{\varepsilon}), \mathbf{d}_{\text{o}}\right) - \ln q_{\boldsymbol{\xi}}\left(f_{\boldsymbol{\xi}}(\boldsymbol{\varepsilon})\right)\right)\right], \tag{1.13}$$

and the ELBO to be optimised using a simple application of the chain rule. With VI again, probabilistic programming frameworks with automatic re-parametrisation and differentiation of the sophisticated proposals described below (and of the joint as in HMC) are an immense practical convenience.

expressivity vs. tractability

The success of VI is largely dependent on the appropriate choice of a family of proposal distributions that balances flexibility (i.e. the ability to accurately capture the desired features of the posterior) with computational efficiency when sampling, evaluating (marginal) posterior properties, and optimising. Assuming an overly restricted family, which cannot faithfully represent $p(\boldsymbol{\psi} \mid \mathbf{d}_{\text{o}})$, can have drastic consequences to the fidelity of inference results, both in terms of parameter constraints and model optimisation through the ELBO (as an evidence proxy). The reason is that, as illustrated by Bishop [49, fig. 10.2], the ELBO is *mode seeking*, i.e. it favours $q_{\boldsymbol{\xi}}(\boldsymbol{\psi})$ with reduced variance when they cannot exactly match the target;[27] which means that the final approximation (even if optimal within the variational family according to eq. (1.11)) may not have the desired coverage properties of the posterior.

mode-seeking vs. mass-covering

mean-field VI

In its genesis [15, 397], VI almost exclusively assumed the so-called *mean-field* limit, in which each parameter ψ_i is *a posteriori* independent of the rest;[28] and the approximation factorises, usually into a product of Gaussians:

$$q_{\boldsymbol{\xi}}(\boldsymbol{\psi}) \to \prod_i q_{\xi_i}(\psi_i) \quad \text{and e.g.} \quad q_{\xi_i}(\psi_i) = \mathcal{N}\left(\psi_i \,\middle|\, \mu_i, \sigma_i^2\right), \tag{1.14}$$

with optimised means and variances: $\boldsymbol{\xi} \to \boldsymbol{\mu}, \boldsymbol{\sigma}^2$ (which are the quantities that marginal inference is ultimately interested in anyway). This, as discussed, can be catastrophic, if the individual parameters are *a posteriori* correlated. Rising one level of sophistication higher, the approximate posterior can adopt a structured covariance matrix reflecting these *a posteriori* dependencies, which can be derived automatically from the *a priori* factorisation (i.e. the graphical representation) of a hierarchical model [e.g. 540]. Since dense matrices are computationally demanding, however, some low-rank approximation has to be made for large hierarchical models: e.g. by including all correlations among global parameters, among the local parameters of each individual object, and between global and local parameters but treating each object as otherwise independent of the rest [267].

density estimation

Nowadays, a plethora of flexible high-dimensional *density estimation* tools have been

[27] In the extreme case of a delta distribution proposal, VI reduces to estimation of the maximum *a posteriori*.
[28] but can still be informed by all the data in a hierarchical model, even if it represents an object-specific property

developed that can add expressivity (e.g. higher-order correlations and moments; i.e., non-Gaussianity) to the proposal. However, to be usable within the framework of ELBO maximisation, a density estimator must allow for both sampling and density (gradient) evaluation, narrowing the choice to the group of models known as *normalising flows (NFs)* [see 396, 55 for reviews]: optimisable nonlinear transformations with a tractable Jacobian applied to a random variable with a simple "latent distribution" (i.e. a glorious manifestation of the re-parametrisation trick). NFs are implemented as neural networks, which allows their training, usually targeting a single $\mathrm{p}(\psi \mid \mathbf{d}_\mathrm{o})$, to be *amortised* over many different $\mathbf{d}$ by "conditioning" the flow [545, 349]: this facilitates simultaneous inference from i.i.d. observations, e.g. for learning $\mathrm{p}(\boldsymbol{\gamma}, \boldsymbol{\lambda}^i \mid \mathbf{d}_\mathrm{o}^i)$, but cannot account for hierarchical relations, i.e. $\mathrm{p}(\boldsymbol{\lambda}^i \mid \{\mathbf{d}_\mathrm{o}^i\})$ or $\mathrm{p}(\boldsymbol{\gamma} \mid \{\mathbf{d}_\mathrm{o}^i\})$.

NF

inference amortisation

Ethics 1.3 The case for likelihood-free inference[29]

"~~All models are wrong but some are useful~~" [61].

summary statistics

Traditional Bayesian methods have—~~or rather had~~—a problem, which is this: they restrict scientists (model builders) to models that are *solvable* rather than models that are *correct*. This may prevent them from extracting the full information content of their observations, e.g. by imposing the use of *summary statistics* as a compressed and supposedly simpler to model representation of the data. The poor scalability of established techniques—especially when the model gradient is unavailable, or HMC fails to auto-tune, or in the presence of strong correlations—might force model builders into unsightly approximations, including, but not limited to: adopting a convenient hierarchical likelihood that allows analytic marginalisation; fixing stochastic components to point estimates in an analysis split in stages; uncertainty quantification through variation (i.e. linear error propagation); and χ^2 fitting.

The scientific insights derived with simplified models are, in general, different from those that practitioners would have *liked* to obtain, had they had appropriate tools; and in many cases, these "insights" will be *wrong*, i.e. will not have the properties of the Bayesian posterior. We will repeated come back to this point throughout this thesis. And while most issues, specifically in SN cosmology, can be addressed by a carefully constructed and patiently sampled BHM, there is one that has no solution within the likelihood-based framework: even though the joint probability density $\mathrm{p}(\psi, \mathbf{d})$ is defined for every model—after all, this is what "model" means—it may not be calculable if no closed-form expressions for $\mathrm{p}(\mathbf{d} \mid \psi)$ and/or $\mathrm{p}(\psi)$ are available.

implicit likelihood

[29] not really free of a sampling distribution (which is always *implicit* in a model) or even free of a likelihood (which some methods end up approximating), but ♫ *freed from reliance* ♫ on evaluating probabilities

Missing: the likelihood. In the most common case, the likelihood may not be tractable[30] when modelling a complicated (usually deterministic) process through which probabilities are impossible[31] to trace. Examples particularly relevant to SN cosmology (see also section 8.3) include the scheduling of telescope observations and their celestial footprint (dependent on weather and balancing diverse scientific goals), tricky instrumental effects, extracting sources and their properties from images, summarising data through model fitting, and the large scale structure and dynamics in the Universe.

Missing: dimensionality. Another challenge for the likelihood framework are what we coin *un-dimensional*®[32] models, in which the *dimensionality* of the parameter and or data space is itself *a priori* uncertain.[33] Situations like this arise in a hierarchical setting when not all objects in a population are observed (see our application to selection effects in chapter 15) or when single observations are influenced by unknown numbers of objects (e.g. when de-blending sources).

Missing: parametrisation. The final source of joint uncalculability is an *implicit prior*,[34] usually of the kind attempting to incorporate *common sense*[35] like "the source in a gravitational lens image must look like a galaxy" [375, 2, 266, 3] or previous empirical findings (e.g. the spectro-temporal flux distribution of SNæ Ia). The challenge in these cases is that (a part of) the model is constructed in reference to external *data* (e.g. images of "galaxies" or spectra of "SNæ Ia", which are used to define these objects) instead of a probability density in a particular parametrisation. *implicit prior*

These hurdles to traditional Bayesian inference, along with the computational issues of likelihood-based techniques that have often prevailed over the desire for scientific rigour, are our motivation for developing an alternative analysis framework based not on the joint probability (density) but merely on samples from it, which we present in the next chapter.

---

30 even if we set aside intractable integrals—that bane of all Bayesians—which are of practical concern if the distributions of nuisance parameters live up to the name or if there is such a multitude of them (e.g. object-specific parameters in a hierarchical model) that the dimensionality is beyond the reach of numerical methods (including high-dimensional sampling)

31 or, if not entirely undoable, then at least, *ludicrous*

32 in reference to terminology of disruption like "un-conference"

33 A likelihood-based method for sampling such a model (or, in fact, any collection of models, each with its own fixed dimensionality), known as *reversible jump* (or *trans-dimensional*) MCMC, was developed by Green [186, see also 187] and involves the construction of proposal distributions across dimensions.

34 One may call this an un-prior™ or un-model® since its foundations lie entirely—or as much as possible—in data and empiricism, rather than in abstract model building.

35 ♫ *My model has no prior, it's got its strong beliefs* ♫

Chapter 2

Neural simulation-based inference

Simulation-based inference (SBI) is a suite of methods in which the model is explicated not by the joint probability distribution but through a *simulator*[36]: a forward process from causes to effects. This has immediate and far-reaching consequences that align inference with the origins of Bayesian reasoning exposed above. Firstly, it restores the symmetry between quantities formerly known as "parameters" and "data": it is up to the scientist to identify which variables in the simulator are of their interest and which have been observed in the real world, recording their (fully stochastic) values as pairs $(\boldsymbol{\theta}, \mathbf{d})$ *defined* as samples from $\mathrm{p}(\boldsymbol{\theta}, \mathbf{d})$. This obviates the need for explicit inference—or even definition—of nuisances: all SBI is *marginal* and performed naturally with samples. Secondly, it introduces the notion of *true values*: accessible for both $\boldsymbol{\theta}$ and $\mathbf{d}$ in computer programs, but only for the latter in the implementation regarded as the real world. Correspondingly, SBI strips "real" data from its special status and regards it as simply another stochastic realisation from the simulator.

simulator

marginal SBI

true values

Moreover, SBI locks model builders into conceptualising the studied phenomenon in the forward fashion, doing away with large swathes of arbitrariness, since each aspect of the model must fit within the simulator's logical flow. This finally permits them to incorporate a variety of processes from the physical world and elements of Bayesian reasoning that elude a straightforward probabilistic description: e.g. data-driven priors and nuisance spaces with arbitrary dimensions and distributions.

[36] in fact, the etymology of "*sim*ulate" itself alludes to together/joint-ness

© The Author(s), under exclusive license to Springer Nature Switzerland AG 2026
K. Karchev, *Supernova Cosmology for the 21st Century*,
https://doi.org/10.1007/978-3-032-15072-1_2

Pre-neural SBI: approximate Bayesian computation

ABC

The idea of using simulations for inference[37] found an early manifestation in *approximate Bayesian computation (ABC)* [425, see also 482, 37], a likelihood-*skimmed*™ technique that implements the conditioning on observed data via sampling. More formally, ABC proposes parameters-of-interest–data pairs from $\mathrm{p}(\boldsymbol{\theta}, \mathbf{d})$ and accepts them with probability that is 1 when $\mathbf{d} \equiv \mathbf{d}_\mathrm{o}$ and 0 when the "distance" between them is much greater than ε. As $\varepsilon \to 0$, and so the acceptance criterion tends to a delta function ($K_\varepsilon(\mathbf{d}, \mathbf{d}_\mathrm{o}) \to \delta(\mathbf{d}, \mathbf{d}_\mathrm{o})$), the marginal distribution of $\boldsymbol{\theta}$, i.e. the collection of $\boldsymbol{\theta}$ that *happen* to produce simulated data equal—or sufficiently similar—to $\mathbf{d}_\mathrm{o}$, tends to the conditional at $\mathbf{d}_\mathrm{o}$, namely, the posterior $\mathrm{p}(\boldsymbol{\theta} \mid \mathbf{d}_\mathrm{o})$. For a finite ε, the result is an average over posteriors from data "close" to $\mathbf{d}_\mathrm{o}$.

Being fully simulation-based, ABC is insensitive to the number of nuisances, which are customarily ignored in simulations and thus marginalised. Still, the original parameter-proposal distribution $\mathrm{p}(\boldsymbol{\theta})$ may be excessively wasteful, especially for highly constraining[38] data that results to a narrow posterior. In such instances, ABC can be implemented through traditional random-walk MCMC methods [350], with the likelihood replaced[39] by $K_\varepsilon(\mathbf{d}, \mathbf{d}_\mathrm{o})$, whose average approximates it: $\mathbb{E}_{\mathrm{p}(\mathbf{d} \mid \boldsymbol{\theta})}[K_\varepsilon(\mathbf{d}, \mathbf{d}_\mathrm{o})] \approx \mathrm{p}(\mathbf{d}_\mathrm{o} \mid \boldsymbol{\theta})$. ABC also greatly benefits from iteratively refining the proposal (starting with $\mathrm{p}(\boldsymbol{\theta})$ and a large ε) so that simulated data is increasingly more similar to $\mathbf{d}_\mathrm{o}$ while progressively reducing ε [481, 9].

distance measure & bandwidth

Crucial to the success of ABC is the selection of suitable *distance measure* in data space, "*bandwidth*" ε, and criterion K that balance simulator efficiency (a major concern for any rejection sampling-like algorithm) and fidelity of the approximated posterior. For some forms of data, the choice may be obvious: e.g. for integers (counts), one may hope to condition *exactly* by retaining parameters that lead to the same number of simulated events as observed. The next simplest candidate is an L2 distance between vectors with an optional arbitrarily specified covariance/metric: $\Delta^2 \equiv \mathrm{Dist}(\mathbf{d}, \mathbf{d}_\mathrm{o}) \to (\mathbf{d} - \mathbf{d}_\mathrm{o})^\top \boldsymbol{\Sigma}^{-1} (\mathbf{d} - \mathbf{d}_\mathrm{o})$, and a sharp cutoff at ε: $K = \mathbb{1}(\Delta \le \varepsilon)$, or a Gaussian kernel $K = \exp(-\Delta^2/\varepsilon^2)$. Many other distance/similarity measures exist and can be used in ABC [50].

regularisation

curse of dimensionality

However, some types of observations—for example, any collection of *a priori* unknown size or order like light curves or population sub-samples—cannot be trivially compared to mock realisations and need to be *regularised* first, adding another layer of arbitrariness. Moreover, since ABC is founded on the idea of data-space locality, a direct application to any high-dimensional data is *cursed* to fail, just like local sampling proposals in high-

[37] as opposed to elevating a given physical model (usually deterministic) from an elementary form, e.g. the law of gravity, to some statement about observables, e.g. two-point correlations of large-scale structure

[38] In the terms of ABC this is reflected in a stringent requirement for a small ε to ensure an accurate approximation of the posterior.

[39] hence the *skimmed* qualifier

dimensional parameter spaces. In this case, the data must first be *summarised* into a low-dimensional (and often interpretable) space before proceeding with ABC. *dimensionality reduction*

However optimised, ABC retains one major flaw: it does not extract information from the rejected examples and is not amortised; i.e. the whole procedure needs to be rerun for different $\mathbf{d}_o$.[40] It also requires a multitude of tunable settings: summary and/or similarity functions, acceptance criterion, and bandwidth, and at best produces an approximation to the posterior. This explains why it is—at best—used as an introduction to more successful SBI methods.

2.1 Flavours of neural SBI

The usability and performance of SBI gained a substantial boost—leading to a rise in popularity[41]—by the introduction[42] of (artificial) *neural networks (NNs)*: universal approximators [see e.g. 24] that can be *trained* via optimisation of a suitably defined *objective functional*. For applications within SBI, this most commonly takes the form of a *Bayesian risk*, i.e. an expectation over $\mathrm{p}(\boldsymbol{\theta}, \mathbf{d})$, which can be approximated by averaging over a simulated *training set* $\{(\boldsymbol{\theta}, \mathbf{d})_i\}_{i=1}^{N_\text{train}}$. In rarer cases, training sets $\{\mathbf{d}_i\}_{i=1}^{N_\text{train}}$ sampled from $\mathrm{p}(\mathbf{d} \mid \boldsymbol{\theta}_\text{fid})$ at fixed *fiducial parameters* may be requested.[43]

NN
objective fn: win™ $\equiv -$loss
Bayesian risk
training set
fiducial parameters

For all its generality and convenience—and philosophical hauteur—, the use of an unconditioned simulator that samples from $\mathrm{p}(\boldsymbol{\theta}, \mathbf{d})$—in contrast to traditional sampling from $\mathrm{p}(\boldsymbol{\theta} \mid \mathbf{d}_o)$—has one seeming drawback: any inference method needs to confront counterfactual data during training. The framework has two responses: on one hand, it incorporates *sequential methods* of restricting/conditioning the simulator so that it generates data that resembles $\mathbf{d}_o$ (subsection 2.2.2); with the other hand it embraces this "weakness" and turns it into a valuable and unique feature: *amortisation*, which, among other uses, allows for verification of the inference procedure from both a Bayesian and a frequentist perspective, as described in section 2.3.

sequential methods
amortisation

---

[40] Certainly, simulations can be saved and compared to the many $\mathbf{d}_o$, but that can only speed up the first round of sequential updates or would require excessively many simulations to fill the data space to within ε.

[41] Consult the automatically collected references at `https://simulation-based-inference.org/` and the curated list of "awesome" applications at `https://github.com/smsharma/awesome-neural-sbi`.

[42] A history and overview of neural networks and their training in practice is beyond the scope of this thesis since they are nowadays taught in mostly any higher-education course and recognised with Nobel Prizes in mostly any discipline.

[43] Such can be obtained via reverse-ABC (CBA™), i.e. retaining only those samples from $\mathrm{p}(\boldsymbol{\theta}, \mathbf{d})$ for which $\boldsymbol{\theta} \approx \boldsymbol{\theta}_\text{fid}$, or by modifying the simulator to always sample $\boldsymbol{\theta}_\text{fid}$, which is usually straightforward (but see section 4.1) since it usually factorises as $\mathrm{p}(\mathbf{d} \mid \boldsymbol{\theta})\, \mathrm{p}(\boldsymbol{\theta})$. However, this somewhat contravenes the essential philosophy of SBI and borders on frequentism!

Table 2.1: Summary and comparison of the principal neural SBI methods

	summarisation		density estimation		density-ratio estimation	
	IMNN	VMIM	NPE	NLE	NRE	
estimator	$\mathbf{s}(\mathbf{d})$		$\mathrm{q}(\boldsymbol{\theta} \mid \mathbf{d})$	$\mathrm{q}(\mathbf{d} \mid \boldsymbol{\theta})$	$\hat{\mathrm{r}}(\boldsymbol{\theta}_I, \boldsymbol{\theta}_{II}, \mathbf{d})$	$\hat{\mathrm{r}}(\boldsymbol{\theta}, \mathbf{d})$
target	—		$\mathrm{p}(\boldsymbol{\theta} \mid \mathbf{d})$	$\mathrm{p}(\mathbf{d} \mid \boldsymbol{\theta})$	$\frac{\mathrm{p}(\mathbf{d} \mid \boldsymbol{\theta}_1)}{\mathrm{p}(\mathbf{d} \mid \boldsymbol{\theta}_2)}$	$\frac{\mathrm{p}(\mathbf{d} \mid \boldsymbol{\theta})}{\mathrm{p}(\mathbf{d})} = \frac{\mathrm{p}(\boldsymbol{\theta} \mid \mathbf{d})}{\mathrm{p}(\boldsymbol{\theta})}$
objective	$\mathbf{F}_{\mathbf{s}(\mathbf{d})}(\boldsymbol{\theta}_{\mathrm{fid}})$	$\mathcal{I}(\boldsymbol{\theta}, \mathbf{s}(\mathbf{d}))$	$\mathbb{E}_{\mathrm{p}(\boldsymbol{\theta}, \mathbf{d})}[\mathrm{q}(\ldots)]$		binary classification	
inference	downstream		direct	traditional* & frequentist		
summaries	explicit: local	explicit: global	optimise $\mathcal{I}(\boldsymbol{\theta}, \mathbf{s}(\mathbf{d}))$	explicit: likelihood (ratio)		optimise $\mathrm{JS}[\mathrm{p}(\boldsymbol{\theta} \mid \mathbf{s}(\mathbf{d})), \mathrm{p}(\boldsymbol{\theta})]$
training prior	$\delta(\boldsymbol{\theta}_{\mathrm{fid}})$	any	$\mathrm{p}(\boldsymbol{\theta})$	any		

*MCMC with the approximate likelihood or re-weighting prior samples

NNs thus conclusively emancipate SBI from the likelihood — and any stand-ins for it, in contrast to ABC — by learning functions of arbitrary $\mathbf{d}$ instead of particular values at $\mathbf{d}_{\mathrm{o}}$: inference is said to be *amortised*. Granted, final scientific conclusions are drawn by evaluating the trained network on $\mathbf{d}_{\mathrm{o}}$, but this is extremely rapid, while training and inference verification (see section 2.3) can be performed even *before* data is available and in parallel, exploiting modern hardware like *graphics processing units (GPUs)*.

amortisation

GPU

Based on what (and how) NNs are taught, their use in SBI falls in three categories: summarisation, density estimation, and density-ratio estimation. The principal methods discussed below are summarised and compared in table 2.1. Cranmer et al. [108] give an in-depth overview and further discussion, while Lueckmann et al. [333] perform a systematic comparison and benchmark.

2.1.1 Neural data summarisation

summary statistics

If a high-dimensional data set is informative of a limited number of parameters, it can be embedded into a lower-dimensional manifold of so-called *summary statistics* ($\mathbf{s}(\mathbf{d})$) that preserve some — hopefully most, or even all — of the relevant information. Such a representation is useful for two main reasons: first, it speeds up repeated evaluation at fixed data $\mathbf{d}_{\mathrm{o}}$, e.g. for (approximate) likelihood-based inference; and second, the distribution of $\mathbf{s}$ is simpler to learn and may be even (assumed to be) known or very well approximated analytically, e.g. by a Gaussian due to the central limit theorem. For some models, convenient summaries can be derived from physical intuition, but, being arbitrarily handcrafted, these may not be optimal (loosely speaking; optimality will be shortly defined in a number of different ways).

As a particular example, consider analysis of large-scale structure (LSS). The "established" summary statistic, the two-point correlation of galaxy counts as a function of spatial separation [see e.g. 401, chapter 19] has been shown to be extremely lossy with respect to the full field-level data [423] and even to other analytic summaries [500]. Hence, large efforts [e.g. 304, 398] are underway to evaluate and characterise the performance of various machine learning (ML)-based compression procedures whose goal is maximising the scientific utility of modern data sets.

The simplest principled summaries take the form of *parameter estimators* $\widehat{\boldsymbol{\theta}}(\mathbf{d})$ and so are exactly as numerous as the parameters of interest. Estimators are derived via minimisation of a given measure of deviation from the true parameters averaged over a given distribution. For example, the *posterior mean*, $\bar{\boldsymbol{\theta}}(\mathbf{d}) \equiv \mathbb{E}_{\mathrm{p}(\boldsymbol{\theta}\,|\,\mathbf{d})}[\boldsymbol{\theta}]$, minimises the Bayesian risk of squared error: $\mathbb{E}_{\mathrm{p}(\boldsymbol{\theta},\mathbf{d})}\left[\|\widehat{\boldsymbol{\theta}} - \boldsymbol{\theta}\|^2\right]$ and so can be directly learnt by a NN from simulations. Similar objectives can be derived also for arbitrary quantiles (e.g. the median) of the posterior and for its higher-order moments [see e.g. 247, 249].

estimator

posterior mean

Also popular, and with a long tradition in frequentist inference, is the maximum likelihood estimator (MLE), which, as is well-known, saturates—but only asymptotically—the Cramér[106](–Aitken-&-Silverstone[5]–Fréchet[159]–Darmois[116])–Radhakrishna-Rao[436] inequality, i.e. has the lowest variance achievable by any unbiased estimator. This precision bound, known as the *Fisher matrix* [151]:

Fisher matrix / information

$$\begin{aligned}\mathbf{F}_{\mathbf{d}}(\boldsymbol{\theta}_{\mathrm{fid}}) &\equiv \mathbb{E}_{\mathrm{p}(\mathbf{d}\,|\,\boldsymbol{\theta}_{\mathrm{fid}})}\left[\left(\nabla_{\boldsymbol{\theta}} \ln \mathrm{p}(\mathbf{d}\,|\,\boldsymbol{\theta})\Big|_{\boldsymbol{\theta}_{\mathrm{fid}}}\right)\left(\nabla_{\boldsymbol{\theta}} \ln \mathrm{p}(\mathbf{d}\,|\,\boldsymbol{\theta})\Big|_{\boldsymbol{\theta}_{\mathrm{fid}}}\right)^{\top}\right] \\ &= -\mathbb{E}_{\mathrm{p}(\mathbf{d}\,|\,\boldsymbol{\theta}_{\mathrm{fid}})}\left[\nabla\nabla^{\top} \ln \mathrm{p}(\mathbf{d}\,|\,\boldsymbol{\theta})\Big|_{\boldsymbol{\theta}_{\mathrm{fid}}}\right], \end{aligned} \tag{2.1}$$

has also widely be used as a measure of information content of summaries. Alsing & Wandelt [7] observed straight from eq. (2.1) that a Fisher-optimal summary is the *score* vector $\nabla_{\boldsymbol{\theta}} \ln \mathrm{p}(\mathbf{d}\,|\,\boldsymbol{\theta})$ itself.[44] This view unifies previous methods specialised to a Gaussian sampling distribution: linear compression with MOPED℗[45] [504, 207] when the sampling covariance is independent of the parameters of interest; and quadratic compression [503],

score

[44] This is a consequence of the *likelihood principle*, which follows from the Fisher[151]–Neyman[382]–Halmos-&-Savage[195] factorisation theorem (which Jaynes [245, section 8.5] calls a "trivial consequence of the product rule of probability theory, [...] no more to be questioned than the multiplication table") and states that the likelihood is a *sufficient summary*.

sufficient summary

[45] Massively optimised Parameter Estimation and Data compression [442], patent US6433710B1 expired

widely used for summarising the cosmic microwave background (CMB), for the converse case that only the covariance depends on parameters.

IMNN

Of course, the score is not available for models with an implicit likelihood, and the MLE cannot be learnt from samples[46] as easily as a mean or median. Nonetheless, the Fisher matrix $\mathbf{F}_{\mathbf{s}(\mathbf{d})}(\boldsymbol{\theta}_{\mathrm{fid}})$ given a tunable nonlinear summary represents a convenient optimisation target to train an *information-maximising neural network (IMNN)* [91]. While this idea might be Platonically appealing, its realisation is more than cumbersome: it requires, at every training step, estimation of the mean and covariance of $\mathrm{p}(\mathbf{s}(\mathbf{d}) \mid \boldsymbol{\theta}_{\mathrm{fid}})$ and a differentiable simulator — or additional simulations at $\boldsymbol{\theta}_{\mathrm{fid}} \pm \delta\boldsymbol{\theta}$ — to estimate the mean's gradient; in addition, the summaries might need to be re-optimised if the initial fiducial parameters are poorly chosen.

mutual information

An alternative criterion is the *mutual information* between parameters of interest $\boldsymbol{\theta}$ and general — non-linear, arbitrary-dimensional[47] — summaries $\mathbf{s}(\mathbf{d})$, which equals the average improvement (KL divergence) of the posterior with respect to the prior:

$$\begin{aligned} \mathcal{I}(\boldsymbol{\theta}, \mathrm{s}) &\equiv \mathrm{KL}[\mathrm{p}(\boldsymbol{\theta}, \mathrm{s}) \parallel \mathrm{p}(\boldsymbol{\theta})\, \mathrm{p}(\mathrm{s})] = \mathbb{E}_{\mathrm{p}(\mathbf{d})}[\mathrm{KL}[\mathrm{p}(\boldsymbol{\theta} \mid \mathrm{s}) \parallel \mathrm{p}(\boldsymbol{\theta})]] \\ &= \mathbb{E}_{\mathrm{p}(\boldsymbol{\theta}, \mathbf{d})}[\ln \mathrm{p}(\boldsymbol{\theta} \mid \mathrm{s})] - \mathcal{H}(\boldsymbol{\theta}), \end{aligned} \tag{2.2}$$

entropy

where $\mathcal{H}(\boldsymbol{\theta})$ is the *entropy* (self-information) in the prior, independent of data. The last form in eq. (2.2) inspires a variational lower bound [34, cf. eq. (1.11)][48]:

$$\mathbb{E}_{\mathrm{p}(\boldsymbol{\theta}, \mathbf{d})}[\ln \mathrm{p}(\boldsymbol{\theta} \mid \mathrm{s})] = \mathbb{E}_{\mathrm{p}(\boldsymbol{\theta}, \mathbf{d})}[\ln \mathrm{q}(\boldsymbol{\theta} \mid \mathrm{s})] + \mathbb{E}_{\mathrm{p}(\mathbf{d})} \underbrace{\mathbb{E}_{\mathrm{p}(\boldsymbol{\theta} \mid \mathbf{d})}[\ln \mathrm{p}(\boldsymbol{\theta} \mid \mathrm{s}) - \ln \mathrm{q}(\boldsymbol{\theta} \mid \mathrm{s})]}_{= \mathrm{KL}[\mathrm{p}(\boldsymbol{\theta} \mid \mathrm{s}) \parallel \mathrm{q}(\boldsymbol{\theta} \mid \mathrm{s})] \geq 0}, \tag{2.3}$$

VMIM

which the method of *variational mutual-information maximisation (VMIM)* [249] optimises with respect to NN-parametrised $\mathbf{s}(\mathbf{d})$. Summaries derived with VMIM can be of any dimension and are relevant across the whole parameter space. This approach itself is also more in line with the spirit of SBI than Fisher maximisation — it only requires samples from $\mathrm{p}(\boldsymbol{\theta}, \mathbf{d})$ — but begs the question: how to produce a good enough approximate posterior $\mathrm{q}(\boldsymbol{\theta} \mid \mathrm{s})$, given a set of possibly lossy summaries, that saturates the variational bound and

---

[46] Makinen et al. [339] described a two-NN setup (Fishnets) that targets the asymptotic MLE by estimating simultaneously a vector and a positive-definite matrix (nominally, the score and Fisher information) and forming an approximation to the single-step Fisher-scoring quasi-MLE: $\boldsymbol{\theta}_{\mathrm{fid}} + \mathbf{F}^{-1}(\boldsymbol{\theta}_{\mathrm{fid}})\, \boldsymbol{\nabla}_{\boldsymbol{\theta}} \ln \mathrm{p}(\mathbf{d} \mid \boldsymbol{\theta})\big|_{\boldsymbol{\theta}_{\mathrm{fid}}}$.

[47] The linear summaries maximising mutual information (equivalent to MOPED) can be derived via canonical correlation analysis (CCA) [229] and are again as numerous as the parameters of interest.

[48] See also Poole et al. [416] for extended discussion of variational mutual-information estimation in the context of representation learning.

can be evaluated for many different pairs $\boldsymbol{\theta}, \mathbf{s}(\mathbf{d})$ produced by the summariser-augmented simulator. This naturally brings us to the next subsection and the quintessence of SBI.[49]

2.1.2 Neural density estimation

SBI is founded in the denial of explicit probabilities; however, scientific conclusions are necessarily concerned with the posterior of different parameter values. Hence, the path of least resistance/effort from simulations to inference is to learn a numeric representation of the joint distribution $\mathrm{p}(\boldsymbol{\theta}, \mathbf{d})$ from the simulated samples and proceed with traditional inference techniques supplied with an explicit low-dimensional marginalised model.

Estimating the density $\mathrm{p}(\mathbf{x})$—which in *(neural) joint estimation (NJE)* represents a concatenation of the parameters of interest and data: $\mathbf{x} \equiv [\boldsymbol{\theta}, \mathbf{d}]$—from samples $\{\mathbf{x}_i\}_{i=1}^{N_{\text{train}}}$—i.e. simulations—is a well-studied statistical task[50] that can be performed with traditional *kernel methods (KDE)* [49, section 2.5.1] (in low dimensions) or with neural techniques like mixture density networks (MDNs) [48] and normalising flows (NFs), already discussed in the context of VI. Density estimation can in fact be regarded as the *opposite* process (samples from p $\rightarrow$ density estimate q) to VI (known density p $\rightarrow$ samples from q) and is consequently achieved through the same objective function, the KL divergence, but with the roles of the two distributions flipped: *NJE* *KDE* *NDE*

$$\mathrm{KL}[\mathrm{p}(\mathbf{x}) \parallel \mathrm{q}(\mathbf{x})] \equiv -\mathbb{E}_{\mathrm{p}(\mathbf{x})}[\ln \mathrm{q}(\mathbf{x})] + \underbrace{\mathbb{E}_{\mathrm{p}(\mathbf{x})}[\ln \mathrm{p}(\mathbf{x})]}_{\text{constant}}, \tag{2.4}$$

which suggests a training procedure that simply maximises the "predicted" probability[51] $\sum_i \ln \mathrm{q}(\mathbf{x}_i)$ across the training set. Of course, the non-negativity bound on the KL divergence relies on the assumption that both p and q are properly normalised probability density functions (PDFs), so optimisation must be constrained by $\int \mathrm{q}(\mathbf{x})\,\mathrm{d}\mathbf{x} = 1$ (otherwise q can be made arbitrarily large), which the particular architecture of NFs as glorified tractable re-parametrisers achieves automatically. Importantly, NFs are also reversible by design, i.e. their Jacobian, and hence $\mathrm{q}(\mathbf{x})$, can be easily evaluated. *NLL*

[49] Still, before proceeding, we must mention a final class of not-supervised summaries: autoencodings [e.g. 92] and contrastive embeddings [e.g. 305]. These general (non-inference-specialised) low-dimensional representations are derived so as to preserve the "fidelity" and "diversity" (broadly and diversely defined) of the data without reference to a specific (e.g. inference) task and just "happen to be" useful as SBI summaries.

[50] In fact, Bishop [49, chapter 2] unorthodoxically (cf. Jaynes [245]) introduces the general notion of a distribution in relation to *parametric* density estimation whose solution is the MLE (cf. eq. (2.4)) of the distribution's parameters.

[51] In the pessimistic and sloppy ML circles, this is equivalently perceived as minimising the negative log-likelihood (NLL) loss. Information theorists, on the other hand, recognise $-\mathbb{E}_{\mathrm{p}(\mathbf{x})}[\ln \mathrm{q}(\mathbf{x})]$ as the *cross-entropy*. *cross-entropy*

diffusion model (DDPM)
forward/reverse processes

Recently, an alternative[52] density-estimation technique, *score-based / denoising diffusion modelling* [486, 487, 224, 488], has taken over ML and canonised the above-mentioned duality between the inference (*forward*) and sampling (*reverse*) processes (cf. also eq. (1.1)). Its centrepiece is the introduction of an auxiliary variable $\tilde{\mathbf{x}}$ with a tractable conditional distribution $\mathrm{p}_\sigma(\tilde{\mathbf{x}} \mid \mathbf{x})$ that reduces to a delta at $\mathbf{x}$ in some limit ($\sigma \to 0 \implies \mathrm{p}_\sigma(\tilde{\mathbf{x}}) \to \mathrm{p}(\mathbf{x})$) and learning its marginal "score"[53] as a function of σ with a NN $\hat{\mathbf{t}}(\tilde{\mathbf{x}}, \sigma)$ trained to optimise:

$$\mathbb{E}_{\mathrm{p}_\sigma(\mathbf{x},\tilde{\mathbf{x}})}\left[\left\|\hat{\mathbf{t}}(\tilde{\mathbf{x}}, \sigma) - \boldsymbol{\nabla}_{\tilde{\mathbf{x}}} \ln \mathrm{p}_\sigma(\tilde{\mathbf{x}})\right\|^2\right] = \mathbb{E}_{\mathrm{p}_\sigma(\mathbf{x},\tilde{\mathbf{x}})}\left[\left\|\hat{\mathbf{t}}(\tilde{\mathbf{x}}, \sigma) - \boldsymbol{\nabla}_{\tilde{\mathbf{x}}} \ln \mathrm{p}_\sigma(\tilde{\mathbf{x}} \mid \mathbf{x})\right\|^2\right] + \text{const.} \tag{2.5}$$

The left-hand side makes it clear that $\hat{\mathbf{t}}$ learns $\boldsymbol{\nabla}_{\tilde{\mathbf{x}}} \ln \mathrm{p}_\sigma(\tilde{\mathbf{x}})$, and so $\hat{\mathbf{t}}(\mathbf{x}, \sigma = 0)$ approximates $\boldsymbol{\nabla}_{\mathbf{x}} \ln \mathrm{p}(\mathbf{x})$, which is (the "score" of) the sought distribution. The—equivalent as shown by Vincent [524]—right-hand side suggests a practical way for the calculation of the objective that only depends on the tractable *conditional* "score", for which a common choice is $\boldsymbol{\nabla}_{\tilde{\mathbf{x}}} \ln \mathrm{p}_\sigma(\tilde{\mathbf{x}} \mid \mathbf{x}) \equiv (\tilde{\mathbf{x}} - \mathbf{x})/\sigma$, i.e. "degradation" with Gaussian noise. In practice, the "de-noiser" (score estimator) of diffusion models is trained across a wide range of σ up to a level at which $\mathrm{p}_\sigma(\tilde{\mathbf{x}}) \approx \mathcal{N}(0, \sigma^2)$. Sampling and "likelihood" (probability density) evaluation can then be performed by following the *probability flow* (with respect to σ) [488] associated to $\hat{\mathbf{t}}$, respectively in the reverse and forward directions by solving a differential equation. Since this is far more computationally intensive than feeding backward through a finite-(usually few-)layer normalising flow, corresponding shortcut discretisation schemes [e.g. 489] for DDPMs have also been devised.

probability flow

Once $\mathrm{q}([\boldsymbol{\theta}, \mathbf{d}])$ is trained to estimate $\mathrm{p}(\boldsymbol{\theta}, \mathbf{d})$, inference can be performed simply by evaluating it at $\mathbf{d}_\mathrm{o}$: $\mathrm{q}([\boldsymbol{\theta}, \mathbf{d}_\mathrm{o}]) \approx \mathrm{p}(\boldsymbol{\theta} \mid \mathbf{d}_\mathrm{o})\,\mathrm{p}(\mathbf{d}_\mathrm{o}) \propto \mathrm{p}(\boldsymbol{\theta} \mid \mathbf{d}_\mathrm{o})$.[54] While this achieves amortisation and allows approximate posteriors—un-normalised, nonetheless—to be derived relatively quickly for many different (real or simulated) data sets, NJE is rarely employed because of its wastefulness in estimating the possibly very high-dimensional density/distribution of $\mathbf{d}$, which is anyway not queried in the end. In practice, a conditional probability—the posterior or sampling distribution—is usually learnt instead.

Neural posterior estimation

NPE
conditioning
context
parameter group

The most popular approach to SBI is to directly target the object of interest, $\mathrm{p}(\boldsymbol{\theta} \mid \mathbf{d})$, through *neural posterior estimation (NPE)* [393], a conditioned counterpart of NDE in which $\mathbf{d}$ is provided to the NN as a *context variable* but only the distribution of the few parameters of interest $\boldsymbol{\theta}$ is learnt—in fact, multiple density estimators can be trained independently and simultaneously for a number of (not necessarily disjoint) *parameter groups*

[52] but only seemingly: de-noising is the continuous limit of a normalising flow

[53] Concordantly with calling the probability density $\mathrm{p}(\mathbf{x})$ a "likelihood", in the literature on diffusion models,

$\boldsymbol{\theta}_g$ (usually, each is one- or two-dimensional) if their marginals are of scientific interest but not the correlations between them.

NPE requires minimal modification from NJE due to the general fact that an expectation $\mathbb{E}_{\mathrm{p}(a,b)}$ over two random variables is optimised by any function $\hat{g}(a, b)$ that optimises the conditional expectation $\mathbb{E}_{\mathrm{p}(a\,|\,b)}$ at almost all b [310, theorem 4.1.1], leading to the objectives

$$\text{"likelihood" maximisation:} \quad \operatorname{argmax}_{\mathrm{q}} \mathbb{E}_{\mathrm{p}(\boldsymbol{\theta},\mathbf{d})}[\ln \mathrm{q}(\boldsymbol{\theta}\,|\,\mathbf{d})] \quad \text{given} \quad \int \mathrm{q}(\boldsymbol{\theta}\,|\,\mathbf{d})\,\mathrm{d}\boldsymbol{\theta} = 1; \tag{2.6}$$

$$\text{score}^{55} \text{ matching:} \quad \operatorname{argmin}_{\hat{\mathbf{t}}} \mathbb{E}_{\mathrm{p}_\sigma(\boldsymbol{\theta},\tilde{\boldsymbol{\theta}},\mathbf{d})}\left[\left\|\hat{\mathbf{t}}(\tilde{\boldsymbol{\theta}}, \mathbf{d}, \sigma) - \boldsymbol{\nabla}_{\tilde{\boldsymbol{\theta}}} \ln \mathrm{p}_\sigma(\tilde{\boldsymbol{\theta}}\,|\,\boldsymbol{\theta})\right\|^2\right]. \tag{2.7}$$

Of the two, eq. (2.6) with explicitly normalised NFs is usually preferred due to its simplicity and sufficient performance in this low-dimensional setup. Moreover, it allows for an arbitrary conditioning context, e.g. a learnt summary representation $\mathbf{d} \to \mathbf{s}(\mathbf{d})$, which can be shared among the inference tasks (groups $\boldsymbol{\theta}_g$) and will be optimised, in terms of mutual information, when training with eq. (2.6): cf. the VMIM objective eq. (2.3). A direct estimate of the posterior density (rather than its gradient) through a NF also enables evaluation of the credibility of *highest posterior density (HPD) regions*[56] in $\boldsymbol{\theta}$-space, which is particularly useful for inference verification (section 2.3). On the flip side, NPE enshrines the prior into the trained density estimator, which presents a challenge to sequential learning (see subsection 2.2.2) and to combining results across data sets [see e.g. 528]. *HPD credible region*

Neural likelihood estimation

Through a similar application of conditional NDE, one can restore the "missing" sampling distribution in an approach correspondingly called *neural sampling-distribution ("likelihood") estimation* [395, 332]: *NLE*

$$\text{likelihood maximisation:} \quad \operatorname{argmax}_{\mathrm{q}} \mathbb{E}_{\mathrm{p}(\boldsymbol{\theta},\mathbf{d})}[\ln \mathrm{q}(\mathbf{d}\,|\,\boldsymbol{\theta})] \quad \text{given} \quad \int \mathrm{q}(\mathbf{d}\,|\,\boldsymbol{\theta})\,\mathrm{d}\mathbf{d} = 1; \tag{2.8}$$

$$\text{"score"}^{57} \text{ matching:} \quad \operatorname{argmin}_{\hat{\mathbf{t}}} \mathbb{E}_{\mathrm{p}_\sigma(\boldsymbol{\theta},\mathbf{d},\tilde{\mathbf{d}})}\left[\left\|\hat{\mathbf{t}}(\boldsymbol{\theta}, \tilde{\mathbf{d}}, \sigma) - \boldsymbol{\nabla}_{\tilde{\mathbf{d}}} \ln \mathrm{p}_\sigma(\tilde{\mathbf{d}}\,|\,\mathbf{d})\right\|^2\right]. \tag{2.9}$$

"score" is incorrectly used to refer to the gradient of its logarithm with respect to the value of the random variable: "score" $\mathbf{t}(\mathbf{x}) \equiv \boldsymbol{\nabla}_{\mathbf{x}} \ln \mathrm{p}(\mathbf{x})$.

[54] Similarly, a trained $\hat{\mathbf{t}}([\boldsymbol{\theta}, \mathbf{d}], 0) \equiv \left[\hat{\mathbf{t}}_{\boldsymbol{\theta}}([\boldsymbol{\theta}, \mathbf{d}]), \hat{\mathbf{t}}_{\mathbf{d}}([\boldsymbol{\theta}, \mathbf{d}])\right]$ evaluated at $\mathbf{d}_\mathrm{o}$ approximates the parameter-gradient of the joint: $\hat{\mathbf{t}}_{\boldsymbol{\theta}}([\boldsymbol{\theta}, \mathbf{d}]) \approx \boldsymbol{\nabla}_{\boldsymbol{\theta}} \ln \mathrm{p}(\boldsymbol{\theta}, \mathbf{d})$.

[55] Sharrock et al. [480], Geffner et al. [168] call this *neural posterior score estimation (NPSE)*, but the score $\boldsymbol{\nabla}_{\boldsymbol{\theta}} \ln \mathrm{p}(\mathbf{d}\,|\,\boldsymbol{\theta})$ can only be recovered if the prior is tractable or estimated as well. *NPSE*

[56] Given a trained NF for $\mathbf{x}$, i.e. a transformation $\mathbf{x} = f(\mathbf{z})$ where the base random variable has a tractable distribution $\mathrm{q}(\mathbf{z})$, the mass enclosed by a given iso-$\mathrm{q}(\mathbf{x}_\mathrm{o})$ contour is equal to that enclosed by the corresponding iso-$\mathrm{q}\big(\mathbf{z}_\mathrm{o} = f^{-1}(\mathbf{x}_\mathrm{o})\big)$, which is easy to calculate by design.

Inference from concrete $\mathbf{d}_o$ is then delegated to traditional methods (e.g. MCMC), with the nuisance variables (i.e. those that are not provided in the conditioning context) already implicitly marginalised during training, which drastically improves performance. If, however, the prior density $p(\boldsymbol{\theta})$ is not explicitly calculable—and hence "likelihood"-based techniques remain inapplicable—, the posterior can still be represented through prior samples *re-weighted* by $w(\boldsymbol{\theta}) \propto p(\mathbf{d}_o \mid \boldsymbol{\theta}) \approx q(\mathbf{d}_o \mid \boldsymbol{\theta})$. Even though this is not as convenient and straightforward as direct sampling from a NPE, it is usually still fast (especially considering the reduced dimensionality $\boldsymbol{\psi} \to \boldsymbol{\theta}$) since NN forward evaluations are extremely quick and massively parallelisable (e.g. over the MCMC chains or prior samples).

weighted samples

As with NJE, however, the high-dimensionality of the learnt distribution is a major hurdle in training, so the vast majority of NLE applications are preceded by a summarisation step. In fact, NLE can be seen as synonymous with data compression due to the likelihood principle.[44] However, owing to the difficulty of density estimation, the summaries are usually explicitly and independently optimised beforehand, using the methods of subsection 2.1.1.

Nevertheless, NLE has one major advantage: it is completely decoupled from the $\boldsymbol{\theta}$ prior, so that the model $p(\boldsymbol{\theta}, \mathbf{d})$ over which expectations are taken in eqs. (2.8) and (2.9) (i.e. the training data) can be replaced with samples from any $p(\mathbf{d} \mid \boldsymbol{\theta})\, \tilde{p}(\boldsymbol{\theta})$, and the same sampling distribution will be learnt. This is in contrast to NPE and allows for absolute flexibility in sequential or active training (subsection 2.2.2), trivial combination of the (approximate) likelihoods from many independent observations, and non-Bayesian inference.

emulation

Lastly, trained NLEs find extensive use as *emulators* of laborious simulators (we will use one in chapter 16) that can reduce the computation time by many orders of magnitude, still providing faithful samples from $p(\mathbf{d} \mid \boldsymbol{\theta})$.[58] In that context, the NN is trained over a wide distribution of input parameters designed specifically to encompass various use cases (e.g. inference with differently constructed priors). Since this approach relies only on samples, it can be used to construct a non-parametric / entirely empirical implicit representation of a phenomenon for which a simulator—and thus, a model—is entirely missing but data is available.

[57] Note that the learnt gradient in this case is with respect to the data, so it is not an estimate of the score, unlike NPSE (sans the prior "score")...

[58] Often, an emulator is trained as a "shortcut" for a deterministic but expensive calculation, i.e. $p(\mathbf{d} \mid \boldsymbol{\theta}) \to \delta(\mathbf{d}(\boldsymbol{\theta}))$. In that case, it is enough to estimate only $\mathbb{E}_{p(\mathbf{d} \mid \boldsymbol{\theta})}[\mathbf{d}]$ as above, reducing emulation to the simplest regression task.

2.1.3 Neural ratio estimation

Density-ratio estimation is a paradigm for distribution learning entirely distinct from density or "score" estimation. Its keystone is the idea of representing the target $\mathrm{p}_1(\mathbf{x}) \to \mathrm{p}(\mathbf{x} \mid l = 1)$ and another tractable $\mathrm{p}_2(\mathbf{x}) \to \mathrm{p}(\mathbf{x} \mid l = 2)$ as *conditional* (sampling) distributions in a model augmented by a categorical *class label* $l \in \{1, 2\}$. Assigning a "uniform" prior $\mathrm{p}(l) = \text{const}$, the posterior probability for $l = 1$ is $\mathrm{p}(l = 1 \mid \mathbf{x}) = \mathrm{p}_1(\mathbf{x}) / (\mathrm{p}_1(\mathbf{x}) + \mathrm{p}_2(\mathbf{x}))$, so learning it allows the target $\mathrm{p}_1(\mathbf{x})$ to be expressed through the tractable $\mathrm{p}_2(\mathbf{x})$ and their *ratio* $\mathrm{r}(\mathbf{x}) \equiv \mathrm{p}_1(\mathbf{x}) / \mathrm{p}_2(\mathbf{x})$, which can be any non-negative function (i.e. the normalisation $\mathbb{E}_{\mathrm{p}_2(\mathbf{x})}[\mathrm{r}(\mathbf{x})] = 1$ is a *consequence* and not a requirement of its definition).

class label

density ratio

This so-called "likelihood-ratio trick" [108] has thus transformed distribution estimation, typically perceived as an *unsupervised problem*,[59] into a *supervised classification* task, which can be solved by any *probabilistic classifier* trained on labeled simulations $l, \mathbf{x} \sim \mathrm{p}(\mathbf{x} \mid l)\, \mathrm{p}(l)$: e.g. a random forest [e.g. 107] or logistic regression [e.g. 509], since the Bayes-optimal decision criterion for label assignment is indeed based on the posterior probability $\mathrm{p}(l \mid \mathbf{x})$ [e.g. 125, theorem 2.1].[60]

(un)supervised learning
classification
probabilistic classifier

Moreover, following the previous discussion on posterior estimation, an approximation for $\mathrm{p}(l \mid \mathbf{x})$—and through it, for $\mathrm{r}(\mathbf{x})$—can be obtained directly by optimising the usual[61] forward KL divergence (eqs. (2.4) and (2.6)):

$$-\mathrm{KL}[\mathrm{p}(l, \mathbf{x}) \parallel \mathrm{q}(l \mid \mathbf{x})\, \mathrm{p}(\mathbf{x})] + \text{const} = \mathbb{E}_{\mathrm{p}(l, \mathbf{x})}[\ln \mathrm{q}(l \mid \mathbf{x})], \tag{2.10}$$

which expands into the *binary cross-entropy (BCE)*:

BCE

$$-\mathrm{BCE} \equiv \mathbb{E}_{\mathrm{p}_1(\mathbf{x})}[\ln \mathrm{q}(l = 1 \mid \mathbf{x})] + \mathbb{E}_{\mathrm{p}_2(\mathbf{x})}[\ln \mathrm{q}(l = 2 \mid \mathbf{x})]. \tag{2.11}$$

Training with this objective is achieved by *simultaneously* maximising $\mathrm{q}(l = 1 \mid \mathbf{x})$ with samples from $\mathrm{p}_1(\mathbf{x})$ while also maximising $\mathrm{q}(l = 2 \mid \mathbf{x})$ with samples from $\mathrm{p}_2(\mathbf{x})$. Given the simplicity of this parameter space, explicit enumeration (that adopts a suggestive parametrisation) is a viable strategy for implementing the estimator in a way that ensures non-negativity and normalisation over l:

$$\mathrm{q}(l = 1 \mid \mathbf{x}) \equiv \sigma(\ln \hat{\mathrm{r}}(x)) = \frac{\hat{\mathrm{r}}(\mathbf{x})}{1 + \hat{\mathrm{r}}(\mathbf{x})} \quad \text{and} \quad \mathrm{q}(l = 2 \mid \mathbf{x}) \equiv \sigma(-\ln \hat{\mathrm{r}}(x)) = \frac{1}{1 + \hat{\mathrm{r}}(\mathbf{x})}, \tag{2.12}$$

where $\sigma(x) \equiv [1 + \exp(-x)]^{-1}$. Thus, *neural ratio estimation (NRE)* [107, 213] allows for

NRE

[59] In eq. (2.4), the target for $\mathrm{q}(\mathbf{x})$ is not explicit, and a normalisation constraint is necessary to form a stationary point.

[60] Devroye et al. [125] straight out *define* the posterior probabilities through this property.

[61] But see Rosasco et al. [457], who expound on the variety of classification objectives that all lead to Bayes-optimal classifiers, and Izbicki et al. [241, and references therein] for alternative—kernel or spectral—methods of approximately computing $\hat{\mathrm{r}}(\mathbf{x})$.

the simplest NN architecture among SBI methods, embedding $\mathbf{x}$ in a single real number: $\ln \hat{\mathrm{r}}(\mathbf{x}) \in (-\infty; \infty)$ (which, as discussed represents an optimal summary), without explicitly imposing any integral constraints. In these respects, it improves upon score matching (a supervised *regression* that models as many NN outputs as the dimension of the random variable), data compression, and flow-based density estimation.

regression

The above formalism has two incarnations useful for inference. One, inspired by the Neyman–Pearson lemma [384] and targeted mostly at frequentist applications, is *likelihood-ratio estimation* [408, 107, 509], whereby the two compared distributions are $\mathrm{p}(\mathbf{d} \mid \boldsymbol{\theta}_I)$ and $\mathrm{p}(\mathbf{d} \mid \boldsymbol{\theta}_{II})$. To amortise the result and allow for parametrised *hypothesis testing* and Bayesian MCMC inference, $\boldsymbol{\theta}_I$ and $\boldsymbol{\theta}_{II}$ are treated as conditioning variables[62] by a NN classifier $\ln \hat{\mathrm{r}}(\boldsymbol{\theta}_I, \boldsymbol{\theta}_{II}, \mathbf{d})$ trained with a BCE averaged across the prior $\mathrm{p}(\boldsymbol{\theta}_I) = \mathrm{p}(\boldsymbol{\theta}_{II}) = \mathrm{p}(\boldsymbol{\theta})$:

likelihood-ratio estimation

hypothesis testing

$$\mathbf{x} \to [\boldsymbol{\theta}_I, \boldsymbol{\theta}_{II}, \mathbf{d}], \quad \left.\begin{matrix} \mathrm{p}_1 \to \mathrm{p}(\boldsymbol{\theta}_I, \mathbf{d})\,\mathrm{p}(\boldsymbol{\theta}_{II}) \\ \mathrm{p}_2 \to \mathrm{p}(\boldsymbol{\theta}_{II}, \mathbf{d})\,\mathrm{p}(\boldsymbol{\theta}_I) \end{matrix}\right\} \quad \Longrightarrow \quad \mathrm{r}(\boldsymbol{\theta}_I, \boldsymbol{\theta}_{II}, \mathbf{d}) \equiv \frac{\mathrm{p}(\mathbf{d} \mid \boldsymbol{\theta}_I)}{\mathrm{p}(\mathbf{d} \mid \boldsymbol{\theta}_{II})}. \tag{2.13}$$

Apart from amortisation, this formulation allows training with two independent samples $(\boldsymbol{\theta}_1, \mathbf{d}_1)$ and $(\boldsymbol{\theta}_2, \mathbf{d}_2)$ from the same black-box simulator $\mathrm{p}(\boldsymbol{\theta}, \mathbf{d})$; then the p_1 and p_2 are represented by

$$[\boldsymbol{\theta}_1, \boldsymbol{\theta}_2, \mathbf{d}_1] \sim \mathrm{p}_1 \quad \text{and} \quad [\boldsymbol{\theta}_2, \boldsymbol{\theta}_1, \mathbf{d}_1] \sim \mathrm{p}_2$$

in a first step and

$$[\boldsymbol{\theta}_2, \boldsymbol{\theta}_1, \mathbf{d}_2] \sim \mathrm{p}_1 \quad \text{and} \quad [\boldsymbol{\theta}_1, \boldsymbol{\theta}_2, \mathbf{d}_2] \sim \mathrm{p}_2$$

in a second step, thus fully utilising both simulator runs.

On the other hand, it is more convenient for Bayesian applications, even simpler in terms of NN inputs and training, and fully adherent to the symmetry between "parameters" and "data"—the founding principle of Bayesianism disrespected by data compression, NPE, NLE, and likelihood-ratio estimation—to target the triumviratio™ suggested by Bayes' theorem:

$$\mathbf{x} \to [\boldsymbol{\theta}, \mathbf{d}], \quad \left.\begin{matrix} \mathrm{p}_1 \to \mathrm{p}(\boldsymbol{\theta}, \mathbf{d}) \\ \mathrm{p}_2 \to \mathrm{p}(\boldsymbol{\theta})\,\mathrm{p}(\mathbf{d}) \end{matrix}\right\} \quad \Longrightarrow \quad \mathrm{r}(\boldsymbol{\theta}, \mathbf{d}) \equiv \frac{\mathrm{p}(\boldsymbol{\theta}, \mathbf{d})}{\mathrm{p}(\boldsymbol{\theta})\,\mathrm{p}(\mathbf{d})} = \frac{\mathrm{p}(\mathbf{d} \mid \boldsymbol{\theta})}{\mathrm{p}(\mathbf{d})} = \frac{\mathrm{p}(\boldsymbol{\theta} \mid \mathbf{d})}{\mathrm{p}(\boldsymbol{\theta})}. \tag{2.14}$$

By the first ratio, in this flavour of SBI (which we will synecdochally refer to as NRE), a classifier network $\ln \hat{\mathrm{r}}(\boldsymbol{\theta}, \mathbf{d})$ is trained with *joint* (dependent) and *marginal* (independent)

joint-to-marginal NRE

[62] Whereas we made a distinction between the random variable and the condition for the NPE $\mathrm{q}(\mathrm{var} \mid \mathrm{cond})$, here the line is blurred since NRE always estimates a density between two distributions for the same random variables. We will capitalise on this shortly.

parameter–data pairs [213], coming, as above, from two independent simulation runs:

$$[\boldsymbol{\theta}_1, \mathbf{d}_1], [\boldsymbol{\theta}_2, \mathbf{d}_2] \sim \mathrm{p}_1 \quad \text{and} \quad [\boldsymbol{\theta}_1, \mathbf{d}_2], [\boldsymbol{\theta}_2, \mathbf{d}_1] \sim \mathrm{p}_2 \,. \tag{2.15}$$

By the second ratio of eq. (2.14), this corresponds to a simple alteration of likelihood-ratio estimation (eq. (2.13)): the replacement of one "hypothesis" with the marginal $\mathrm{p}(\mathbf{d})$, which leads to increased accuracy of the approximation across the parameter space because the "reference" distribution ($\mathrm{p}_2 \to \mathrm{p}(\boldsymbol{\theta})\,\mathrm{p}(\mathbf{d})$) always covers the target $\mathrm{p}(\boldsymbol{\theta}, \mathbf{d})$ [213, fig. 2]. Miller et al. [363] extend this concept to multiple *contrastive*[63] $\boldsymbol{\theta}$ examples all drawn from the prior, explicitly teaching the network the parameter values that are *un*likely to have produced given data.

contrastive NRE

Finally, a joint-to-marginal ratio estimator provides a direct approximation (up to a normalisation) of the likelihood/sampling probability $\mathrm{p}(\mathbf{d} \mid \boldsymbol{\theta}) \propto \mathrm{r}(\boldsymbol{\theta}, \mathbf{d}) \approx \hat{\mathrm{r}}(\boldsymbol{\theta}, \mathbf{d})$. Consequently, NRE bears many resemblances to NLE: training examples can be generated from any convenient alternative prior $\tilde{\mathrm{p}}(\boldsymbol{\theta})$; and many i.i.d. data can be jointly analysed simply by summing the respective $\ln \hat{\mathrm{r}}\left(\boldsymbol{\theta}, \mathbf{d}_{\mathrm{o}}^{(i)}\right)$ (but see section 4.2). Marginal inference is still performed via traditional techniques or by simply re-weighting prior samples by $w(\boldsymbol{\theta}) = \mathrm{r}(\boldsymbol{\theta}, \mathbf{d}) \approx \hat{\mathrm{r}}(\boldsymbol{\theta}, \mathbf{d})$ (as per the last equality in eq. (2.14)).

NEW Automatic summarisation within NRE

A common design choice for $\ln \hat{\mathrm{r}}(\boldsymbol{\theta}, \mathbf{d})$, in which parameters and data enter on equal footing, is to include a pure-data compression component $\mathbf{d} \to \mathbf{s}(\mathbf{d})$ (cf. NPE contextualisation discussed above), to twofold benefit: firstly, this significantly reduces the time needed for repeated evaluations at the same $\mathbf{d}$ (at $\mathbf{d}_{\mathrm{o}}$ when evaluating the posterior and/or when re-using the summary across inference tasks, i.e. for different groups $\boldsymbol{\theta}_g$); and secondly, $\mathbf{s}(\mathbf{d})$ themselves are interpretable representations of the data, directly optimised for constraining power. Concretely, a NRE of the form $\ln \hat{\mathrm{r}}(\boldsymbol{\theta}, \mathbf{s})$, with $\mathbf{s}(\mathbf{d})$ a learnable function, maximises the averaged *Jensen–Shannon (JS) divergence* [323] between the prior and the posterior resulting from the given compression:

JS divergence

$$\begin{aligned}
\mathrm{JS}[\mathrm{p}(\boldsymbol{\theta} \mid \mathbf{s}), \mathrm{p}(\boldsymbol{\theta})] &\equiv \mathrm{KL}\left[\mathrm{p}(\boldsymbol{\theta} \mid \mathbf{s}) \,\|\, \frac{\mathrm{p}(\boldsymbol{\theta} \mid \mathbf{s}) + \mathrm{p}(\boldsymbol{\theta})}{2}\right] + \mathrm{KL}\left[\mathrm{p}(\boldsymbol{\theta}) \,\|\, \frac{\mathrm{p}(\boldsymbol{\theta} \mid \mathbf{s}) + \mathrm{p}(\boldsymbol{\theta})}{2}\right] \\
&= \mathbb{E}_{\mathrm{p}(\boldsymbol{\theta} \mid \mathbf{s})}\left[\ln \frac{2\,\mathrm{p}(\boldsymbol{\theta} \mid \mathbf{s})}{\mathrm{p}(\boldsymbol{\theta} \mid \mathbf{s}) + \mathrm{p}(\boldsymbol{\theta})}\right] + \mathbb{E}_{\mathrm{p}(\boldsymbol{\theta})}\left[\ln \frac{2\,\mathrm{p}(\boldsymbol{\theta})}{\mathrm{p}(\boldsymbol{\theta} \mid \mathbf{s}) + \mathrm{p}(\boldsymbol{\theta})}\right] \\
\implies \mathbb{E}_{\mathrm{p}(\mathbf{d})}[\mathrm{JS}] &= -\,\mathrm{BCE} + \mathrm{const.}
\end{aligned} \tag{2.16}$$

[63] Previously, Durkan et al. [134] had "unified" in a contrastive framework (that essentially enumerates/discretises the parameter space) likelihood-ratio and posterior estimation, but their focus on the latter spawned an unwieldy normalisation term that Miller et al. [363] cancelled via marginal examples.

This connection and the fact that the same objective (JS $\leftrightarrow$ −BCE) is simultaneously maximised by a powerful classifier (ratio estimator) and minimised by a target distribution (posterior) that matches the reference (prior) is the mathematical foundation for the formerly immensely popular generative adversarial network (GAN) models [179].

Autoregressive NRE

Estimating high-dimensional[64] density ratios from a limited set of training examples and/or with a finitely expressive network is prone to inaccuracies. Moreover, it may be beneficial, for the purposes of truncation / sequential inference (subsection 2.2.2), to place constraints on any nuisance parameters correlated with those of interest, so as to reduce training data variability.

ARNRE

Thus motivated, Anau Montel et al. [14] introduce *autoregressive neural ratio estimation (ARNRE)*, in analogy with the eponymous NF architectures [291, 394, 234], which reduces the complexity of inferring jointly all parameters in $\boldsymbol{\theta}$ to an ordered series of N_g low-dimensional ratio estimators that each learn a small group of parameters $\boldsymbol{\theta}_g$ (or even a single parameter) conditioned on data *and* the values of the "previously" inferred $\boldsymbol{\theta}_{<g}$:

$$\mathrm{p}(\boldsymbol{\theta} \mid \mathbf{d}) = \prod_g \mathrm{p}(\boldsymbol{\theta}_g \mid \boldsymbol{\theta}_{<g}, \mathbf{d}) = \prod_g \frac{\mathrm{p}(\boldsymbol{\theta}_g)}{\mathrm{p}(\boldsymbol{\theta}_g)} \frac{\mathrm{p}(\boldsymbol{\theta}_g, \boldsymbol{\theta}_{<g}, \mathbf{d})}{\mathrm{p}(\boldsymbol{\theta}_{<g}, \mathbf{d})} = \prod_g \mathrm{r}(\boldsymbol{\theta}_g, [\boldsymbol{\theta}_{<g}, \mathbf{d}])\, \mathrm{p}(\boldsymbol{\theta}_g), \tag{2.17}$$

partition

where $[\boldsymbol{\theta}_g]_{g=1}^{N_g}$ form a *partition* of the parameters of interest, i.e. their union is $\boldsymbol{\theta}$, and each pair is disjoint: $\cup_g \boldsymbol{\theta}_g = \boldsymbol{\theta}$ and $\boldsymbol{\theta}_{g_1} \cap \boldsymbol{\theta}_{g_2} = \emptyset \iff g_1 \neq g_2$. Estimators $\{\hat{\mathrm{r}}_g\}$ for all ratios $\{\mathrm{r}(\boldsymbol{\theta}_g, [\boldsymbol{\theta}_{<g}, \mathbf{d}])\}$ can be trained simultaneously (and naturally, any data summarisers $\mathbf{d} \to \mathbf{s}(\mathbf{d})$ shared among them) with the same black-box-simulated examples $\boldsymbol{\theta}, \mathbf{d} \to \{\boldsymbol{\theta}_g\}, \mathbf{d}$ simply by *interpreting* differently which random variables are considered "parameters of interest" (unknown: $\boldsymbol{\theta}_g$) and which "data" (given: $\boldsymbol{\theta}_{<g}, \mathbf{d}$). Still, the choice of autoregressive ordering—dictating how much information is provided through $\boldsymbol{\theta}_{<g}$ to the estimator of $\boldsymbol{\theta}_g$—is arbitrary, i.e. any factorisation is in principle valid; however, Anau Montel et al. [14] argue for placing "first" (i.e. with least conditioning) the parameters that can be inferred best solely from data; and moreover, the choice can be automated (at least somewhat) through model introspection [e.g. 540].

joint vs. marginal prior

It is important to recognise that the originally defined joint-to-marginal ratio $\mathrm{r}(\boldsymbol{\theta}, \mathbf{d}) \neq \prod_g \mathrm{r}(\boldsymbol{\theta}_g, [\boldsymbol{\theta}_{<g}, \mathbf{d}])$, which appears in eq. (2.17): their ratio is that between the *joint prior*

[64] Even though NRE learns from high-dimensional distributions $\mathrm{p}(\boldsymbol{\theta}, \mathbf{d})$ and $\mathrm{p}(\boldsymbol{\theta})\,\mathrm{p}(\mathbf{d})$, its performance is bounded by the size of the smaller among $\boldsymbol{\theta}$ and $\mathbf{d}$ (usually the former) because of eq. (2.14).

$p(\boldsymbol{\theta})$ and the *product of marginals* $\prod_g p(\boldsymbol{\theta}_g)$. Naturally, this can be learnt[65] by another NN, but as already discussed, any prior can be readily used for NRE inference[66] and the $\{p(\boldsymbol{\theta}_g)\}$ are easily accessible from simulations; even their densities, being low-dimensional, can be robustly estimated with traditional techniques—if not already explicitly available due to the prior structure of the model.

2.2 Inside the black box

The discussion of the preceding section adheres strictly to the principle of *sample*-based learning, assuming that the simulator is a black box that can only stochastically draw from $p(\boldsymbol{\theta}, \mathbf{d})$ (or possibly from $p(\mathbf{d} \mid \boldsymbol{\theta}_{\text{fid}})$). Pure SBI methods, as generally described, are applicable even in the absence of understanding of the model—or of the scientific goals of inference—and in the Platonic setting of infinite network expressivity, training data, and practitioner patience.

In this section, we aim to aid the machine learner cope with sums over limited examples and gradient descent of weights and biases in lieu of expectations and functional variation; and to re-attach them to the physical problem they are working on by unveiling the simulator's internals, extracting additional information from it beyond the values of stochastic variables, modifying the distributions from which any or all of them are sampled, and exploiting—or at least referring to—the model's particular conditional and/or hierarchical structure.

2.2.1 Augmented training

While methods for learning purely from $(\boldsymbol{\theta}, \mathbf{d})_i \sim p(\boldsymbol{\theta}, \mathbf{d})$ are appealing, liberating, and general, they can be complemented by additional information extracted from the simulator/model if such is available, increasing the learning efficiency, i.e. the approximation quality of a NN trained with a given finite number of simulator runs.

Outside of SBI, the joint probability $p(\boldsymbol{\psi}, \mathbf{d}) = p(\mathbf{d} \mid \boldsymbol{\theta}, \boldsymbol{\nu})\, p(\boldsymbol{\theta}, \boldsymbol{\nu})$ and its gradient are the *only* available means for inference, and albeit often inconvenient, they are still tractable for vast classes of models. Moreover, frameworks for probabilistic programming with automatic differentiation (treated more comprehensively in chapter 9) allow a forward imple-

[65] This is essentially the approach adopted by Anau Montel et al. [14, appendix A] through an auxiliary variable that "switches" between autoregressive inference of the joint posterior and the joint prior with the same network.

[66] ARNRE corresponds to a particular choice of alternative prior $\tilde{p}(\boldsymbol{\theta}) = \prod_g p(\boldsymbol{\theta}_g)$ for which the (original-) posterior-to-prior ratio does factorise as $\tilde{r}(\boldsymbol{\theta}, \mathbf{d}) = p(\boldsymbol{\theta} \mid \mathbf{d}) / \tilde{p}(\boldsymbol{\theta}) = \prod_g r(\boldsymbol{\theta}_g, [\boldsymbol{\theta}_{<g}, \mathbf{d}])$. From this formulation, it is also evident that ARNRE is invariant to permuting the groups.

mentation of the model (the simulator) to double as likelihood (joint probability) and score (joint gradient) evaluator. Brehmer et al. [68, 67, 69] first detailed how these additional quantities "mined"[67] from the simulator can be used to construct conventional supervised objective functionals that enhance training of summarisers, density, or ratio estimators.

Augmented training makes use of the theorem [310, theorem 4.2.3; see also 69, eq. (6)] that estimating a joint quantity by minimising the mean squared error (MSE) results in its conditional expectation:

$$g^*(b) \equiv \operatorname{argmin}_{\hat{g}} \mathbb{E}_{\mathrm{p}(a,b)}\left[|\hat{g}(b) - g(a,b)|^2\right] = \mathbb{E}_{\mathrm{p}(a\,|\,b)}[g(a,b)], \tag{2.18}$$

of which we already saw two examples.[68] Identifying $a, b \to \mathbf{v}, (\mathbf{d}, \boldsymbol{\theta})$ allows us to approximate the marginal score[69] and the marginal likelihood ratio (eq. (2.13)) from their counterparts calculated jointly with the nuisance parameters $\mathbf{v}$:

$$g(a,b) \to \boldsymbol{\nabla}_{\boldsymbol{\theta}} \ln \mathrm{p}(\mathbf{v}, \mathbf{d} \,|\, \boldsymbol{\theta}) \quad \Longrightarrow \quad g^* \to \boldsymbol{\nabla}_{\boldsymbol{\theta}} \ln \mathrm{p}(\mathbf{d} \,|\, \boldsymbol{\theta}), \tag{2.20}$$

$$g(a,b) \to \mathrm{r}(\tilde{\boldsymbol{\theta}}, \boldsymbol{\theta}, [\mathbf{v}, \mathbf{d}]) \quad \Longrightarrow \quad g^* \to \mathrm{r}(\tilde{\boldsymbol{\theta}}, \boldsymbol{\theta}, \mathbf{d}). \tag{2.21}$$

Training objectives with eqs. (2.20) and (2.21) substituted in eq. (2.18) can be added in arbitrary proportion (i.e. with a scaling hyperparameter) to the previously introduced expressions: "predicted likelihood" (eq. (2.6)) and the BCE (eq. (2.11)), respectively, increasing the benefits reaped from any given simulator run.

Notice, finally, that both opportunities for augmentation are in a sense *differential*: they allow learning a function at $(\boldsymbol{\theta}, \mathbf{d})$ from nearby points $(\boldsymbol{\theta} + \delta\boldsymbol{\theta}, \mathbf{d})$ by calculating the explicit probability for a "trajectory" connecting them through $\mathbf{v}$. While joint-to-marginal NRE is less amenable to augmentation because the "joint evidence"™ $\mathrm{p}(\mathbf{v}, \mathbf{d})$ cannot be evaluated, it already implements a "globalisation" measure to a similar effect by utilising the marginals as reference distributions.

---

[67] The biggest gold nuggets that can be mined from [107, 68, 67, 499, 69] are the abbreviations CARL, RASCAL, CASCAL, SCANDAL, ALICE(S), SALLY, & SALLINO.

[68] namely, the posterior mean estimator $\mathbb{E}_{\mathrm{p}(\boldsymbol{\theta}\,|\,\mathbf{d})}[\boldsymbol{\theta}]$ discussed above ($a, b \to \boldsymbol{\theta}, \mathbf{d}$ and $g \to \boldsymbol{\theta}$) and the "score"-matching objective in eq. (2.5), where $a, b \to \mathbf{x}, \tilde{\mathbf{x}}$, and

$$g(\mathbf{x}, \tilde{\mathbf{x}}) \to \boldsymbol{\nabla}_{\tilde{\mathbf{x}}} \ln \mathrm{p}(\tilde{\mathbf{x}} \,|\, \mathbf{x}) \quad \underset{\text{theorem}}{\overset{\text{Bayes'}}{\Longrightarrow}} \quad g^*(\tilde{\mathbf{x}}) = \mathbb{E}_{\mathrm{p}(\mathbf{x}\,|\,\tilde{\mathbf{x}})}[\cancel{\boldsymbol{\nabla}_{\tilde{\mathbf{x}}} \ln \mathrm{p}(\mathbf{x} \,|\, \tilde{\mathbf{x}})} + \boldsymbol{\nabla}_{\tilde{\mathbf{x}}} \ln \mathrm{p}(\tilde{\mathbf{x}})], \tag{2.19}$$

since the first term famously vanishes as the data-averaged score, and the second does not depend on the random variable $\mathbf{x}$, so its expectation is trivial; cf. also the more general theorem 4.1.1 in [310] that we applied to arrive at eq. (2.6)

[69] This is equivalent to the "score"-matching eq. (2.19), but there an auxiliary variable ($\tilde{\mathbf{x}} \leftrightarrow \mathbf{d}$) must be introduced to establish the necessary conditional structure.

2.2.2 Sequential training

Inference amortisation—i.e. training with diverse simulated data sets—comes at the price of precision in the results for any one example, typically manifesting in weaker[70] than optimal constraints. This issue is similar to that of rejection-ABC: ε must balance simulator efficiency against the mixing of results from $\mathbf{d}_o + \delta\mathbf{d}$; in modern SBI methods, a similar role is assumed by the network's interpolation capability and the density of training examples. First introduced exactly in the context of ABC [481], *sequential (targeted) training* attempts to improve the fidelity of the posterior approximation by focusing the simulator on examples from the vicinity of the observed $\mathbf{d}_o$. *sequential SBI*

In practice, often the only handle of control over the simulator is the sampling of $\boldsymbol{\theta}$, and so sequential methods amount to choosing a *proposal prior* $\tilde{p}(\boldsymbol{\theta})$ so that the resulting distribution of mock data $\tilde{p}(\mathbf{d}) \equiv \int p(\mathbf{d} \mid \boldsymbol{\theta})\, \tilde{p}(\boldsymbol{\theta})\, d\boldsymbol{\theta}$ assign a larger density[71] to $\mathbf{d}_o$ than the original $p(\mathbf{d})$, i.e. a larger fraction of the training data be similar to $\mathbf{d}_o$. At the same time, $\tilde{p}(\boldsymbol{\theta})$ must ensure that it also faithfully represents the $\boldsymbol{\theta}$ values associated with $\mathbf{d}_o$, so the usual choice for a parameter proposal is the current posterior estimate: $\tilde{p}(\boldsymbol{\theta}) \leftarrow q(\boldsymbol{\theta} \mid \mathbf{d})$. Given the new proposal, the estimator is iteratively retrained with a new training set from $\tilde{p}(\boldsymbol{\theta}, \mathbf{d}) \equiv p(\mathbf{d} \mid \boldsymbol{\theta})\, \tilde{p}(\boldsymbol{\theta})$ until some convergence criterion is met (e.g. the estimate does not change/improve noticeably).[72] *proposal prior*

As already discussed, SBI methods that estimate the likelihood (NLE and NRE) can train with samples from any $\tilde{p}(\boldsymbol{\theta}, \mathbf{d}) \equiv p(\mathbf{d} \mid \boldsymbol{\theta})\, \tilde{p}(\boldsymbol{\theta})$, and their result (provided idealised convergence and matching support between $p(\boldsymbol{\theta})$ and $\tilde{p}(\boldsymbol{\theta})$) will not change. Of course, when performing inference on any particular example $\mathbf{d}_o$, the originally intended $p(\boldsymbol{\theta})$ must be used.[73] In contrast, methods for direct posterior estimation ingrain the distribution used for training into the final result, so sequential NPE needs to re-weight the resulting posterior

[70] That both NPE and NRE tend to be conservative can be understood informally by considering their learning objectives eqs. (2.6) and (2.14): for a randomly initialised network, there is no preference for assigning high probability to any one given $\boldsymbol{\theta}$, or to classifying a $(\boldsymbol{\theta}, \mathbf{d})$ pair as either joint or marginal particularly confidently, and so untrained estimators tend to represent the prior, i.e. they are unable to extract any useful information at first and learn to do so increasingly well with training.

[71] The largest $\tilde{p}(\mathbf{d}_o)$ is achieved when the proposal consists only of the MLE: $\tilde{p}(\boldsymbol{\theta}) \to \delta(\boldsymbol{\theta} - \operatorname{argmax}_{\boldsymbol{\theta}} p(\mathbf{d}_o \mid \boldsymbol{\theta}))$. Moreover, and more importantly in high-dimensions, it is also desirable that $\mathbf{d}_o$ is in the *typical set* of $\tilde{p}(\mathbf{d})$ (containing the examples that the NN can learn best), which usually does not include the single highest-probability values. We discuss and demonstrate this further in appendix 17. *typical set*

[72] Sequential SBI methods thus hover on the border of improper posterior-as-prior reuse in the same analysis, i.e. "double counting" the data—and cross it multiple times in both directions.

[73] But compare with ARNRE, which modifies the inferred likelihood/sampling distribution: $p(\mathbf{d} \mid \boldsymbol{\theta}) \to \prod_g p(\boldsymbol{\theta}_{<g}, \mathbf{d} \mid \boldsymbol{\theta}_g)$, and so also requires a modified prior $\prod_g p(\boldsymbol{\theta}_g)$ to be used in inference.

$\tilde{p}(\boldsymbol{\theta} \mid \mathbf{d})$ to arrive at the correct

$$p(\boldsymbol{\theta} \mid \mathbf{d}_o) = \tilde{p}(\boldsymbol{\theta} \mid \mathbf{d}) \frac{p(\boldsymbol{\theta})\, \tilde{p}(\mathbf{d}_o)}{\tilde{p}(\boldsymbol{\theta})\, p(\mathbf{d}_o)} \quad \Longrightarrow \quad q(\boldsymbol{\theta} \mid \mathbf{d}_o) \equiv \tilde{q}(\boldsymbol{\theta} \mid \mathbf{d}_o) \frac{p(\boldsymbol{\theta})\, \tilde{p}(\mathbf{d}_o)}{\tilde{p}(\boldsymbol{\theta})\, p(\mathbf{d}_o)}, \tag{2.22}$$

where $\tilde{q}(\boldsymbol{\theta} \mid \mathbf{d}_o)$ is the NPE trained with $\tilde{p}(\boldsymbol{\theta}, \mathbf{d})$. This correction invalidates the simple procedure[56] for evaluating HPD credibilities and requires at least $p(\boldsymbol{\theta}) / \tilde{p}(\boldsymbol{\theta})$ to be taken into account when evaluating the posterior, by multiplying a numerical $\tilde{q}(\boldsymbol{\theta} \mid \mathbf{d}_o)$ or re-weighting samples (the evidence ratio $\tilde{p}(\mathbf{d}_o) / p(\mathbf{d}_o)$ can be treated as a proportionality constant). Alternatively, the correction can be applied during training via a modified objective

$$\operatorname{argmax}_q \mathbb{E}_{\tilde{p}(\boldsymbol{\theta}, \mathbf{d})} \left[\frac{p(\boldsymbol{\theta})}{\tilde{p}(\boldsymbol{\theta})} \ln q(\boldsymbol{\theta} \mid \mathbf{d}) \right] \quad \text{given} \quad \int q(\boldsymbol{\theta} \mid \mathbf{d})\, d\boldsymbol{\theta} = 1. \tag{2.23}$$

The optimal $q(\boldsymbol{\theta} \mid \mathbf{d})$ in this case is again the correct $p(\boldsymbol{\theta} \mid \mathbf{d})$, restoring all its original convenience. Moreover, training with a finite-sample approximation of eq. (2.23) can have reduced variance with respect to eq. (2.6) if $\tilde{p}(\boldsymbol{\theta})$ is chosen to be largest in the same regions as $p(\boldsymbol{\theta}) \ln p(\boldsymbol{\theta} \mid \mathbf{d})$, as is well known from the theory of *importance sampling* [294], justifying the use of $\tilde{p}(\boldsymbol{\theta}) \leftarrow p(\boldsymbol{\theta} \mid \mathbf{d}_o)$. Yet, the variance of eq. (2.23) may still be high since a simple transformation $p(\boldsymbol{\theta}) \rightarrow \tilde{p}(\boldsymbol{\theta})$ cannot in general be optimal for all $\mathbf{d}$.

importance sampling

Prior truncation

De-biasing sequential inference through importance re-weighting (either in training or during evaluation) requires that the prior and proposal (or at least their ratio[74]) be tractable. A simpler, yet often as effective, technique is *prior truncation* [361, 122][75]:

truncated prior

$$\tilde{p}_{T(\mathbf{d}_o)}(\boldsymbol{\theta}) = \begin{cases} p(\boldsymbol{\theta})/c & \text{if } \boldsymbol{\theta} \in T(\mathbf{d}_o), \\ 0 & \text{otherwise;} \end{cases} \quad \text{with} \quad \boldsymbol{\theta} \in T(\mathbf{d}_o) \iff \begin{cases} p(\mathbf{d}_o \mid \boldsymbol{\theta}) \approx 0, \\ p(\boldsymbol{\theta} \mid \mathbf{d}_o) \approx 0, \end{cases} \tag{2.24}$$

and the normalisation $c \equiv \int_{T(\mathbf{d}_o)} p(\boldsymbol{\theta})\, d\boldsymbol{\theta}$. As opposed to prior transformation, truncation keeps $p(\boldsymbol{\theta}) / \tilde{p}(\boldsymbol{\theta})$ constant while still restricting $\boldsymbol{\theta}$ samples to the region $T(\mathbf{d}_o)$ of parameter space that is relevant to a given observation. It corresponds to slicing through the joint as illustrated in fig. 2.1, thus naturally enhancing the importance of retained parameters and reducing the variance of simulated mock data. The effect on posteriors—perpendicular slices through the joint—is minimal: they are simply re-scaled by the factor of the retained mass inside $T(\mathbf{d}_o)$; for $\mathbf{d}_o$, this is by design close to unity but can be arbitrary for other $\tilde{\mathbf{d}} \neq \mathbf{d}_o$. Nevertheless, since likelihood ratios are, naturally, retained, inference is said to still be *locally* amortised, and thus, validation with credible regions can still be per-

truncation region

local amortisation

[74] that can, of course, be estimated from samples using NRE

[75] introduced first in the context of NRE and then co-opted essentially without modification in NPE

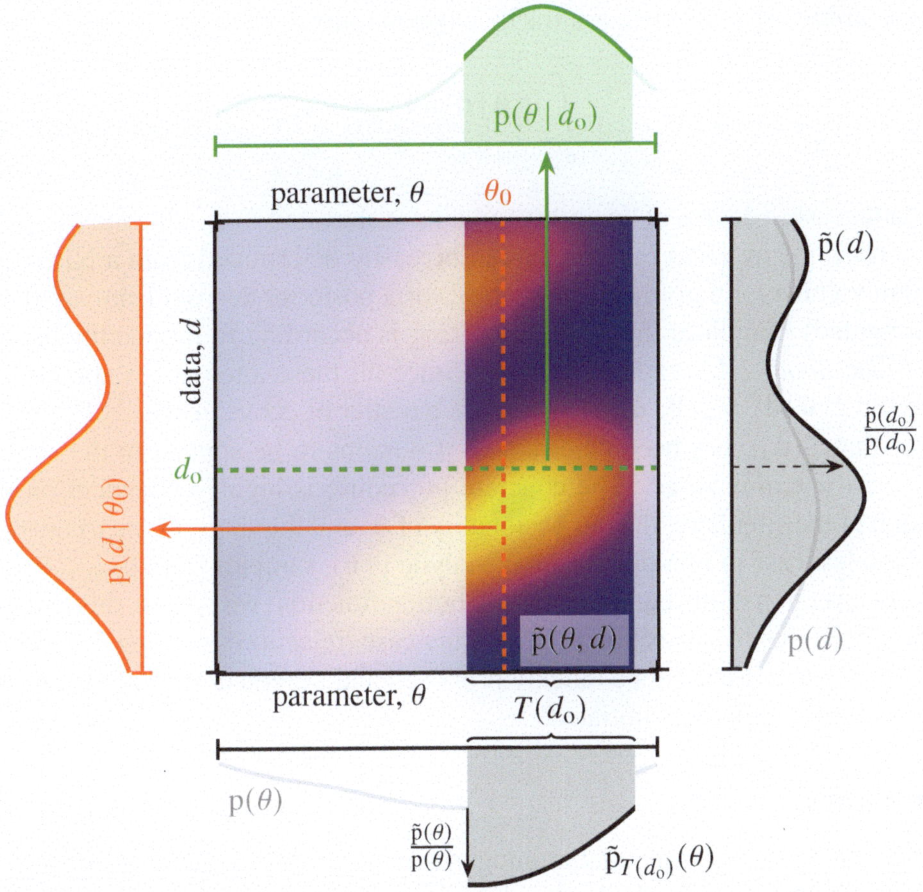

Figure 2.1: Elements of truncated Bayesian reasoning (cf. fig. 1.1).

formed, provided they fall fully within $T(\mathbf{d}_o)$. However, the marginal data distribution is arbitrarily distorted, depending on the discarded posterior mass for $\tilde{\mathbf{d}}$; this can have dire undesired consequences, especially for high-dimensional parameter spaces, which we discuss in appendix 17.

Defined in terms of the posterior and likelihood, which in general may not fully decrease to 0, the truncation region is in practice constructed from the—assumed imperfect—best neural estimates thereof. The approximate bound that is used either controls the likelihood (relative to the maximum at $\boldsymbol{\theta}_{\mathrm{MLE}}$) inside $T(\mathbf{d})$ to be higher than a pre-defined ε:

$$\boldsymbol{\theta} \in T(\mathbf{d}_o) \iff \varepsilon \leq \begin{cases} \hat{\mathrm{r}}(\boldsymbol{\theta}, \boldsymbol{\theta}_{\mathrm{MLE}}, \mathbf{d}_o), \\ \hat{\mathrm{r}}(\boldsymbol{\theta}, \mathbf{d}_o) / \hat{\mathrm{r}}(\boldsymbol{\theta}_{\mathrm{MLE}}, \mathbf{d}_o); \end{cases} \tag{2.25}$$

or the credibility of $T(\mathbf{d})$ to be arbitrarily close to unity:

$$\int_{T(\mathbf{d}_o)} \mathrm{q}(\boldsymbol{\theta} \mid \mathbf{d})\, \mathrm{d}\boldsymbol{\theta} = 1 - \varepsilon. \tag{2.26}$$

In this latter case, a degree of flexibility remains in choosing $T(\mathbf{d}_o)$; the most popular choice is the corresponding HPD region that can be easily determined from a NPE and has the minimum volume for a given ε. In contrast, for a posterior derived from a ratio estimator (via (weighted) sampling) the natural ordering is according to the likelihood, leading to *highest-likelihood (HL) credible regions*. Since all these alternatives should lead to the same result $\tilde{\mathrm{p}}(\boldsymbol{\theta} \mid \mathbf{d}) \approx \mathrm{p}(\boldsymbol{\theta} \mid \mathbf{d})$, the choice is a matter of convenience.

HL credible region

Now that $T(\mathbf{d}_o)$, has been determined, it remains to be applied to the simulator in a successive *truncation stage*. The concrete procedure is highly dependent on the implementation of the simulator, the dimensionality of $\boldsymbol{\theta}$, and the shape of the truncation region. In the simplest case of one-dimensional analytic $\mathrm{p}(\boldsymbol{\theta})$, samples from $\tilde{\mathrm{p}}_{T(\mathbf{d}_o)}(\boldsymbol{\theta})$ can be obtained via inversion of the cumulative distribution function; we discuss this and other more general procedures in subsection 9.1.2. In any case, rejection sampling is a viable option, especially if parameters can be quickly proposed and accepted/rejected before the expensive part of the simulator is initiated.

truncation stage

Active learning

A general alternative to sequential training is active learning, a method of *Bayesian optimisation* and decision-making [166] that does not target distillation in parameter and/or data space and does not proceed in stages but rather continuously and deliberately proposes the $\boldsymbol{\theta}$ values for which the result is sub-optimal. Since this results in a complicated and intractable effective $\tilde{\mathrm{p}}(\boldsymbol{\theta})$, active learning is generally applied for NLE, with two classes of *acquisition rules* considered: local (targeting inference from a particular $\mathbf{d}_o$) and global (optimising emulation across the parameter space). Similarly to summarisation methods (cf. eqs. (2.1) and (2.2)), these aim to maximise, respectively, precision and mutual information [243, 332]. The optimisation in this case is with respect to the parameter value to be sampled and needs to consider the *uncertainty* in the unconverged NN result, calculated either using an *ensemble* or *Bayesian neural network (BNN)*.

Bayesian optimisation

acquisition rule

ensemble / BNN

2.3 Verification of amortised SBI

Besides scalability and flexibility, the other foundational advantage of SBI—in most of its forms—over likelihood-based techniques is *amortisation*: the ability, after the upfront cost of training, to rapidly analyse different data assumed to follow the same generative model. Put otherwise, training in SBI *teaches a procedure* for performing inference[76] that can later be applied to i.i.d. observations from the real world or to different *simulated* data in a variety of ways that can improve the final results from the target $\mathbf{d}_o$. This is particularly appealing to space scientists, who have long lamented their inability to perform controlled experiments: they can on mock data.[77]

amortisation

2.3.1 Training diagnostics

It is standard practice in deep ML to diagnose training using a *validation set*, explicitly extracted as a random sub-sample of the training data. This is crucial if simulation is computationally intense or the training set is externally produced—or a set of "labelled" real-world observations is used—and only limited samples from $p(\boldsymbol{\theta}, \mathbf{d})$ are available. During optimisation, the same objective functional is evaluated separately with these "held-out" examples to test for convergence and prevent overfitting. When, conversely, the simulator is sufficiently "cheap" as to be callable *online* during training, i.e. at will when requested by the optimiser, a permanent training set is never established, so a separate validation step is not necessary.

validation set

online training

Particularly to NRE, two other diagnostics are available, corresponding to global properties of the optimal classifier that are not strictly enforced; namely, following its definition in eq. (2.14), the joint-to-marginal ratio satisfies

$$\text{posterior normalisation:} \quad \mathbb{E}_{p(\boldsymbol{\theta})}[r(\boldsymbol{\theta}, \mathbf{d})] = 1, \quad \forall \mathbf{d}, \tag{2.27}$$

$$\text{likelihood normalisation:} \quad \mathbb{E}_{p(\mathbf{d})}[r(\boldsymbol{\theta}, \mathbf{d})] = 1, \quad \forall \boldsymbol{\theta}. \tag{2.28}$$

While these are necessary criteria for a NN approximation $\hat{r}$ to be optimal, they are not sufficient, and even the trivial classifier $\hat{r}(\boldsymbol{\theta}, \mathbf{d}) = 1$ (which does not extract any information and returns the prior) fulfills them. Lastly, compliance with these normalisation conditions can be explicitly promoted during training through a *balancing objective* [123], e.g.

balanced NRE

[76] hence our insistence on calling the learning objective a function*al*: because the NN represents a function from data to some final result (likelihood, posterior, ratio) rather than the result itself, which is the goal of likelihood-based methods

[77] Of course, mock data cannot lead to scientific discovery like experimentation in the physical world. This section will, therefore, remain firmly rooted in the assumption that the underlying model is a veracious representation of the real Universe, and all performance *verification* methods discussed will be applied with respect to that given model, leaving the question of its possible *mis-specification* to future work.

model mis-specification

$$\operatorname{argmin}_{\hat{r}} \left(\mathbb{E}_{p(\boldsymbol{\theta},\mathbf{d})} \left[\frac{\hat{r}(\boldsymbol{\theta},\mathbf{d})}{1+\hat{r}(\boldsymbol{\theta},\mathbf{d})} \right] + \mathbb{E}_{p(\boldsymbol{\theta})\,p(\mathbf{d})} \left[\frac{\hat{r}(\boldsymbol{\theta},\mathbf{d})}{1+\hat{r}(\boldsymbol{\theta},\mathbf{d})} \right] - 1 \right)^2, \tag{2.29}$$

which penalises extreme ratios, i.e. overconfident results, while preserving the Bayes optimal classifier as the global solution.

SICRET

2.3.2 Coverage tests (P–P plots)

After training is complete, amortisation[78] allows us to verify and enforce certain properties of the inference procedure learnt by the NN so as to comply with the established (Bayesian and frequentist) interpretations of "probability". Specifically, we can examine empirically (i.e. by analysing test data $\mathbf{d}_t$ with known "true" parameters $\boldsymbol{\theta}_t$ simulated according to $\tilde{p}(\boldsymbol{\theta}_t, \mathbf{d}_t)$) the *nominal credibilities* under the approximate posterior $q(\boldsymbol{\theta} \mid \mathbf{d}_t)$[79]:

credibility (nominal)

$$\gamma_q(\boldsymbol{\theta}_t, \mathbf{d}_t) \equiv \int_{\Gamma(\boldsymbol{\theta}_t,\mathbf{d}_t)} q(\boldsymbol{\theta} \mid \mathbf{d}_t)\, d\boldsymbol{\theta}, \tag{2.30}$$

of given regions Γ in parameter space that include $\boldsymbol{\theta}_t$. As with truncation, the definition of Γ is a matter of convenience, and most commonly, the HPD or HL region bounded by $\boldsymbol{\theta}_t$:

$$\boldsymbol{\theta} \in \Gamma^{\mathrm{HPD}}(\boldsymbol{\theta}_t, \mathbf{d}_t) \iff q(\boldsymbol{\theta} \mid \mathbf{d}_t) \geq q(\boldsymbol{\theta}_t \mid \mathbf{d}_t), \tag{2.31}$$

$$\boldsymbol{\theta} \in \Gamma^{\mathrm{HL}}(\boldsymbol{\theta}_t, \mathbf{d}_t) \iff q(\boldsymbol{\theta} \mid \mathbf{d}_t)/\,p(\boldsymbol{\theta}) \geq q(\boldsymbol{\theta}_t \mid \mathbf{d}_t)/\,p(\boldsymbol{\theta}_t), \tag{2.32}$$

is chosen for NPE or NLE/NRE, respectively.

test statistic

Note that γ is a deterministic scalar-valued function of $\boldsymbol{\theta}_t$ and $\mathbf{d}_t$, i.e. a *test statistic*[80] that encodes the inference procedure. Therefore, its cumulative distribution function (CDF), resulting from the sampling of verification examples according to $\tilde{p}(\boldsymbol{\theta}_t, \mathbf{d}_t)$, can be interpreted as the *coverage frequency* of regions with a given nominal credibility γ^*:

coverage frequency (empirical)

$$F_{\tilde{p}}(\gamma^*) \equiv \int_0^{\gamma^*} \mathbb{E}_{\tilde{p}(\boldsymbol{\theta}_t,\mathbf{d}_t)} \left[\delta(\gamma' \mid \gamma_q(\boldsymbol{\theta}_t, \mathbf{d}_t)) \right] d\gamma' = \mathbb{E}_{\tilde{p}(\boldsymbol{\theta}_t,\mathbf{d}_t)} \left[\mathbb{1}(\gamma^* > \gamma_q(\boldsymbol{\theta}_t, \mathbf{d}_t)) \right]. \tag{2.33}$$

P–P plot

Thus, the aptly named *probability–probability (P–P) plot* [see e.g. 173, section 4.8]

[78] In principle, the same procedures can be implemented for any non-amortised likelihood-based inference technique as well, but at an immense computational cost.

[79] represented as $q(\mathbf{d}_t \mid \boldsymbol{\theta})\, p(\boldsymbol{\theta})$ or $\hat{r}(\boldsymbol{\theta}, \mathbf{d}_t)\, p(\boldsymbol{\theta})$ for NLE or NRE, respectively; in any case, the procedures we discuss here can be implemented with weighted samples (in sufficiently low dimensions so that sampling is reasonably dense), so they are independent of the particular inference approach

[80] A *test* statistic is like a *summary* $\mathbf{s}(\mathbf{d})$ but is also allowed to depend on the model (i.e. sampling-distribution) "label" $\boldsymbol{\theta}$.

depicting $F(\gamma)$ can be used to confront the nominal credibility (posterior probability) assigned by approximate inference with its empirical properties. If $F(\gamma) > \gamma$, inference is said to be conservative: results cover the "true" values more frequently than expected; and if $F(\gamma) < \gamma$, on the other hand, it is said to be undercovering. Moreover, these statements can be made in both Bayesian (F_B) and frequentist (F_f) contexts by choosing appropriate $\tilde{p}(\boldsymbol{\theta}_t, \mathbf{d}_t)$.

Bayesian validation [105, 502, 214] A Bayesian interpretation of the empirical frequencies can be made with validation data (and true parameters) simulated from the original model: $\tilde{p} \to p(\boldsymbol{\theta}_t, \mathbf{d}_t)$. Then, after defining the inverse mapping $\gamma_q^{-1}(\gamma^*, \mathbf{d}_t)$ as the $\Gamma^{HPD/HL}$ that has nominal credibility γ^*, given $\mathbf{d}_t$, so that

$$\boldsymbol{\theta}_t \in \gamma_q^{-1}(\gamma^*, \mathbf{d}_t) \iff \gamma^* > \gamma_q(\boldsymbol{\theta}_t, \mathbf{d}_t), \tag{2.34}$$

eq. (2.33) can be rewritten as

$$F_B(\gamma) \equiv \mathbb{E}_{p(\mathbf{d}_t)}\left[\mathbb{E}_{p(\boldsymbol{\theta}_t \mid \mathbf{d}_t)}\left[\mathbb{1}\left(\boldsymbol{\theta}_t \in \gamma_q^{-1}(\gamma, \mathbf{d}_t)\right)\right]\right]. \tag{2.35}$$

That is, a Bayesian P–P plot depicts the *true credibility* (mass of the true posterior), averaged over all possible data, in regions of *nominal* γ^*. While an exact posterior estimate ($q(\boldsymbol{\theta} \mid \mathbf{d}) \to p(\boldsymbol{\theta} \mid \mathbf{d})$) naturally leads to a perfectly diagonal $F_B(\gamma) \to \gamma$, a variety of other distributions have exactly the same property as well due to the mixing of the "analysis results" of different data, with different true parameters, within the expectation over $p(\mathbf{d}_t)$.[81] A trivial example is the prior, for which: *true credibility*

$$q(\boldsymbol{\theta} \mid \mathbf{d}) \to p(\boldsymbol{\theta}) \implies F_B(\gamma) \to \mathbb{E}_{p(\boldsymbol{\theta}_t)}\left[\mathbb{1}\left(\boldsymbol{\theta}_t \in \gamma_{p(\boldsymbol{\theta})}^{-1}(\gamma)\right)\right] = \gamma. \tag{2.36}$$

Consequently, a Bayesian P–P plot can serve as a diagnostic tool (cf. normalisation(s) of a NRE) to identify improper learning but does not provide a direct means for rectifying it, and so Bayesian inference remains approximate. Still, If the credible regions are constructed according to the distance to random point (DRP) method [318], a diagonal P–P plot can be made a sufficient condition for $q(\boldsymbol{\theta} \mid \mathbf{d}) = p(\boldsymbol{\theta} \mid \mathbf{d})$, but we will not explore this possibility.

Frequentist calibration Exact inference from possibly approximate results can be achieved by re-defining inference. Instead of (posterior) probability densities, frequentist statistics aims to produce regions C in parameter space with *prescribed confidence* (empir- *confidence (prescribed)*

[81] Delaunoy et al. [123] illustrate this point exactly by referring to ABC losing constraining power as the bandwidth $\varepsilon \to \infty$.

ical frequency) F of covering *fixed* true parameters $\boldsymbol{\theta}_*$. Since the latter are unknown, and C needs to be constructed solely from $\mathbf{d}_\mathrm{t}$, the procedure needs to satisfy

$$\mathbb{E}_{\mathrm{p}(\mathbf{d}_\mathrm{t} \mid \boldsymbol{\theta}_*)}[\mathbb{1}(\boldsymbol{\theta}_* \in C_F(\mathbf{d}_\mathrm{t}))] = F \quad \text{for all } \boldsymbol{\theta}_* \text{ and } F. \tag{2.37}$$

This is a more constraining property that credible regions $\Gamma = \gamma_\mathrm{q}^{-1}(F, \mathbf{d}_\mathrm{t})$ are not guaranteed to have, even when the true posterior is used since the expectation is taken over distributions of data instead of parameters (and is not a probability mass like $\mathrm{F_B}$)—unless $\mathrm{p}(\boldsymbol{\theta} \mid \mathbf{d})$ and $\mathrm{p}(\mathbf{d} \mid \boldsymbol{\theta})$ are trivially related, i.e. the prior is uniform.

Instead, confidence regions are usually derived via the Neyman construction [383] after identifying a test statistic: usually the likelihood[82] [114, 115, 353], which leads to smallest C as per the Neyman–Pearson lemma [384]. The above discussion presents another opportunity: using the approximate credibilities of HPD/HL regions instead. If the prior of $\boldsymbol{\theta}$ is uniform—and the approximate posterior is exact—, the two choices are equivalent since the nominal credibilities are, in that case, monotonic with the likelihood. Otherwise, the procedure we describe below takes into account the effect that the prior has on posterior credibility and corrects for it implicitly. It also does away with concerns over the inaccuracy in $\mathrm{q}(\boldsymbol{\theta} \mid \mathbf{d})$—i.e. incorrect coverage of credible regions—since any Neyman-constructed C has *guaranteed exact* confidence.

Strategy 1. Building a confidence set from an approximate posterior:strategy 1

1. first, calculate a series of frequentist P–P plots

$$\mathrm{F_f}(\gamma \mid \boldsymbol{\theta}_*) \equiv \mathbb{E}_{\mathrm{p}(\mathbf{d}_\mathrm{t} \mid \boldsymbol{\theta}_*)}\left[\mathbb{1}\left(\boldsymbol{\theta}_* \in \gamma_\mathrm{q}^{-1}(\gamma, \mathbf{d}_\mathrm{t})\right)\right] \tag{2.38}$$

for different fixed parameters $\boldsymbol{\theta}_*$ across the support of the prior (i.e. the parameter space); to this end, either explicitly generate test data $\{\mathbf{d}_\mathrm{t}\}$ from a conditioned simulator $\mathrm{p}(\mathbf{d}_\mathrm{t} \mid \boldsymbol{\theta}_*)$ for each of a set of pre-defined values for $\boldsymbol{\theta}_*$ (e.g. a grid), which requires a very large number of simulations; or obtain a smoothed kernel-based approximation:

$$\mathbb{E}_{\mathrm{p}(\mathbf{d}_\mathrm{t} \mid \boldsymbol{\theta}_*)}[f(\boldsymbol{\theta}_*, \mathbf{d}_\mathrm{t})] = \lim_{K \to \delta} \mathbb{E}_{\mathrm{p}(\mathbf{d}_\mathrm{t} \mid \boldsymbol{\theta}_\mathrm{t})\mathcal{U}(\boldsymbol{\theta}_\mathrm{t})}[K(\boldsymbol{\theta}_\mathrm{t} \mid \boldsymbol{\theta}_*) f(\boldsymbol{\theta}_\mathrm{t}, \mathbf{d}_\mathrm{t})],$$

using a more reasonable number of random test examples $\boldsymbol{\theta}_\mathrm{t}, \mathbf{d}_\mathrm{t} \sim \mathrm{p}(\mathbf{d}_\mathrm{t} \mid \boldsymbol{\theta}_\mathrm{t})\mathcal{U}(\boldsymbol{\theta}_\mathrm{t})$, each of which contributes to the P–P plots of all $\boldsymbol{\theta}_*$ near the sampled $\boldsymbol{\theta}_\mathrm{t}$;

profile likelihood

[82] Frequentist inference considers the maximal (*profiled*) likelihood over the nuisance parameters, in keeping with the Neyman–Pearson lemma, whereas we always mean to marginalise them, or rather, avoid defining them in the first place, relying on black-box simulation.

2. then, given a confidence level F, derive the map of "threshold" credibilities $\hat{\gamma}(\boldsymbol{\theta}_*, F)$ defined to satisfy $\mathrm{F_f}(\hat{\gamma} \mid \boldsymbol{\theta}_*) = F$ (in practice, simply intersect each frequentist P–P plot with F); this corresponds to defining a lower bound for the approximate posterior density/likelihood (when using HPD/HL regions);

3. finally, for any given data, e.g. $\mathbf{d}_\mathrm{o}$, collect into the confidence set those $\boldsymbol{\theta}$ that bound a nominal credibility smaller (i.e. have posterior density/likelihood larger) than the "threshold":

$$C_F(\mathbf{d}_\mathrm{o}) \equiv \left\{\boldsymbol{\theta} : \gamma_\mathrm{q}(\boldsymbol{\theta}, \mathbf{d}_\mathrm{o}) \leq \hat{\gamma}(\boldsymbol{\theta}, F)\right\}, \tag{2.39}$$

with the interpretation[83] that this *construction* $\mathbf{d}_\mathrm{t} \to C_F(\mathbf{d}_\mathrm{t})$—and not this *particular* $C_F(\mathbf{d}_\mathrm{o})$—includes with frequency F over $\mathbf{d}_\mathrm{t} \sim \mathrm{p}(\mathbf{d}_\mathrm{t} \mid \boldsymbol{\theta}_\mathrm{o})$ the **true** parameters $\boldsymbol{\theta}_\mathrm{o}$ that generated $\mathbf{d}_\mathrm{o}$, regardless of the **true** value $\boldsymbol{\theta}_\mathrm{o}$.

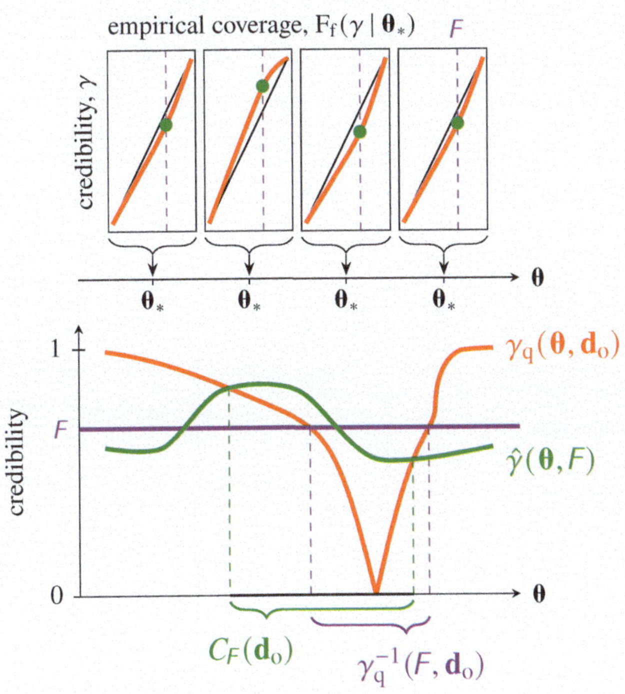

Figure 2.2: (SICRET) Procedure for obtaining frequentist confidence regions with exact coverage from an approximate amortized Bayesian posterior.
Top: by repeated draws of $\mathbf{d}_\mathrm{t}$ at fixed $\boldsymbol{\theta}_*$, we obtain samples for $\gamma(\boldsymbol{\theta}_*, \mathbf{d}_\mathrm{t})$, from which we build its empirical cumulative distribution, $\mathrm{F_f}(\gamma \mid \boldsymbol{\theta}_*)$ (red lines). The "threshold" $\hat{\gamma}(\boldsymbol{\theta}_*, F)$ for regions to cover $\boldsymbol{\theta}_*$ exactly with a given frequency (confidence) F are indicated with green dots and determined as the Fth quantiles of $\mathrm{F_f}$. The green line in the bottom panel connects the dots at all $\boldsymbol{\theta}_*$.
Bottom: $C_F(\mathbf{d}_\mathrm{o})$, the region with confidence level F constructed from the observed data $\mathbf{d}_\mathrm{o}$, is that in which the approximate credibility bounded by $\boldsymbol{\theta}$: $\gamma_\mathrm{q}(\boldsymbol{\theta}, \mathbf{d}_\mathrm{o})$ (red), is lower than the threshold: $\hat{\gamma}(\boldsymbol{\theta}, F)$ (green). For comparison, if $\mathrm{q}(\boldsymbol{\theta} \mid \mathbf{d})$ matches the true posterior—and a uniform prior is used—$\hat{\gamma}(\boldsymbol{\theta}, F)$ is constant across all $\boldsymbol{\theta}$ and equal to the target confidence level (purple line), so C matches the region $\gamma_\mathrm{q}^{-1}(F, \mathbf{d}_\mathrm{o})$ with *credibility* F.

[83] that is apparently useful or sensible to *frequentists*... but consider that parameters and repeatability/exchangeability are mathematical constructs, and transferring them to the "real" world (by designating **true** values) does them as much a disservice as the projection of real objects onto the walls of Plato's cave

frequentist statistics

Chapter 3

Neural simulation-based model selection

3.1 Bayesian model selection

Bayesian model selection[84] is — simply — marginal inference of a *super*global *categorical* hierarchical parameter $\mathcal{M} \in \{M_m\}_{m=1}^{N_\mathrm{mod}}$, and as such, it aims to derive the posterior probability *distribution* $\mathrm{p}(\mathcal{M} \mid \mathbf{d}_\mathrm{o})$ given observations.

categorical (discrete) variable

As in any Bayesian context, before confronting the data, a model selector must adopt some *model prior* $\mathrm{p}(\mathcal{M})$, but since the domain of $\mathcal{M}$ is discrete, a uniform prior $\mathrm{p}(M_m) = 1/N_\mathrm{mod}$ is assumed in practically all cases; the alternative, of course, is to (re-)use a non-trivial model posterior $\mathrm{p}\big(\mathcal{M} \,\big|\, \mathbf{d}_\mathrm{prev}\big)$ from a previous analysis.

model posterior

model prior

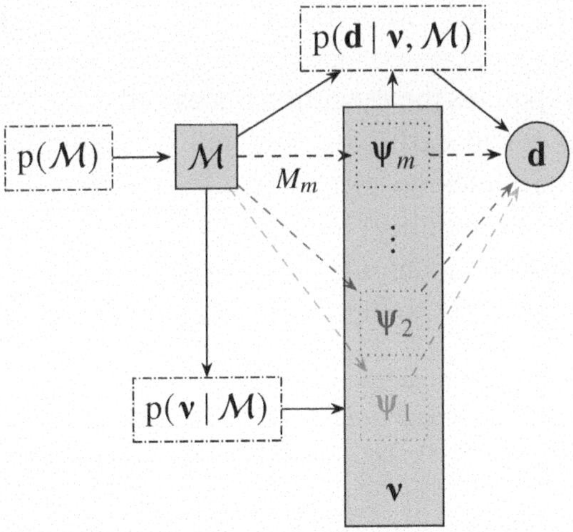

Figure 3.1: Model selection as marginal superglobal inference with a Bayesian supermodel. Note that the $\boldsymbol{\psi}_m$ cannot be regarded as parameters in themselves: they represent the relevant subsets of $\boldsymbol{\psi} = \cup_m \boldsymbol{\psi}_m$ that influence the sampling distribution, given *a particular value* $\mathcal{M} = M_m$ (rather than *any* value of the $\boldsymbol{\gamma}$ in conventional BHMs); as a consequence, the dashed lines indicate different *execution paths*, rather than conditional structure.

[84] We will use this terminology for its visual/aural qualities (in juxtaposition with the terms "simulation" and "supernova") but will be interested in inference over all models and not in simply selecting one.

© The Author(s), under exclusive license to Springer Nature Switzerland AG 2026

K. Karchev, *Supernova Cosmology for the 21st Century*,
https://doi.org/10.1007/978-3-032-15072-1_3

model evidence

The central object of model selection, thus, is the *model evidence*[85] $\mathrm{p}(\mathbf{d}_\mathrm{o} \mid \mathcal{M})$. Evaluated for a single given M_m, however, it is meaningless: its value can be made arbitrary by re-parametrisation of **d**. Instead, at least two models need to be *compared* through the model-evidence ratio: $\mathrm{p}(\mathbf{d}_\mathrm{o} \mid M_1) / \mathrm{p}(\mathbf{d}_\mathrm{o} \mid M_2)$, also known as *Bayes factor*, which is invariant to the chosen data representation. Interpretation of Bayes factors — or *posterior odds* $\mathrm{p}(M_1 \mid \mathbf{d}_\mathrm{o})/\mathrm{p}(M_2 \mid \mathbf{d}_\mathrm{o})$ in case a non-uniform model prior is used — is notoriously subjective and usually qualified, following Jeffreys [250], on a scale from "not worth more than a bare mention" to "decisive". Such labels, however, are not necessary: the Bayes factor is quantitatively the ratio of the probabilities that (the forward models associated to) each model generate data $\mathbf{d} \approx \mathbf{d}_\mathrm{o}$.

Bayes factor

posterior odds

Naturally, each M_m may be parametrised by $\boldsymbol{\psi}_m$, which can, in general, be different in size and/or composition and not necessarily disjoint.[86] For the purposes of model selection, however, they are all *nuisances*: $\cup_m \boldsymbol{\psi}_m \to \mathbf{v}$, and the problem can be familiarly restated as inference of a single parameter of interest $\mathcal{M} \to \boldsymbol{\theta}$ in a *Bayesian supermodel™*, as depicted in fig. 3.1, with a superjoint probability

Bayesian supermodel™

$$\mathrm{p}(\mathcal{M}, \{\boldsymbol{\psi}_m\}, \mathbf{d}) = \mathrm{p}(\mathbf{d} \,|\, \{\boldsymbol{\psi}_m\}, \mathcal{M})\, \mathrm{p}(\{\boldsymbol{\psi}_m\} \,|\, \mathcal{M})\, \mathrm{p}(\mathcal{M}). \tag{3.1}$$

trans-dimensional MCMC

This model can be sampled, but its peculiar and possibly *trans-dimensional*[33] structure requires specialised — and very sophisticated — MCMC proposals [186, see also 16, 187].[87] Instead, the discreteness of M_m — which otherwise precludes the use of gradient-based techniques — is usually exploited through enumeration, i.e. explicit independent calculation of

$$\mathrm{p}(\mathbf{d}_\mathrm{o} \mid M_m) = \int \mathrm{p}(\mathbf{d}_\mathrm{o} \,|\, \boldsymbol{\psi}_m, M_m)\, \mathrm{p}(\boldsymbol{\psi}_m \,|\, M_m)\, \mathrm{d}\boldsymbol{\psi}_m\,, \tag{3.2}$$

for each M_m, typically via nested sampling. This separates the execution traces and presents a well-defined set of nuisances to be integrated out in each case–but, unlike sampling, focuses on individually meaningless values and provides no mechanism for steering the computational effort to the relevant (high-posterior-probability) models.

Whichever the technique employed, evidence-based model selection echos the deficiencies of likelihood-based inference: disastrous scalability[88] to high-dimensional parameter

[85] Note that this is not *the* evidence $\mathrm{p}(\mathbf{d}_\mathrm{o}) = \sum_{m=1}^{N_\mathrm{mod}} \mathrm{p}(\mathbf{d}_\mathrm{o} \mid M_m)\, \mathrm{p}(M_m)$, which is, as always, ignorable.

model-averaged posterior

[86] In case a certain parameter $\boldsymbol{\theta} \in \cap_m \boldsymbol{\psi}_m$ figures in *all* models, its *model-averaged posterior* $\sum_{m=1}^{N_\mathrm{mod}} \mathrm{p}(\boldsymbol{\theta} \mid \mathbf{d}_\mathrm{o}, M_m)\, \mathrm{p}(M_m \mid \mathbf{d}_\mathrm{o})$ might be of scientific interest [e.g. 564]. This is nothing more than marginal nuisance-parameter inference.

[87] Of course, when the evidences of a large number N_mod of models are magically tractable, $\mathcal{M}$ can easily be marginally sampled [e.g. 471].

[88] State-of-the-art NS methods of evidence calculation boast [e.g. 199, fig. 7] applicability to "hundreds" of parameters, which can easily be exceeded even in non-hierarchical settings.

spaces (especially for hierarchical models of numerous i.i.d. objects) and a reliance on the tractability of the likelihood(s) and prior(s) and on explicit model parametrisations.

3.2 Simulation-based model selection

SimSIMS

Model selection is a (discrete-)parameter inference task that can be accomplished with the usual SBI tools[89] trained on examples $\{(\mathcal{M}, \mathbf{d})_k\}_{k=1}^{N_{\text{train}}}$ simulated from the $\text{p}(\mathcal{M}, \mathbf{d})$ that corresponds to eq. (3.1). In practice, this involves picking a model $M_m \sim \text{p}(\mathcal{M})$ at random from the model prior and then generating a $\mathbf{d}$ from the respective *supersampling™ distribution* $\text{p}(\mathbf{d} \mid M_m)$.

supersampling™ distribution

Recall, now, that we already discussed model selection! Replacing $l \to \mathcal{M}$ in eq. (2.10) and letting it extend over $\{M_m\}$ instead of only $\{1, 2\}$, we can approximate the model posterior with a $\text{q}(\mathcal{M} \mid \mathbf{d})$ that optimises

$$\mathbb{E}_{\text{p}(\mathcal{M},\mathbf{d})}[\ln \text{q}(\mathcal{M} \mid \mathbf{d})] = \sum_{m=1}^{N_{\text{mod}}} \text{p}(M_m) \times \mathbb{E}_{\text{p}(\mathbf{d} \mid M_m)}[\ln \text{q}(M_m \mid \mathbf{d})]. \tag{3.3}$$

(The prior factors $\text{p}(M_m)$ are implicitly encoded in the *abundance* of training data from each model.) The $\text{q}(\mathcal{M} \mid \mathbf{d})$ can be any valid categorical distribution: e.g. Radev et al. [434] use a NN-parametrised regularised[90] Dirichlet; more commonly throughout the ML literature (and advocated by Radev et al. [434], Elsemüller et al. [143] for simulation-based model selection), the NN is designed as a *multi-classifier* that directly outputs the (unnormalised log-)probabilities for all models: $[\ln \hat{\text{r}}_m(\mathbf{d})]_{m=1}^{N_{\text{mod}}}$, as unconstrained real numbers, which are then again explicitly normalised:

multi-classifier

$$\text{q}(M_m \mid \mathbf{d}) \equiv \hat{\text{r}}_m(\mathbf{d}) \Big/ \sum_{k=1}^{N_{\text{mod}}} \hat{\text{r}}_k(\mathbf{d}) \, . \tag{3.4}$$

This parametrisation differs very slightly from a ratio estimator — which has $N_{\text{mod}} - 1$ outputs (e.g. eq. (2.12) for $N_{\text{mod}} = 2$), and one distribution is considered the "basis" of comparison — in a way that does not matter to infinitely expressive networks trained on limitless data sets. Nevertheless, many further ML tricks and alternative objective functionals, recounted by e.g. Rosasco et al. [457], Jeffrey & Wandelt [248], can be employed to (attempt to) improve performance, should the need arise in practice.

---

[89] Alternative ML methods for evidence estimation based on posterior samples have been developed by e.g. Heavens et al. [209], Jia & Seljak [255], McEwen et al. [355], Srinivasan et al. [493].

[90] Their aim is to penalise confident model posteriors for data that is *not* representative of the simulator. We will never consider such possibility... in this thesis.

3.2.1 Verification of amortised model selection

Just like amortised (continuous-)parameter inference, properties of the approximate model posterior $q(\mathcal{M} \mid \mathbf{d})$ can be tested and verified using validation data unseen during training. Since the model space is discrete, we do not need to define "credible" or "confidence" regions and can directly examine instead the probability (mass) apportioned to the *true model* from which mock data has been simulated. This motivates two diagnostic tools distinct from the ones presented in section 2.3: refinedness and reliability diagrams [120].

true model

Refinedness is the *average* (approximate) *posterior* derived from data simulated according to a given model:

average posterior

$$\text{Refinedness}_{q}(\mathcal{M} \mid M_m) \equiv \mathbb{E}_{p(\mathbf{d}_t \mid M_m)}[q(\mathcal{M} \mid \mathbf{d})]. \tag{3.5}$$

A *refined* (well trained) classifier would assign the most probability to the "correct" model, leading to a pronounced diagonal. But unlike in usual ML applications,[91] in which there is usually a clear distinction between classes, i.e. the supports of each $p(\mathbf{d} \mid M_m)$ are assumed to not overlap—the "cat" model never produces an image that can be mistaken for a "dog"—, Bayesian comparison of scientific models is in general expected to assign non-zero posterior probability to all M_m (i.e. produce non-zero off-diagonal entries), especially in the presence of significant noise. Concretely, the refinedness of the true model posterior,

$$\text{Refinedness}_{p}(\mathcal{M} \mid M_m) \equiv \mathbb{E}_{p(\mathbf{d} \mid M_m)}[p(\mathcal{M} \mid \mathbf{d})] = \frac{\mathbb{E}_{p(\mathbf{d})}[p(M_m \mid \mathbf{d})\, p(\mathcal{M} \mid \mathbf{d})]}{p(M_m)}, \tag{3.6}$$

is the "covariance" of the posterior probabilities, considered as functions of data. The prominence of the diagonal, therefore, depends both on how well $q(\mathcal{M} \mid \mathbf{d})$ approximates $p(\mathcal{M} \mid \mathbf{d})$ and on how powerful the data inherently is in distinguishing the models.

Lastly, note from eq. (3.6) that $\text{Refinedness}_{p}(\mathcal{M} \mid M_m) \times p(M_m)$ is symmetric and sums to the model prior both across rows ($\sum_{M_m} \to p(\mathcal{M})$) and columns ($\sum_{\mathcal{M}} \to p(M_m)$), presenting two additional summary diagnostics that a converged $q(\boldsymbol{\theta} \mid \mathbf{d})$ must satisfy (the row-wise normalisation is always ensured by the explicit normalisation).

Reliability is a neither-Bayesian-nor-frequentist P–P plot. It specifies a fixed "region" in "parameter" space, i.e. a model M_m, and a $p_* \in [0; 1]$ and considers only those test

confusion matrix

[91] Classification performance in ML is often illustrated through a *confusion matrix*, which is equivalent to a refinedness diagram but the posterior estimator is first degraded into a delta distribution at the top-ranked, i.e. most probable model.

examples $\mathcal{M}, \mathbf{d} \sim \mathrm{p}(\boldsymbol{\theta}, \mathbf{d})$ that lead to an approximate posterior probability of M_m equal to[92] p_*; it then determines the *fraction* among them that came from $\mathcal{M} = M_m$:

fraction (empirical)

$$\text{Reliability}_{\mathrm{q}}(M_m, p_*) \equiv \mathrm{p}(\mathcal{M} = M_m \,|\, \mathrm{q}(M_m \,|\, \mathbf{d}) = p_*). \tag{3.7}$$

Thus, $\mathrm{q}(M_m \,|\, \mathbf{d})$ is treated as a deterministic function, corresponding to the credibilities $\gamma(\boldsymbol{\theta}, \mathbf{d})$ of section 2.3, but with $\Gamma(\boldsymbol{\theta}, \mathbf{d}) \to M_m$. One peculiar consequence of this is that the reliability diagram is not guaranteed to span the whole range $p_* \in [0; 1]$: consider e.g. completely uninformative data, for which the true model posterior is always equal to the prior, leaving the reliability undefined except for at $p_* = \mathrm{p}(M_m)$. Where defined, however, a theorem[93] [119] assures that a "well-calibrated", i.e. exact, approximate posterior $\mathrm{q}(\mathcal{M} \,|\, \mathbf{d}) \to \mathrm{p}(\mathcal{M} \,|\, \mathbf{d})$ produces diagonal reliability plots: $\text{Reliability}_{\mathrm{p}}(M_m, p_*) = p_*$, for any M_m, i.e. any fixed "region" in "parameter" space. Indeed, this discussion resonates with the natural interpretation of posterior probability.

3.2.2 Occam's razor

Aristotle's razor—the principle of ontological parsimony—professes the superiority of inference (logical demonstration) from "fewer postulates, hypotheses, or premises" [17], which has been interpreted, by the "orthodox" [245] school of statistics, as referring to the multiplicity of model *parameters*.[94] The motivation for this reading comes from the so-called *Bayesian information criterion (BIC)* [472], which is related to the evidence of a BHM with N_Υ parameters, Gaussian marginal likelihood (in the sense of eq. (1.8); alternatively, any BHM with $N_{\text{obj}} \to \infty$), and uniform prior (in the vicinity of the MLE):

BIC

$$\mathrm{p}(\mathbf{d}_\mathrm{o} \,|\, M_m) \to \mathrm{p}\big(\mathbf{d}_\mathrm{o} \,\big|\, \Upsilon_{\text{MLE}}\big) \times \left(\frac{2\pi}{N_{\text{obj}}}\right)^{N_\Upsilon} \times \ldots \tag{3.8}$$

[92] in the vicinity of, since reliability diagrams are primarily a practical tool, and exact conditioning is impossible on (non-categorical) samples

[93] A theorem by Dawid [119] states that if an infinite number of independent events $\{\mathcal{M}_\mathrm{t} = M_m\}_\mathrm{t}^N$ are considered, each of which having a certain (possibly different) probability, then the overall fraction of "successes" equals the average probability:

$$\mathrm{q}_\mathrm{t} \equiv \mathrm{q}(\mathcal{M}_\mathrm{t} = M_m \,|\, \mathbf{d}_\mathrm{t}) \quad \Longrightarrow \quad \tfrac{1}{N} \textstyle\sum_\mathrm{t} \mathbb{1}(\mathcal{M}_\mathrm{t} = M_m) \to \tfrac{1}{N} \textstyle\sum_\mathrm{t} \mathrm{q}_\mathrm{t} \quad \text{as} \quad N \to \infty,$$

as long as the choice of events is independent of their outcome (lest the theorem be foiled by Maxwell's dæ mon). The reliability diagram is a special case where the selection is based on $\mathrm{q}_\mathrm{t} = p_*$ (note this does not imply knowledge of the realisation of $\mathcal{M}_\mathrm{t}$). To a Bayesian—even a "well-calibrated" [119] one—, this must sound absurd (and obvious): there are no realisations and fractions, just indicator variables that take deterministically the values $\in \{0, 1\}$ and expectations.

[94] Jeffreys [250] places "the onus of proof [...] on the advocate of the more complicated hypothesis," meaning one that introduces "new parameters in laws".

(ignoring terms independent of data and its size), indicating that models with fewer N_γ that can still assign as much probability to the observed data are preferred in the Bayesian sense.

In less asymptotic scenarios, the principle attains a fundamentally different—and, arguably, more significant—form. In fact, consider the evidence of any model, as in eq. (3.2): values with vanishing likelihood but considerable prior probability still contribute to the integration by reducing $\mathrm{p}(\psi_m)$ in regions where the ψ_m support $\mathbf{d}_\mathrm{o}$. The prior mass in the latter (arbitrarily delineated) is called the *Occam factor* [190, 337, fig. 28.5], and the corresponding razor shaves off not parameters *per se* but superfluous parameter *values*—more precisely, their prior probabilities.

Occam factor

But alas, recall that parameters are auxiliary constructs introduced to ensure tractability of likelihood-based calculations. To SBI (essentially marginal), their *existence* is not necessary, and exactly in this light do we—diSBIdents™—read William of Ockham's nominalist philosophy: "There is no universal [$\mathbf{v}$ or ψ] outside the mind really existing in individual substances [$\mathrm{p}(\mathbf{d}_\mathrm{o} \mid \gamma)$ or $\mathrm{p}(\mathbf{d}_\mathrm{o} \mid \mathcal{M})$] or in the essences of things [$\mathrm{p}(\mathbf{d} \mid \gamma)$ or $\mathrm{p}(\mathbf{d} \mid \mathcal{M})$]," [273].

Instead, Bayesian model selection is directly concerned with the marginal sampling distribution of the *data* variable, $\mathrm{p}(\mathbf{d} \mid \mathcal{M})$, and automatically enforces parsimony on it; for a model that generates *a priori* a large variety of data—regardless of whether by means of a wide range of allowed parameter values or a large multiplicity thereof—cannot assign a too-high probability to any of them and so will be renounced in favour of another more specialised one [337, fig. 28.3]—provided that the latter, by chance or design, has specialised exactly onto the observed realisation $\mathbf{d}_\mathrm{o}$.[95]

Still, for the "mind", i.e. simulator/model/model-*builder*, parameters are an essential concept, and the amortisation of simulation-based model selection offers a unique perspective on their impact and a motivates handle of control on Occam factors, as we describe next.

[95] Curiously, this may seem to contradict the principle that Aristotle [17] establishes immediately before originally formulating Occam's razor in his *Posterior Analytics*: "universal demonstration is superior to particular", i.e. a model that can predict a greater variety of observations should be preferred. However, to Aristotle, "universal" concepts in fact bear the greatest specificity since they are defined in the purest/most precise terms. That is, if our data = "the sum of this figure's interior angles equal two right angles" and the models are that it is either a "triangle" or an "isosceles triangle", we should prefer the former since its "sampling distribution" predicts little else than what has been observed, just as Aristotle would have us.

Visualisation with amortisation

The discussion of Occam factors is particularly relevant to comparisons of *nested models*[96] ($M_1 \subset M_2$), which share the same sampling distribution but represent different parameter spaces, one a subset of the other. Often, for example, the "simplified" M_1 *fixes*[97] certain parameters $\boldsymbol{\theta}_\mathrm{f}$ that e.g. control the strength of effects only considered in M_2 to a specific value $\boldsymbol{\theta}_\mathrm{f}^*$ (e.g. zero) instead of *floating* (i.e. inferring from data) them.

nested models

Now, the data generated by M_1 will in general be much more concentrated than that by M_2 (see fig. 3.2), and any realisation from the latter that gets scattered due to noise within the high-density regions of $\mathrm{p}(\mathbf{d} \mid M_1)$ will lead to an erroneous preference for the former—a so-called type II error: the failure to reject the null hypothesis (M_1). Minimising the prevalence of such instances is the goal of *experimental design*, and so the extent $\boldsymbol{\varepsilon}$ around $\boldsymbol{\theta}_\mathrm{f}^*$ of the region in which this effect is expected is an important quantity, interpretable as a *detection limit*. It can be determined by balancing two factors: the amount of noise, i.e. the overlap between $\mathrm{p}\left(\mathbf{d} \,\middle|\, \boldsymbol{\theta}_\mathrm{f}^*, M_2\right)$ and $\mathrm{p}\left(\mathbf{d} \,\middle|\, \boldsymbol{\theta}_\mathrm{f}^* + \boldsymbol{\varepsilon}, M_2\right)$, causing the data confusion[98]; and the amount by which $\mathrm{p}(\mathbf{d} \mid M_1) = \mathrm{p}\left(\mathbf{d} \,\middle|\, \boldsymbol{\theta}_\mathrm{f} = \boldsymbol{\theta}_\mathrm{f}^*, M_2\right)$ dominates the marginal $\mathrm{p}(\mathbf{d} \mid M_2)$, which depends on the *additional parameter space* considered by the latter: again, see fig. 3.2.

experimental design

detection limit

While detection limits are difficult to calculate from first principles for all but the simplest models, they can be obtained by applying model selection to mock examples from M_2 with known true $\boldsymbol{\theta}_\mathrm{f}$. Importantly, since the mapping $\boldsymbol{\theta}_\mathrm{f} \to \mathbf{d}$ is non-deterministic, the posterior model probabilities and the Bayes factor cannot be expressed or depicted directly as a function of $\boldsymbol{\theta}_\mathrm{f}$; instead, the *average* trend (i.e. the weighted mean of the arrows in fig. 3.2) needs to be extracted from analyses of numerous mock examples, ruling out the possibility of likelihood-based (e.g. NS) evaluation.[99] With amortised simulation-based model selection, on the other hand, such exploration is trivial and requires nothing more than plotting the validation results—which are anyway calculated during / at the end of training as part of good ML practice. We show examples from our application to SNæ in fig. 13.2.

---

[96] In fact, some authors [e.g. 330, 549] consider *ab initio* the Occam factor as the contribution to the Bayes factor of nested models formed by the ratio of their prior volumes.

[97] One might rightfully, in this case, consider the models to have a different number of parameters altogether.

[98] Conversely, data with true $\boldsymbol{\theta}_\mathrm{f} = \boldsymbol{\theta}_\mathrm{f}^*$ from *any* of the two models may assign significant posterior probability to various $\boldsymbol{\theta}_\mathrm{f}$ under model M_2. This has given rise to a deplorable approach to nested-model comparison via the posterior probability of $\boldsymbol{\theta}_\mathrm{f}^*$ under the augmented model. Crucially, this does not take into consideration the Occam factor in any way—but is also, consequently, insensitive to the considerations presented below regarding the amount of extra space.

[99] An extremely laborious—and still approximate—poor man's attempt at examining using likelihood-based methods the distribution of Bayes factors in (hyper)parameter space is present in Benito et al. [38, fig. 10]

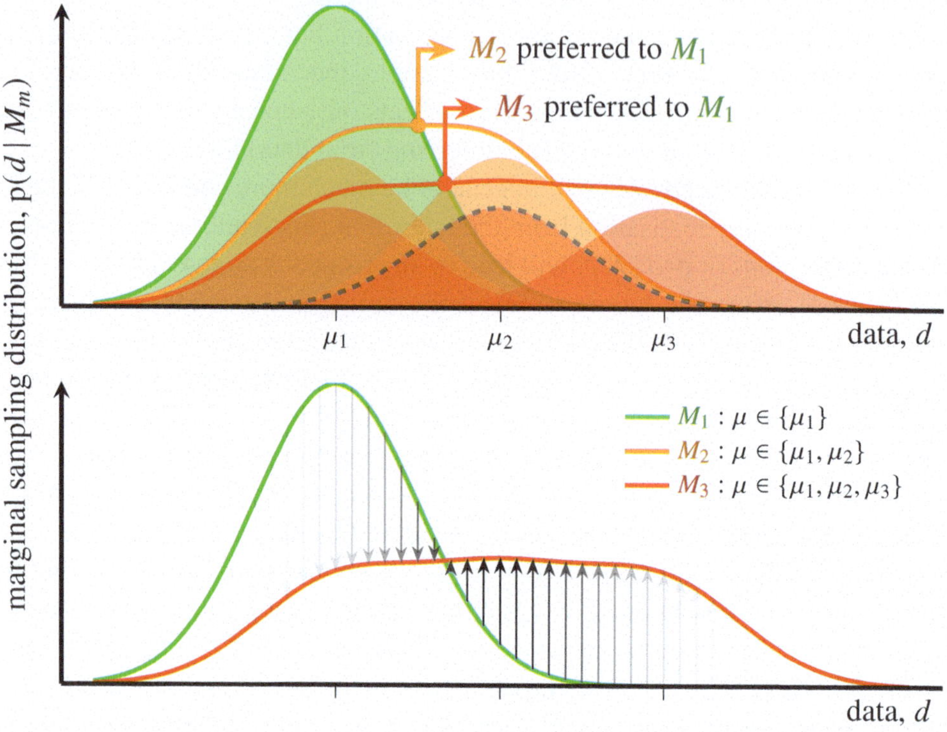

Figure 3.2: Marginal sampling distributions / evidences of three nested models $M_1 \subset M_2 \subset M_3$ as a function of the data variable. The models are mixtures of up to three distributions at different locations: $\mu \in \{\mu_1, \mu_2, \mu_3\}$, which can be viewed as a simplified *parameter space*, different portions of which are included in each model. *Top*: a given model is preferred over another when its $p(d \mid M_m)$ is higher. The additional parameter/data space in M_3 with respect to M_2 makes it *less* preferred in regions around μ_1, in which the two models are essentially the same—up to the normalisation—, "pushing" the decision boundary further in parameter/data space from μ_1. *Bottom*: Arrows depict the possible Bayes factors $p(d \mid M_3) / p(d \mid M_1)$ inferred from data sampled from $\mu = \mu_2$ (dashed in the top plot; the arrow opacity follows the density/prominence of data at d). Thus, even at "fixed parameter", i.e. mixture component, model comparison results may greatly vary depending on the "noise" realisation. Nevertheless, the general trend is that the further in parameter space the data is sampled from (e.g. at μ_3), the less likely it is that the Bayes factor will prefer the restricted model.

Occam's axe™: truncation for model selection

Similarly, we previously aimed our truncation axe exactly at the same parameter-space regions with exactly the same goal: increasing the marginal sampling density at (evidence at) $\mathbf{d}_o$, remarking that this does not modify posterior parameter inference (from $\mathbf{d}_o$)—provided truncation is at the highest[104] hierarchical level of interest. Even though this is not the case for $\boldsymbol{\psi}_m$ within model selection, now our goal *is* to modify inference (of the top-level parameter $\mathcal{M}$) in a desirable/interpretable way, and furthermore, our concerns with middle-layer truncation (appendix 17) mainly refer to arguments of typicality in settings containing i.i.d./exchangeable variables (object-specific parameters and/or data). Therefore, $\boldsymbol{\psi}_m$-space truncation can be considered a viable procedure to counteract the effects (with regards to the Occam factor) of the arbitrariness of prior selection when it is applied to *distinguishable* parameters, i.e. the globals $\boldsymbol{\gamma}$, which are not expected to suffer the caveats of appendix 17.

Concretely, we propose

Strategy 2. Truncated inference and trustworthy simulation-parsimonious model selection:

1. Perform truncated *parameter inference* separately with each model, possibly deriving model-averaged[86] posteriors and truncation regions for any shared parameters.

2. Train a *model-posterior* estimator *re-using* the simulations from the last truncation step, which represent only the regions in each model's parameter space that are consistent with the data at least to a minimal degree.

Although we will not apply this strategy in the present thesis, we are currently preparing a demonstration (on real data) in the ever-enticing (for obvious reasons) field of exoplanet analysis, concretely to atmospheric "retrievals" [see e.g. 152], which allow for almost indefinite—but arguably unwarranted—model sophistication.

Chapter 4

Developments in hierarchical SBI

In this chapter, we develop methodologies that combine and build upon the concepts described above, specifically targeting inference of Bayesian hierarchical models (BHMs) with *truncated marginal neural ratio estimation (TMNRE)* [361], which will prove useful in the context of — and we will later apply to — supernova cosmology. *TMNRE*

4.1 Complete hierarchical TMNRE

SICRET

Significant role in motivating the development of SBI methods plays the inconvenience — and frequently, impossibility — of marginal likelihood-based inference of global/population parameters in the presence of numerous object-specific nuisances. That is, often one identifies in a hierarchical model $\boldsymbol{\gamma} \to \boldsymbol{\theta}$, $\{\boldsymbol{\lambda}^i\} \to \boldsymbol{\nu}$. Here, we are also interested in the converse: inferring marginally the many $\boldsymbol{\lambda}^i \to \boldsymbol{\theta}_g$, with $\boldsymbol{\gamma} \to \boldsymbol{\nu}$ marginalised. In principle, there should be no distinction when employing marginal SBI: a brute-force approach can train *bespoke* posterior/ratio estimators for each $\boldsymbol{\lambda}^i$ (of course, one or many subsets $\boldsymbol{\lambda}^i_g$ of the parameters of each object may instead be considered), conditioned on *the full data set* $\{\mathbf{d}^i_\mathrm{o}\}$. The conditional independence of the objects on $\boldsymbol{\gamma}$ (cf. section 1.1 and eq. (1.6)), however, leads to two simplifications.

Consider the marginal posterior(s) being approximated:

$$\mathrm{p}\big(\boldsymbol{\lambda}^i \,\big|\, \{\mathbf{d}^j\}\big) = \frac{\displaystyle\int \mathrm{p}\big(\boldsymbol{\gamma}, \boldsymbol{\lambda}^i, \mathbf{d}^i \,\big|\, \{\mathbf{d}^{j\neq i}\}\big)\,\mathrm{d}\boldsymbol{\gamma}}{\mathrm{p}(\mathbf{d}^i)} = \frac{\displaystyle\int \mathrm{p}\Big(\boldsymbol{\lambda}^i, \mathbf{d}^i \,\Big|\, \boldsymbol{\gamma}, \cancel{\{\mathbf{d}^{j\neq i}\}}\Big)\,\mathrm{p}\big(\boldsymbol{\gamma} \,\big|\, \{\mathbf{d}^{j\neq i}\}\big)\,\mathrm{d}\boldsymbol{\gamma}}{\mathrm{p}(\mathbf{d}^i)}. \tag{4.1}$$

By defining an alternative global-parameter prior $\tilde{\mathrm{p}}_i(\boldsymbol{\gamma}) \leftarrow \mathrm{p}\big(\boldsymbol{\gamma} \,\big|\, \{\mathbf{d}^{j\neq i}\}\big)$, inference of $\boldsymbol{\lambda}^i$ can therefore be performed by conditioning only on data for the respective object, $\mathbf{d}^i$, rather

© The Author(s), under exclusive license to Springer Nature Switzerland AG 2026
K. Karchev, *Supernova Cosmology for the 21st Century*,
https://doi.org/10.1007/978-3-032-15072-1_4

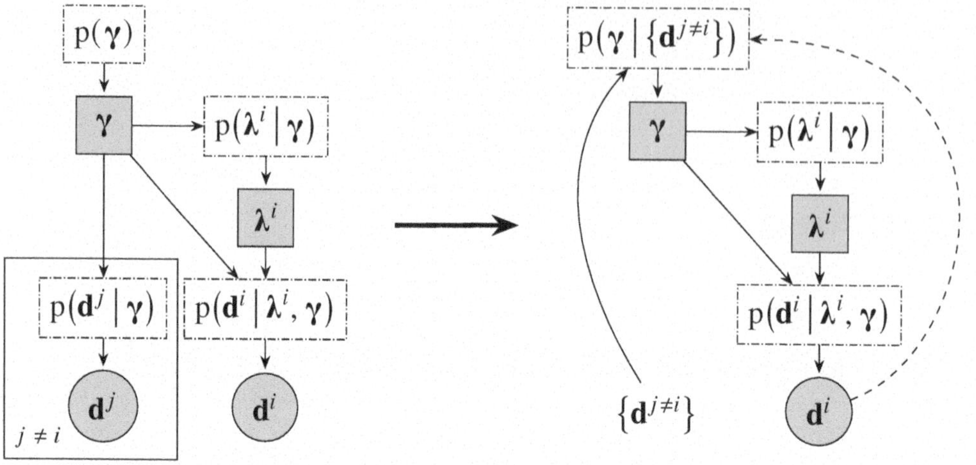

Figure 4.1: Re-configuration of a BHM: when inferring marginally the object-specific parameters $\boldsymbol{\lambda}^i$, data on other objects, $\{\mathbf{d}^{j\neq i}\}$, can be considered given/input and the model re-stated to use the respective global-parameter posterior $\mathrm{p}(\boldsymbol{\gamma}\,|\,\{\mathbf{d}^{j\neq i}\})$ as an alternative prior. Moreover, for large data sets, adding also $\mathbf{d}^i$ to the conditioning (the dashed line, which technically creates a forbidden cycle) modifies the posterior very mildly, justifying the use of the full $\mathrm{p}(\boldsymbol{\gamma}\,|\,\{\mathbf{d}^j\})$ instead. This approach, convenient with SBI, is contrary to traditional hierarchical inference that, conceptually, derives first constraints on local parameters $\{\boldsymbol{\lambda}^i\}$ and subsequently pools them to infer the $\boldsymbol{\gamma}$.

than the full set. In fact, this re-configuration, illustrated in fig. 4.1, corresponds to the original — literal — interpretation of the prior as representing information from independent "previous" observations, i.e. of objects other than i, and is a general statement about BHMs. Importantly, all parameters that ensure conditional independence must be jointly inferred.

When, furthermore, the number of observed objects is large, each one individually contributes relatively little to constraining the global/population parameters[100] — indeed, pooling the information across the data set is the main reason for introducing hierarchical modelling in the first place. Therefore, we can approximate $\mathrm{p}(\boldsymbol{\gamma}\,|\,\{\mathbf{d}^{j\neq i}\}) \approx \mathrm{p}(\boldsymbol{\gamma}\,|\,\{\mathbf{d}^i\}) \to \tilde{\mathrm{p}}(\boldsymbol{\gamma})$, discarding the unwieldy i-dependence of the alternative prior.

exchangeability If the objects are furthermore identically distributed (i.e. are i.i.d.) and hence, *exchangeable*, local quantities (parameters and data) for different objects can be considered realisations of the *the same* random variables[101]: $\boldsymbol{\lambda}^i \to \boldsymbol{\lambda}$ and $\mathbf{d}^i \to \mathbf{d}$. In case the BHM contains

[100] Concretely, $\mathrm{p}(\boldsymbol{\gamma}\,|\,\{\mathbf{d}^j\}) / \mathrm{p}(\boldsymbol{\gamma}\,|\,\{\mathbf{d}^{j\neq i}\}) \propto \mathrm{p}(\mathbf{d}^i\,|\,\boldsymbol{\gamma})$ is approximately constant across the high-density regions of $\mathrm{p}(\boldsymbol{\gamma}\,|\,\{\mathbf{d}^j\})$.

[101] Note, accordingly, the lack of a plate in the re-configured graph in fig. 4.1, making the use of i unnecessary.

object-specific settings $\mathbf{a}^i$ (see eq. (1.9) and fig. 1.2) that modify the data-sampling distribution (e.g. measurement noise, constraints from external analyses, or even the label $i \to \mathbf{a}^i$), exchangeability is still realised on the level of $\{(\boldsymbol{\lambda}, \mathbf{d}, \mathbf{a})^i\}$, since $\mathrm{p}(\mathbf{d}^i \,|\, \mathbf{a}^i, \boldsymbol{\lambda}^i, \boldsymbol{\gamma})$ is the same for two objects if their corresponding parameters and *metadata*[102] match. *metadata*

Thus, every run of the simulator results in N_{obj} samples from

$$\mathrm{p}(\boldsymbol{\lambda}, \mathbf{a}, \mathbf{d} \,|\, \boldsymbol{\gamma}) = \mathrm{p}(\mathbf{d} \,|\, \mathbf{a}, \boldsymbol{\lambda}, \boldsymbol{\gamma})\, \mathrm{p}(\boldsymbol{\lambda} \,|\, \boldsymbol{\gamma})\, \mathrm{p}(\mathbf{a}), \tag{4.2}$$

where $\mathrm{p}(\mathbf{a}) = \sum_i \delta(\mathbf{a}^i) / N_{\mathrm{obj}}$ in accordance with eq. (1.9) and is represented by the (fixed/given) collection $\{\mathbf{a}^i\}$. In accordance with eq. (4.1), marginalising over the alternative global-parameters prior $\tilde{\mathrm{p}}(\boldsymbol{\gamma}) \leftarrow \mathrm{p}(\boldsymbol{\gamma} \,|\, \{\mathbf{d}^i_{\mathrm{o}}\})$ enables training *a single* posterior/ratio estimator $\mathrm{q}(\boldsymbol{\lambda} \,|\, [\mathbf{d}, \mathbf{a}])$ or $\hat{\mathrm{r}}(\boldsymbol{\lambda}, [\mathbf{d}, \mathbf{a}])$ that can be used to evaluate *all* object-specific marginal posteriors, given only the particular $[\mathbf{d}^i, \mathbf{a}^i]$. Note that the full data set $\{\mathbf{d}^i\}$ need not be specifically provided to the NN since the (global) information contained in is still extracted and encoded during training.

A final simplification can be made when the observational noise is much smaller than the population scatter.[103] Then, local constraints are object-, as opposed to population-, driven, and precise globally aggregated information in the form of $\mathrm{p}(\boldsymbol{\gamma} \,|\, \{\mathbf{d}^i\})$ is not necessary. Therefore, credible results can still be obtained by using simply the original $\mathrm{p}(\boldsymbol{\gamma})$ or—to a better approximation and ease of implementation—the *truncated* global-parameters prior $\tilde{\mathrm{p}}_{T(\{\mathbf{d}^i_{\mathrm{o}}\})}(\boldsymbol{\gamma})$. This setup is particularly convenient for *simultaneous* global and local SBI since the same simulations can be used for both tasks, and it does not require joint inference of all $\boldsymbol{\gamma}$, i.e. truncation can be marginal over smaller groups $\boldsymbol{\gamma}_g$.

Still, often only some object-specific parameters can be precisely estimated from $\mathbf{d}^i$ alone while others require hierarchical pooling, i.e. only their population can be reliably inferred (see chapter 12). In such cases, the information discarded when degrading $\mathrm{p}(\boldsymbol{\gamma} \,|\, \{\mathbf{d}^i_{\mathrm{o}}\}) \to \tilde{\mathrm{p}}_{T(\{\mathbf{d}^i_{\mathrm{o}}\})}(\boldsymbol{\gamma})$ (or even to the original $\mathrm{p}(\boldsymbol{\gamma})$) can be restored by explicitly providing $\{\mathbf{d}^i\}$—or a *global summary* $\mathbf{s}(\{\mathbf{d}^i\})$—to the local-parameters posterior/ratio estimator: $[\mathbf{d}, \mathbf{a}] \to [\mathbf{d}, \mathbf{a}, \mathbf{s}(\{\mathbf{d}^i\})]$. When trained as part of a combined global+local inference NN, the summary assumes the structural role of the global-parameters posterior and is the natural conditioning context / data-input of the respective estimator, $\mathrm{q}(\boldsymbol{\gamma} \,|\, \mathbf{s}(\{\mathbf{d}^i\}))$ or $\hat{\mathrm{r}}(\boldsymbol{\gamma}, \mathbf{s}(\{\mathbf{d}^i\}))$, as illustrated in fig. 4.2. *global summary*

102 So-called because it always appears as conditioning of probability distributions, i.e. is fixed. In fact, since $\{\mathbf{a}^i\}$ often represent (*auxiliary*) measurements that describe the particular realisation of the observing run, their treatment as stochastically produced data (possibly resulting from an adjacent hierarchical process) is an obvious opportunity for extending the model. In this thesis, such is the case of the SN redshifts, which start off fixed and, after a series of improvements, end up explicitly inferred from external observations as part of a unified analysis.

103 more formally, when the object-specific $\mathrm{p}(\boldsymbol{\lambda} \,|\, \mathbf{d}, \boldsymbol{\gamma})$ is significantly more concentrated than $\mathrm{p}(\boldsymbol{\lambda} \,|\, \boldsymbol{\gamma})$

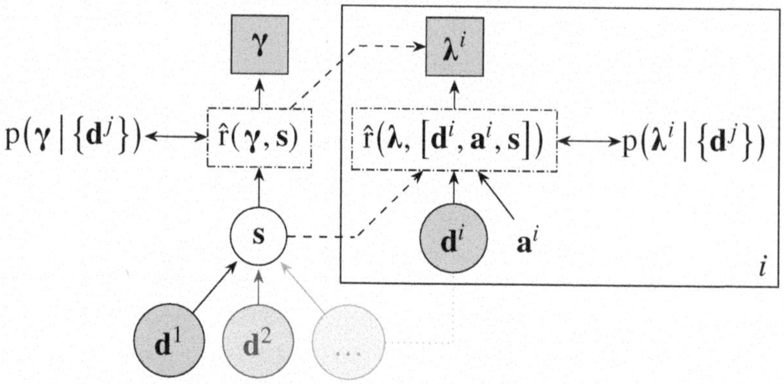

Figure 4.2: Abstract structure of a combined global+local inference network (joint-to-marginal ratio estimator as in our applications in part IV). The dashed arrows present two alternatives for including global information in inferring local parameters: either explicitly conditioning the posterior/ratio estimator, or simulating training data according to the estimated global posterior. Compare with the specific NNs used throughout part IV and with the re-configured BHM in fig. 4.1.

Strategy 3. Complete truncated[104] marginal SBI of a hierarchical model:

1. Infer the global parameters from the complete data set, truncating their prior so as to tightly encompass the posterior $\mathrm{p}(\boldsymbol{\gamma} \mid \{\mathbf{d}^i_{\mathrm{o}}\}) \to \tilde{\mathrm{p}}_{T(\{\mathbf{d}^i_{\mathrm{o}}\})}(\boldsymbol{\gamma})$. Inference and truncation may be marginal: $\tilde{\mathrm{p}}_{T(\{\mathbf{d}^i_{\mathrm{o}}\})}(\boldsymbol{\gamma}) \to \prod_g \tilde{\mathrm{p}}_{T(\{\mathbf{d}^i_{\mathrm{o}}\})}(\boldsymbol{\gamma}_g)$, but must include all global parameters, $\boldsymbol{\gamma} = \cup_g \boldsymbol{\gamma}_g$, so as to ensure conditional independence.

2. In the last truncation stage, record also any object-specific parameters of interest $\boldsymbol{\lambda}^i_g$ for all objects in the simulations, collecting thus $N_{\mathrm{obj}} \times N_{\mathrm{train}}$ samples

$$\left\{\left\{\left(\boldsymbol{\lambda}^{i,k}_g, \mathbf{d}^{i,k}, \mathbf{a}^i\right)\right\}_{i=1}^{N_{\mathrm{obj}}}\right\}_{k=1}^{N_{\mathrm{train}}} \quad \text{that represent} \quad \mathrm{p}(\boldsymbol{\lambda}_g, \mathbf{d}, \mathbf{a} \mid \{\mathbf{d}^i_{\mathrm{o}}\}) \tag{4.3}$$

and use them for marginal SBI of $\mathrm{p}(\boldsymbol{\lambda} \mid \mathbf{a}, \mathbf{d}, \{\mathbf{d}^i_{\mathrm{o}}\})$. The posterior for the parameters of object i is obtained by evaluating the estimator given the relevant $\mathbf{d}^i_{\mathrm{o}}$ and $\mathbf{a}^i$.

In general, the full data set (possibly summarised through $\mathbf{s}(\{\mathbf{d}^i\})$ as part of global inference) must also be supplied to the NN: $\mathrm{q}(\boldsymbol{\lambda} \mid [\mathbf{d}, \mathbf{a}, \{\mathbf{d}^i\}])$ or $\hat{\mathrm{r}}(\boldsymbol{\lambda}, [\mathbf{d}, \mathbf{a}, \{\mathbf{d}^i\}])$. It can be omitted in two cases:

[104] Truncation here refers to the global parameters. The extension/converse application of local truncation for global inference—which proves significantly more challenging—is discussed in appendix 17; it also relates to strategy 2, where we were much more blasé about it.

a) either the training data for local inference is re-generated with $\boldsymbol{\gamma} \sim \mathrm{p}(\boldsymbol{\gamma} \mid \{\mathbf{d}^i_\mathrm{o}\})$ (in fact, the estimate thereof) or corresponding weights are used in training;

b) or the particular local parameter(s) $\boldsymbol{\lambda}_g$ is/are deemed well enough constrained solely from the respective object-specific data $\mathbf{d}^i$ and the truncated prior.[105]

A note on NRE for local parameters: In evaluating the posterior using a local-parameters ratio estimator $\hat{\mathrm{r}}(\boldsymbol{\lambda}, [\mathbf{d}_\mathrm{o}, \mathbf{a}, \ldots])$ trained as above, reference is made to the prior $\mathrm{p}(\boldsymbol{\lambda})$. As always, this is the marginal from the original model, which for a BHM is the compound distribution $\int \mathrm{p}(\boldsymbol{\lambda} \mid \boldsymbol{\gamma})\, \mathrm{p}(\boldsymbol{\gamma})\, \mathrm{d}\boldsymbol{\gamma}$, regardless of the training prior (i.e. $\int \mathrm{p}(\boldsymbol{\lambda} \mid \boldsymbol{\gamma})\, \tilde{\mathrm{p}}(\boldsymbol{\gamma})\, \mathrm{d}\boldsymbol{\gamma}$). Since the $\mathrm{p}(\boldsymbol{\lambda})$ is thus often intractable, simulating according to the truncated prior instead of the (estimate of the) posterior $\mathrm{p}(\boldsymbol{\gamma} \mid \{\mathbf{d}^i\})$ has the additional advantage that the training/validation samples can be used as the base distribution to be re-weighted by $\hat{\mathrm{r}}$ for representing $\mathrm{p}(\boldsymbol{\lambda} \mid \{\mathbf{d}^i\})$.

4.2 Catalogue-based NRE

RESSET

In section 1.1, we presented the simplest ontology of a BHM: in which the data on each object is identically sampled (conditioned on some given object-specific settings). However, this description crucially relies on the definition of an "object", which we left abstract. In fact, it is up to the data collector (observer) to define the available measurements, and they may or may not align with the notion of an "object" as used within a BHM. For instance, some astronomical data comes as an aggregate of signals from multiple sources that are not immediately distinguishable. For this case, Anau Montel & Weniger [13] developed a TMNRE methodology that defines and identifies individual "sources" only for the purposes of truncation; they then need to *correct* the aggregate-based results for the effects of this modelling choice.

More commonly, the objects *are* identifiable entities, but data is available only for a *non-random*[106] subset of the population determined by a given procedure that, in principle, represents three distinct processes: identifying the existence of an object (*detection*), obtaining data for it (*measurement*), and including it (*selection*) in the collection analysed in a particular study. Assuming that their outcome is binary[107] (i.e. they always either fail

detection
measurement
selection

[105] Under this hypothesis, including $\{\mathbf{d}^i\}$ or $\mathbf{s}(\{\mathbf{d}^i\})$ as a NN input can improve constraints but/or contribute noise to the training procedure due to the additional network parameters needed to extract relatively little additional information with respect to $\mathbf{d}^i$.

[106] If the subset were random, it would be indistinguishable from a smaller complete population.

[107] We will briefly touch upon the possibility of partially missing data, i.e. the availability of distinct *multi-modal* data on each object, in section 8.4.

multi-modal

or completely succeed), detection + measurement + selection can be represented through a single random variable S indicating whether an object has ultimately been included ("detected": $S^i = \mathfrak{s}$ for $i = 1, \dots, N_{\text{sel}}$) or not ("missed": $S^j = \mathfrak{m}$ for $j = N_{\text{sel}} + 1, \dots, N_{\text{tot}}$) in a released *catalogue*.

catalogue

The indicator variables for all objects—*observed or missed*—thus have well-defined values: $\{\mathfrak{s}^i\}$ and $\{\mathfrak{m}^j\}$, and must be considered given, i.e. part of the data alongside the usual object-specific measurements $\{\mathbf{d}^i\}$ (limited, of course, to observed objects i), with likelihood / sampling distribution $p(S^k \,|\, \mathbf{d}^k, \gamma)$, also called *efficiency*. In this picture (fig. 4.3), inference with missing data corresponds to marginalisation of

efficiency

marginal efficiency

1. the un-observed $\{\mathbf{d}^j\}$, defining the *marginal efficiency*

$$p(S \,|\, \gamma) \equiv \int p(S \,|\, \mathbf{d}, \gamma)\, p(\mathbf{d} \,|\, \gamma)\, \mathrm{d}\mathbf{d}\,; \tag{4.4}$$

   and since all objects are assumed to have the same sampling distribution[108] $p(\mathbf{d} \,|\, \gamma)$, the indices $\{i\}$ and $\{j\}$ are once again *exchangeable*, so a further combinatorial factor $\binom{N_{\text{sel}}}{N_{\text{tot}}}$ must be taken into account when "selecting" the observed sample:

$$\begin{aligned} p(\{\mathbf{d}^i\}, \{\mathfrak{s}^i\}, \{\mathfrak{m}^j\} \,|\, N_{\text{tot}}, \gamma) &= \binom{N_{\text{sel}}}{N_{\text{tot}}} \int \prod_k^{N_{\text{tot}}} p\left(S^k \,\middle|\, \mathbf{d}^k, \gamma\right) p\left(\mathbf{d}^k \,\middle|\, \gamma\right) \mathrm{d}\{\mathbf{d}^{j > N_{\text{sel}}}\} \\ &= \binom{N_{\text{sel}}}{N_{\text{tot}}} \prod_{i=1}^{N_{\text{sel}}} \underbrace{p(\mathfrak{s}^i \,|\, \mathbf{d}^i, \gamma)\, p(\mathbf{d}^i \,|\, \gamma)}_{= p(\mathbf{d}^i \,|\, \mathfrak{s}^i, \gamma)\, p(\mathfrak{s}^i \,|\, \gamma)} \prod_{j = N_{\text{sel}}+1}^{N_{\text{tot}}} p(\mathfrak{m}^j \,|\, \gamma) \\ &= \prod_{i=1}^{N_{\text{sel}}} p(\mathbf{d}^i \,|\, \mathfrak{s}^i, \gamma) \times \underbrace{\mathrm{Binom}(N_{\text{sel}} \,|\, N_{\text{tot}}, p(\mathfrak{s} \,|\, \gamma))}_{p(\{\mathfrak{s}^i\}, \{\mathfrak{m}^j\} \,|\, N_{\text{tot}}, \gamma) \,\equiv\, p(N_{\text{sel}} \,|\, \gamma)}; \end{aligned} \tag{4.5}$$

2. the uncertain—*a priori* and *a posteriori*—total size of the population, N_{tot}, which can meaningfully depend on parameters of interest through a $p(N_{\text{tot}} \,|\, \gamma)$:

$$p(N_{\text{sel}} \,|\, \gamma) = \int \mathrm{Binom}(N_{\text{sel}} \,|\, N_{\text{tot}}, p(\mathfrak{s} \,|\, \gamma))\, p(N_{\text{tot}} \,|\, \gamma)\, \mathrm{d}N_{\text{tot}}\,. \tag{4.6}$$

population size (total)

Like any other *a priori* distribution, $p(N_{tot} \,|\, \gamma)$ is to an extent a modelling choice. How-

[108] This is, in fact, not a trivial statement: it means that objects j *could* have potentially been observed but were not, owing to the particular realisation of their latent properties and observational noise.

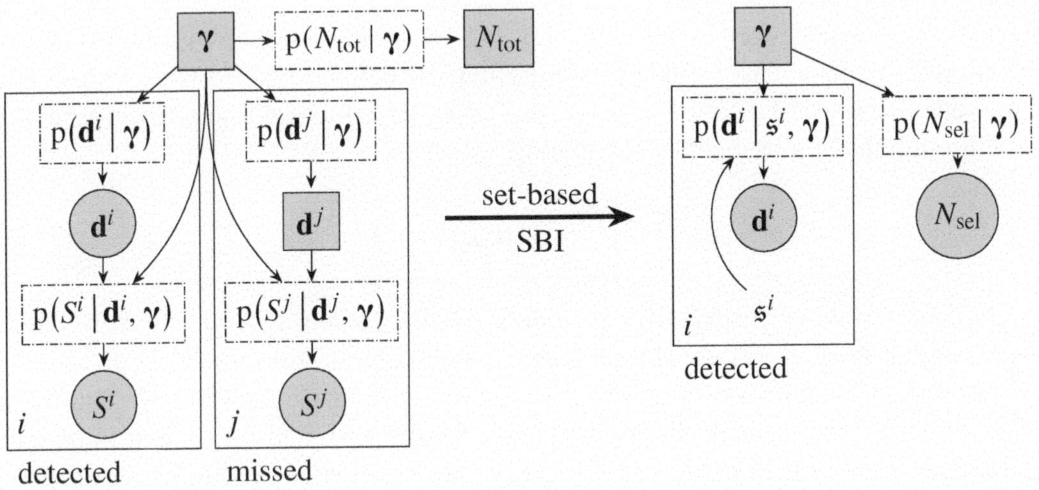

Figure 4.3: Re-configuration of a BHM for population inference from a non-random subsample of the objects. In the original formulation (left: straightforward to implement as a simulator), both the population size (N_{tot}) and the data realisations of missed objects ($\{\mathbf{d}^j\}$) are unobserved and must be marginalised, making it intractable for likelihood-based inference. Set-based SBI, corresponding to the simpler probabilistic model on the right, is instead trained directly on simulated sets of data for "detected" objects and their stochastic count.

ever, often concrete physical models are available for the expected abundance or rate of occurrence of the category of objects considered, i.e. for the size of the population. They usually express a deterministic relationship $\langle N_{\mathrm{tot}}\rangle(\boldsymbol{\gamma})$ that serves as the rate parameter of a Poisson distribution, simplifying the integral

$$\begin{aligned} &\mathrm{p}(N_{\mathrm{tot}} \mid \boldsymbol{\gamma}) \to \mathrm{Pois}[\langle N_{\mathrm{tot}}\rangle(\boldsymbol{\gamma})] \\ \Longrightarrow\ &\mathrm{p}(N_{\mathrm{sel}} \mid \boldsymbol{\gamma}) \to \mathrm{Pois}[\langle N_{\mathrm{tot}}\rangle(\boldsymbol{\gamma}) \times \mathrm{p}(\mathfrak{s} \mid \boldsymbol{\gamma})]. \end{aligned} \tag{4.7}$$

All in all, inference can be reduced to the superficially simple likelihood / sampling distribution illustrated in fig. 4.3:

$$\mathrm{p}(\mathbf{D} \mid \boldsymbol{\gamma}) = \prod_{i=1}^{N_{\mathrm{sel}}} \mathrm{p}(\mathbf{d}^i \mid \mathfrak{s}^i, \boldsymbol{\gamma}) \times \mathrm{p}(N_{\mathrm{sel}} \mid \boldsymbol{\gamma}), \tag{4.8}$$

where by $\mathbf{D}$ we label the observed data set $\{\mathbf{d}^i\}$ *and* its information-carrying cardinality N_{sel}. Since both terms are wrought of intractable integration, they are extremely hard *data set* to calculate in practice, even if the efficiency $\mathrm{p}(S \mid \mathbf{d}, \boldsymbol{\gamma})$ is known.[109] Furthermore, even in the simplest realistic cases, this probability may be defined instead with respect to the unobserved object-specific parameters, i.e. $\mathrm{p}(S \mid \mathbf{d}, \boldsymbol{\lambda}, \boldsymbol{\gamma})$, which adds another layer of complexity through the requirement of calculating

$$\mathrm{p}(S \mid \mathbf{d}, \boldsymbol{\gamma}) \equiv \int \mathrm{p}(S \mid \mathbf{d}, \boldsymbol{\lambda}, \boldsymbol{\gamma})\, \mathrm{p}(\boldsymbol{\lambda} \mid \boldsymbol{\gamma})\, \mathrm{d}\boldsymbol{\lambda}\,. \tag{4.9}$$

Very often, moreover, and especially for complicated data resulting from an involved detection and measurement procedure and subjected to more or less arbitrary selection criteria—e.g. light curves used for SN cosmology—, no numeric expression for the efficiency at any level is available, and instead, the integrals eqs. (4.4), (4.6), and (4.9) need to be calculated stochastically from simulations. Crucially, these estimates need to be performed for the proposed $\boldsymbol{\gamma}$ at every step of MCMC sampling if a likelihood-based approach is chosen, Clearly, then, direct simulation-based inference—in which all marginalisations are implicit—is the superior method to account for non-random sample selection.

4.2.1 Apologia of SBI with stochastic cardinality

"what I do not know I do not think I know either" [Plato, Apology 21d]

MLP

The direct SBI approach to solving eq. (4.8) and fig. 4.1, simulating mock $\mathbf{D}$ and using them to train a posterior or ratio estimator, presents a peculiar challenge in that each example may in principle have different cardinality, whereas traditional NNs, i.e. *multi-layer perceptrons (MLPs)*, require that their input size be pre-determined and fixed during training and evaluation. To circumvent this, a number of methods have been devised to work either with individual objects, or to condition the simulator on the observed N_{sel} (corresponding with the different re-configurations of the original BHM), both enabling the use of fixed-input-size NNs. Below, we survey them briefly, highlighting their crippling deficiencies as motivation for stochastic-cardinality SBI with set-based networks, which we introduce at the end.

SBI from individual objects [528, 168, 339] One obvious idea is to learn the likelihood $\mathrm{p}(\mathbf{d} \mid \mathfrak{s}, \boldsymbol{\gamma})$ from a single *observed* object[110] and then evaluate and combine it across the data set $\{\mathbf{d}_{\mathrm{o}}^{i}\}$. This has numerous perceived advantages: it requires a simpler and more

[109] The simplest case often employed in toy models—that of collecting the data that surpass a parameter-independent detectability threshold, i.e. a signal-to-noise ratio (SNR) cut—is rarely realised in practice but often used in simplistic analyses [e.g. 172, 163].

[110] or, alternatively, the single-*observed*-object posterior $\mathrm{p}(\boldsymbol{\gamma} \mid \mathbf{d}, \mathfrak{s}) \propto \mathrm{p}(\mathbf{d} \mid \mathfrak{s}, \boldsymbol{\gamma})\, \mathrm{p}(\boldsymbol{\gamma})$ via e.g. NPE and then take care to divide out the prior $\mathrm{p}(\boldsymbol{\gamma})$ since it is still the individual likelihoods that compose; [see e.g. 528, eq. (10) or 168, subsection 3.1]

streamlined NN that inputs only a single object, less training (the task is—seemingly!—simpler in proportion to the data), and $O(N_{\text{sel}})$ times less memory (during training), while allowing inference from future additions to the data (newly observed objects) without re-training.

Nevertheless, this strategy does not address the information contained in the observed counts ($\mathrm{p}(N_{\text{sel}} \mid \gamma)$) and, crucially, is only applicable when the observations are i.i.d. conditioned on the inferred parameters, which is often not the case with marginal inference and may thus enforce learning a much higher-dimensional likelihood than needed. Moreover, combining $O(N_{\text{sel}})$ independent results is prone to accumulation of the inevitable approximation errors—minor on the level of individual likelihoods/posteriors but magnified thus $O(N_{\text{sel}})$ times. For example, Wagner-Carena et al. [528, fig. 8] observed a bias when combining inference from 1000 objects, while Makinen et al. [339] were able to scale to (exactly) 10^4 only by training an additional fixed-input-size network to aggregate the likelihood estimates, correcting the inaccuracy.

SBI from fixed-size collections [172, 163] Inference (and training) should then be performed from the entire observed set, whose size is in general *a priori* unknown due to the stochasticity of both population size and sample selection. Observations collapse uncertainty in the latter, fixing N_{sel}. In contrast, N_{tot} remains uncertain also *a posteriori* and makes conditioning the simulator (which follows the pre-configured (left-hand) variant in fig. 4.1) on N_{sel}—so that all training examples have the same ("correct") cardinality, and a fixed-input-size NN can be used—difficult and resource-intensive. For the only option to implement this conditioning is usually rejection sampling,[111] whose efficiency reduces with the spread of *a priori* plausible data set sizes, i.e. as $1/\sqrt{N_{\text{sel}}}$, dropping prohibitively low for large populations. The problem can be mitigated by allowing a range (cf. ε in ABC) of acceptable N_{sel}—especially if $\mathrm{p}(N_{\text{sel}} \mid \gamma)$ is not extremely informative—and re-sampling to exactly N_{sel} for input into the network, but this comes with all the previously discussed caveats.

SBI training with unobserved data [13] The complete population, as opposed to non-random subsets of it, is trivial to simulate from, once N_{tot} is determined: it is simply a BHM with $N_{\text{obj}} \to N_{\text{tot}}$. This leads to the final alternative SBI strategy for eq. (4.8): learning from *random* sets of fixed size N_{sel} without reference to whether objects are observed or missed and correcting the results *post factum*:

[111] Note that rejection sampling at least naturally implements $\mathrm{p}(N_{\text{sel}} \mid \gamma)$ by only accepting γ that lead to N_{sel}, i.e. following $\mathrm{p}(\gamma \mid N_{\text{sel}})$.

$$\underbrace{\prod_{i=1}^{N_{\text{sel}}} \mathrm{p}(\mathbf{d}^i \mid \boldsymbol{\gamma})}_{\substack{\text{conventional}\\ \text{inference}}} \times \underbrace{\prod_{i=1}^{N_{\text{sel}}} \frac{\mathrm{p}(\mathfrak{s}^i \mid \mathbf{d}^i, \boldsymbol{\gamma})}{\mathrm{p}(\mathfrak{s}^i \mid \boldsymbol{\gamma})}}_{\substack{\text{correction for}\\ \text{selection effects}}} \tag{4.10}$$

In practice, this involves first solving the usual inference task, for which NLE/NPE/NRE can be trained using sub-samples from the full simulation output $\{\mathbf{d}^i\} \cup \{\mathbf{d}^j\}$. To correct this to eq. (4.8), eq. (4.10) accounts for *selection effects*[112] through the ratio

selection effects

$$\prod_{i=1}^{N_{\text{sel}}} \frac{\mathrm{p}(\mathfrak{s}^i \mid \mathbf{d}^i, \boldsymbol{\gamma})}{\mathrm{p}(\mathfrak{s}^i \mid \boldsymbol{\gamma})} = \prod_{i=1}^{N_{\text{sel}}} \frac{\mathrm{p}(\mathbf{d}^i \mid \mathfrak{s}^i, \boldsymbol{\gamma})}{\mathrm{p}(\mathbf{d}^i \mid \boldsymbol{\gamma})} = \prod_{i=1}^{N_{\text{sel}}} \frac{\mathrm{p}(\mathbf{d}^i, \mathfrak{s}^i \mid \boldsymbol{\gamma})}{\mathrm{p}(\mathbf{d}^i \mid \boldsymbol{\gamma})\, \mathrm{p}(\mathfrak{s}^i \mid \boldsymbol{\gamma})}. \tag{4.11}$$

Noticing the similarity between this and eq. (2.14) and working within the usual framework of SNR-cut detection akin to eq. (2.25), Anau Montel & Weniger [13] declare "detection is truncation" and proceed to train a NRE for eq. (4.11) using the simulated detection labels of mock populations scaled to size N_{sel}: $\left(\boldsymbol{\gamma}, \{(\mathbf{d}, S)^k\}_{k=1}^{N_{\text{sel}}}\right)$.

While easy to simulate training examples for, this approach has the major disadvantage that the observed[113] data $\{\mathbf{d}_{\text{o}}^i\}, \{\mathfrak{s}^i\}$ cannot be regarded as a *typical sample*[114] of the simulator, and so will fall outside the region in data space in which the inference networks have been trained, leading to undefined—and often very biased—results. While this is subtler to see for the data $\{\mathbf{d}_{\text{o}}^i\}$,[112] it is clear that the probability of sampling a large N_{sel} number of objects at random from the population and having all of them end up detected is vanishingly small. And even if training data were made somewhat representative through e.g. a truncation procedure, combining the two terms in eq. (4.10) requires extreme precision in the tails of each, for if selection effects are important, they strongly bias the first term; and while the combination may vary reasonably over the extent of the final posterior, the two range of each separate term is proportional to N_{sel} and can thus reach many orders of magnitude. An illustration of this effect for a Gaussian toy model is shown in fig. 4.4, while appendix A in RESSET contains the equivalent for the simple BHM for SN Ia cosmology we adopt in chapter 15.

typical set

[112] the difference between the sampling distributions of observed objects and the total population or, equivalently, between the posterior derived with and without regard for the probability of detection + measurement + selection; concrete examples are discussed in subsection 8.3.6 and chapter 15

[113] both in the sense of "real" and containing exclusively detected, etc. objects

[114] This problem, discussed in appendix 17, is similar to that of hierarchical truncation: **d** plays the role of a *local parameter* for the observation of S, whose distribution is, as Anau Montel & Weniger [13] exclaim, being "truncated".

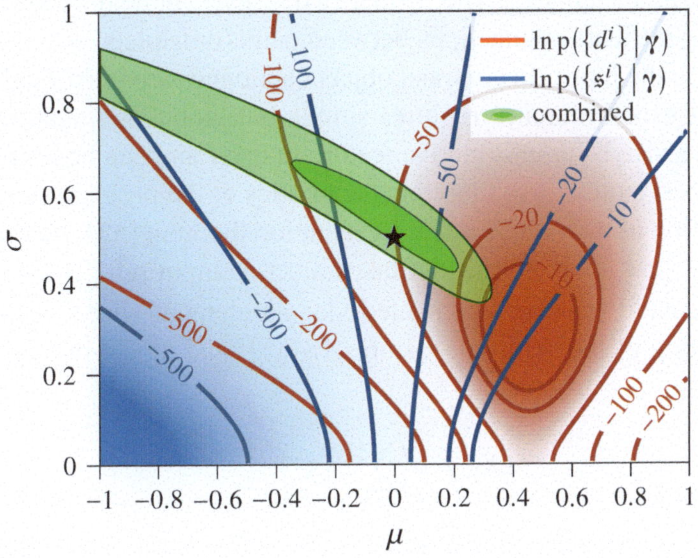

Figure 4.4: (RESSET) Illustration of catastrophic cancellation for the Gaussian toy model. The data comprise 100 selected* objects ($d^i > 0$) drawn from $\mathcal{N}(\mu, \sigma^2)$ indicated by a star and with added "measurement" noise $\varepsilon = 0.2$. Depicted as blue and red gradients and iso-log-likelihood contours are the two competing terms** from eq. (4.10) (normalised to their respective maximum values within the prior range). The setup is engineered so that they nearly cancel, leading to a combined likelihood $\mathrm{p}(\{d^i\} \mid \gamma)/\mathrm{p}(\{\mathfrak{s}^i\} \mid \gamma)$ (illustrated as green filled 1- and 2-σ contours) that peaks in the tail of both. Notice the magnitude of the cancellation: across the 2-σ region, where the combined likelihood is within e^{-2} of the maximum, both terms individually change by a factor $\sim \mathrm{e}^{100}$. Therefore, approximating each separately and then combining them could lead to large numerical inaccuracies. See RESSET, fig. 8 for an equivalent demonstration for cosmological inference from 2000 SNæ Ia with selection effects (cf. chapter 15).

*realised through rejection sampling (the population size is considered independent of μ and σ)

**Note that in this example, $\mathrm{p}(\mathfrak{s}^i \mid d^i, \gamma) \equiv 1$ because the selection is deterministic.

SBI with stochastic cardinality [455, 435, 143, 82, 210] Having established the need for training with stochastic cardinality when dealing with a model for object catalogues, we now need to select among the (small) variety of NN architectures that admit such arbitrary-sized input: namely, between recurrent neural networks (RNNs), attention-based (transformer) models, and deep-sets. The former two[115] excel at dealing with *ordered* sequences and capturing intricate connections between pairs (or triplets, etc. as their depth is increased) of objects in the input. In contrast, object catalogues are intrinsically *unordered* collections whose Bayesian inference requires simple but global aggregation. Moreover, their *a priori* unknown size is an informative feature in itself and can be immense.[116]

deep set

The architecture best suited[117] to those characteristics of the problem is Zaheer et al.'s *deep set*: a manifestation of their universal representation theorem [554, theorem 2] stating that any function $f : \{x \in \mathbb{R}^d\} \rightarrow \mathbb{R}^n$ that takes a set as input (and is thus invariant to the order of its elements)[118] can be represented via: an element-wise transformation $\phi : \mathbb{R}^d \rightarrow \mathbb{R}^m$, a summation of the resulting features $\{\phi(x)\}$, and a post-processing function $\rho : \mathbb{R}^m \rightarrow \mathbb{R}^n$; i.e.

$$f(X) = \rho\Big(\sum_{x \in X} \phi(x)\Big). \tag{4.12}$$

Learning $f(X)$ then consists of optimising two conventional, i.e. *fixed-size/order-input*, NNs $\hat{\phi}$ and $\hat{\rho}$ to approximate the featuriser and post-processor.

The simplest application of a deep set in SBI is to derive a fixed-size summary $\mathbf{s}(\mathbf{D}) \leftarrow \hat{\rho}\big(\sum_{\mathbf{d} \in \mathbf{D}} \hat{\phi}(\mathbf{d})\big)$ and use it conventionally in downstream inference. This approach forces the NN to extract the complete information from the set (i.e. encode the full posterior at all $\boldsymbol{\theta}$ into $\mathbf{s}$), straining the network capacity and requiring long training with a large number of simulations. Nevertheless, it is the only option if relying on NLE or NPE for inference, which consider the data separately from the parameters (either for density estimation or as

[115] An important application of RNNs and transformers (with positional embedding) is natural-language processing (NLP)—that has recently experienced great research interest and generated wide societal impact—, in which, too, the data (sentences/text) is composed of an arbitrary number of units (words/tokens). We will revisit them in the similar context of SN light-curve analysis in part II.

[116] Only recently has the context window of large language models (LLMs) extended beyond 100 000 tokens—a pessimistic estimate for the number of SNæ we hope to have the opportunity to analyse with SBI.

[117] Over its main competitor, the set transformer [306] (used e.g. by Campeau-Poirier et al. [82], Heinrich et al. [210]), the deep set architecture boasts a more favourable linear memory and computational complexity and better information aggregation from large sets [see e.g. 210, fig. 2].

[118] The concrete statement concerns only sets of elements from a "countable universe", which excludes the real numbers. However, in the digital universe™, *everything* is discretised and hence, countable, so we will not concerns ourselves with mathematical subtleties. Nevertheless, we point the interested reader to Wagstaff et al. [530] for extended discussion on the requirements for applicability (in worst-case scenarios) of Zaheer et al.'s conjecture for varying-sized sets of real-number arrays ("vectors").

a conditioning context). In contrast, NRE combines the data and parameters at the onset, allowing for a much more efficient construct, the (parameter-)*conditioned deep set*, *conditioned deep set*

$$\ln \hat{\mathrm{r}}(\boldsymbol{\theta}, \mathbf{D}) = \hat{\rho}\Big(\boldsymbol{\theta}, \sum_{\mathbf{d}\in\mathbf{D}} \hat{\phi}(\boldsymbol{\theta}, \mathbf{d})\Big). \tag{4.13}$$

In the simplest scenario, in which the data are independent, conditional on $\boldsymbol{\theta}$, this form can trivially represent the *optimal* summary:

$$\ln \mathrm{p}(\mathbf{D} \mid \boldsymbol{\theta}) = \sum_{\mathbf{d}\in\mathbf{D}} \ln \mathrm{p}(\mathbf{d} \mid \boldsymbol{\theta}), \quad \text{given } \mathbf{d}^i \perp\!\!\!\perp \mathbf{d}^j \mid \boldsymbol{\theta}, \tag{4.14}$$

which cannot as straightforwardly be implemented as a $\mathbf{s}(\mathbf{D})$ that does not reference $\boldsymbol{\theta}$. Importantly, a similar expression can be learnt[119] even in the presence of global nuisances, but $\hat{\phi}$ will not, in general represent exactly the individual-object sampling probability.

[119] Capitalising on the connection between universal representation of sets and the equivalence of exchangeability (permutation-invariance) and *conditional* independence, i.e. de Finetti's theorem [563], Zaheer et al. [554, eq. (3)] derive explicitly the expression for a deep-Finetti-set™, which implements the marginal likelihood for the special but widely-applicable case of nuisances with a likelihood from the exponential family and a conjugate prior.

Part II

Supernova cosmology

Chapter 5

Supernova cosmology for philosophers

Stripped of astrophysical jargon and millennial tradition, supernova cosmology can be seen as a purely epistemic field that considers its objects (section 5.1) only insofar as the concordance of their properties makes them *useful* for inference of background quantities (section 5.2). That is, it is not interested in the *essence* of "supernovæ" but is content with employing sufficiently faithful *representations* thereof, i.e. abstract models that re-fashion *espistemic ignorance* as prior uncertainty. Since the field is, therefore, still not completely agnostic of the particular real-world ontology it deals with, in this chapter, we briefly review the conception/conceptualisation of supernova cosmology before elaborating its various statistical descriptions in chapters 6 and 8.

utility

essence vs. representation

ignorance as a prior

5.1 A brief history of novælty

Novælty™ is a highly contentious topic, "part and parcel of a philosophical revolt against the overweening pretensions of science," i.e. determinism [494]. In an astronomical context, the name *nova* was first used by Brahe [64] to refer to a *new* "star" that had appeared on the night sky.[120] About a year later—promptly, on cosmic scales—it disappeared, leaving behind a dim remnant (discovered 380 years later [198]) and the conclusive refutation of that most basic form of determinism: Aristotle's static Universe. Only, it was no star at all, since those take hundreds of thousands of years to "come into existence" and millions to

[120] The appearance of a "guest" or "temporary" star, often also visible during the day, had always been a notable event: Chinese accounts, for example, take note of the explosions that left behind supernova remnants RCW 86 [526, 544] and G347.3-0.5 [532, but see 149] as far back as 185 and 393 AD, respectively; they even describe the brightness evolution and give a multi-band characterisation: "it was as big as half a mat; it was multicoloured, and it fluctuated. It gradually became smaller and disappeared," [498].

© The Author(s), under exclusive license to Springer Nature Switzerland AG 2026

K. Karchev, *Supernova Cosmology for the 21st Century*,
https://doi.org/10.1007/978-3-032-15072-1_5

subsequently "vanish" [356] (so as to fool Aristotle), whereas Tycho's nova had transpired on much shorter scales: it was an example of an *astronomical transient*![121]

transient

As more transients were observed, it became evident [30, 561, 364] that they can be sorted into observational *classes*, hopefully signifying a similar categorisation of underlying physical phenomena. At first, a broad distinction was made based on the total power and duration of the events. It was later found that the less energetic and shorter transients, also called *cataclysmic* (cleansing) variables [539], indeed originate from similar binary systems involving a *white dwarf (WD)* star accreting material from its companion. A *dwarf* or *classical* nova then corresponds to a sudden increase in emission from the accretion disk due to overheating or runaway thermonuclear fusion on the WD's surface, respectively. The released energy then leads to a negative feedback that quenches accretion, leaving the system intact and prone to repeated in-nova-tion™.

classification

white dwarf

In contrast, the more powerful events become *catastrophic*: they destroy their progenitor system either entirely or to such an extent as to make replication impossible.[122] While the initial classification of these so-called *super*novæ (SNæ) again referred to their observational—this time *spectroscopic*—characteristics, i.e. the presence (type II) or absence (type I) of hydrogen lines, the division has since shifted to the physical scenarios they represent [see e.g. 562, 99, fig. 1]. Concretely, on one hand, *core-collapse supernovæ (CC SNæ)* [e.g. 76] result from the final disintegration of a (nominally, isolated) main-sequence star following the exhaustion of its usable nuclear fuel and consequent gravitational collapse of its core into a neutron star or a black hole. On the other hand, a thermonuclear *type Ia supernova (SN Ia)* occurs when an accreting white dwarf starts fusing carbon *internally* and explodes [232, 18, 389, 52] *before* reaching the critical Chandrasekhar mass. The heavy elements synthesised in this rapid process then gradually decay (Ni → Co → Fe) and release energy into the ejecta, leading to a sustained late-time brightness and even a secondary peak in infrared wavelengths [19, 354].

supernova

spectroscopy

CC SN

SN Ia

While this overarching physical description of SNæ Ia has generally consolidated, certain aspects of the explosion mechanism remain uncertain, e.g. the role of rotation and anisotropic detonation [147, 148] and the amount of nuclear fuel burnt and of radioactive material created [for more details, see e.g. 507, 218]. Moreover, the exact nature of the white dwarf's companion—whether it is an "ordinary" star or another white dwarf, respectively labelled a single- (SD) or double-degenerate (DD) *progenitor scenario*[123]—is still a heavily debated topic, which can have a significant reflection on the stellar environments

progenitor

---

[121] different from *transits* (e.g. of an exoplanet in front of its host), *periodically variable* stars, and *wanderers* (e.g. planets or comets) in our Solar System, all examples of *predictable* celestial mutability, i.e. determinism

[122] Their existence, then, is truly *transient*, implying, in turn, unique and genuinely new.

[123] Of course, considering possibilities beyond these two, namely triple- and higher-order systems [438], and "solitary" WDs in dense environments like stellar clusters is also a worthwhile endeavour.

and populations in which SNæ Ia are likely to occur, as well as on their overall rate [426]. This all has led to the tentative institution of *sub-classes* within the Ia type[124]: subluminous ("1991bg"- and "2002cx"-like, also called type Iax [156])) and *superluminous* ("1991T"-like) [see 66, and references therein].

sub-classification SLSN

Despite the few outstanding challenges in their physical modelling and the mild evidence for irregularities, SNæ Ia remain an "extremely homogeneous" [364] class, owing to the specificity of their formation scenario, and exhibit in practice an *intrinsic* diversity around peak of no more than a few magnitudes.[125] In combination with their tremendous power[128] and ubiquity [100], this makes them ideal *standard candles*[129]: indicators of cosmological distances via the relation of absolute and apparent brightnesses (eq. (5.2)).

standard candle

---

124 again defined with respect to *observational* (spectral) features rather than intrinsic properties

125 The traditional system of measuring "brightnesses"—(spectral) irradiance/flux density—in astronomy was introduced by Hipparchus of Nicæa, fleshed out by Ptolemy, and—after remaining more or less arbitrary for the next 2000 years—systematised in its current form by Pogson [359]. It derives from the usual Ancient Greek "*exhaustive*" *categorisation* of all stars (visible to them) into five roughly logarithmically-spaced (as per the general Weber–Fechner law of human perception [145]) "classes". Pogson's quantitative version, instead, assigns an overall ratio of exactly 100 across any five "degrees of magnitude" (or $100^{1/5} \approx 2.512$ per magnitude) and selects a particular value f_0 to serve as reference (null magnitude) when measuring a physical quantity f:

magnitudes: apparent & absolute

$$m\,[\mathrm{mag}] \equiv -2.5\log_{10}(f/f_0). \tag{5.1}$$

For further details, more rigour, and practical considerations, refer to section 7.1.

Moreover, distinction is made between the *apparent* magnitude m of an object to a given observer and the *absolute*/intrinsic brightness M, defined as that which it would attain were it in Euclidean space at a distance of 10 pc (parsec)[126] from the observer. At other distances d (and in other spaces), its (spectral) flux density, i.e. brightness, will be attenuated over a (hyper-)spherical sector, whose surface $S(d) \sim d^2$ in our everyday spatially flat three-dimensional Universe. Thus, the *apparent* magnitude m is related to M through the *distance modulus*:

distance modulus

$$\mu \equiv m - M \equiv 2.5\log_{10}[S(d)/S(10\,\mathrm{pc})] = 2.5\log_{10}[(d/10\,\mathrm{pc})^2]. \tag{5.2}$$

In more exotic space (see below), the last equality has often been treated as reverse equivalence to *define* d in terms of observed and intrinsic brightness.

126 1 pc, in turn, equals $180 \times 60 \times 60/\pi \approx 206\,265$ times the semi-major axis of the Earth's orbit (the *astronomical unit*[127]) [136, page 342*] or ≈3.26 ly (light-years)… The blame for these particular ratios is shared between the Greeks, the Babylonians [246], and the number of days in a year.

127 Yes, astronomers would use any system of units but the metric…

128 The fusion energy released in each *needs* to exceed the gravitational self-potential of a white dwarf: 5–6 × 10^{43} J [507]; and since the nucleosynthesis in SNæ Ia is more *rapid* than beta decay, the excess is dissipated entirely by photons (and velocity of the ejecta) and not by neutrinos as in a CC SN.

129 *À propos*, a candela (cd) is the SI unit of luminous intensity equal to the subjective brightness of a standard wax candle.

a crandle™ (non-standard)

5.2 A crash course in cosmo~~logy~~graphy

the night sky

To appreciate the importance of SNæ Ia, one first needs to gaze at *the night sky* and consider: "Why is it dark?" For, Olbers — and Digges, Kepler, Halley, and Chesaux [201] — thought, in a homogeneous universe, in which locations are not special (merely spatial), the abundance of light emitters balances exactly the attenuation of light due to distance (in fact, light is extinguished precisely *by* its spread in space: eq. (5.2)), and as a consequence, in an infinity of space and time (so as to allow light — proven finite in speed[130] — to reach an observer from all locations), implies that the night — and day — sky should be infinitely bright, which it is not: a *paradox* (but actually simply a contraction between an assumption and an observation). The resolution is, of course, to reject the hypothesis: the Universe *is not* infinite (either in time and/or space) and/or our position (in space and/or time) *is* special.

cosmology

The study of alternatives to the Aristotelian world-order — of the history and future of the Universe, its composition, and our place within and whenin™ it — is called *cosmology*. For the purposes of this thesis, it suffices to refer the reader to Peebles [401] and here summarise the two relevant Principles of physical cosmology:

FLRW
curvature
(spatial) scale factor

1. Einstein's general theory of relativity (GTR) [140] holds, and the Universe is homogeneous and isotropic on large-enough spatial scales. This implies that it is described by the *Friedman(n)–Lemaître–Robertson–Walker (FLRW)* metric, with *spatial curvature* $k \in (-\infty; \infty)$ and arbitrary time evolution encoded in a *scale factor* $a(t)$; in hyperspherical[131] coordinates:

$$\mathrm{d}s^2 = \mathrm{c}^2\,\mathrm{d}t^2 - a^2(t)\left[\mathrm{d}r^2 + S_k(r)\,\mathrm{d}\Omega^2\right] \quad \text{with} \quad S_k(r) = \left[r\,\mathrm{sinc}\left(\sqrt{kr^2}\right)\right]^2, \tag{5.4}$$

where Ω measures solid angles, r radial distance, t time, and c the speed of light. We will take a unitful k of dimension inverse squared length so as to make the scale factor unitless.

light echo

130 Incidentally, this also allowed Brahe's "new star" to be conclusively confirmed as a regular SN Ia based on observations of its reflection in interstellar dust (a *light echo*) [443, 296] more than four centuries after its initial transpiration. Another kind of re-(super)nova-tion™ enabled by the finite speed of light (and causality/determinism): multiple imaging through a strong gravitational lens, has recently (re-)ignited h0pes of a determination of the Hubble constant independent of the distance ladder [441, see also 178].

131 The use and choice of the *same* factor S as in eq. (5.2) is intentional for the elegance of the overall presentation. To the same effect, we take advantage of the continuity of complex trigonometric functions:

$$\mathrm{sinc}(x) \equiv \frac{\sin(x)}{x} = \frac{\sinh(\mathrm{i}x)}{\mathrm{i}x} \to 1 \text{ as } x \to 0, \tag{5.3}$$

to handle gracefully all possible (negative, positive, or vanishing) curvatures.

Redshift In such a universe, egocentrism is afforded to everyone: i.e. the natural choice of *coordinate origin* ($r = 0$, $t = 0$) is the observation event: the location and present of the observer. It is also the observer who sets the scale of the Universe through the units of r and t, which fix $a(0) = 1$. An important corollary concept is the *cosmological redshift*[132]

coordinate origin
cosmological redshift

$$z_c \equiv a^{-1} - 1, \tag{5.5}$$

which can by this *definition* be used as a *label for time*.[133] It is paramount to understand that z_c is, therefore, not equivalent to the *observed (total) redshift* of an object [see e.g. 117]:

total redshift

$$(1 + z) = (1 + z_c) \times \left(1 + z_{pec}\right) \times \left(1 + z_{grav}\right), \tag{5.6}$$

which for a supernova combines the effects of

- cosmological expansion,
- the total *peculiar velocity* of the source with respect to the observer, comprising the motions of the SN within its galaxy, of the host and of the Milky Way with respect to the CMB (taken as a cosmic frame of reference), and of the Sun and the Earth (the latter three are usually explicitly corrected since they are precisely known [413, 414], and a z_{CMB} in the CMB frame is reported),
- *gravitational redshift* due to inevitable small-scale inhomogeneities of cosmic structure and, possibly, surface gravity of the emitter.

peculiar velocity
eppur si muove
gravitational redshift

Distances The metric eq. (5.4) is used for all purposes that concern distances or time intervals: *cosmography*. In the spirit of egocentrism, this is usually[134] reduced to assigning labels r and t to observed events (assigning Ω, i.e. *angular/sky coordinates*, is unaffected by cosmology since the Cosmos is assumed isotropic), which is significantly simplified by the fact that the two events (happening and observation) are by definition connected by a light-like path ($\mathrm{d}s^2 = 0 \implies c\,\mathrm{d}t = a(t)\,\mathrm{d}r$). Then, the spatial and temporal coordinates of an event, also called the (radial) *comoving distance* and *lookback time* to it, are

cosmography
sky location
comoving distance (radial)
lookback time

$$D_c(z_c) = \int \mathrm{d}r = \int \frac{\mathrm{c}\,\mathrm{d}t}{a(t)} = \mathrm{c}\int \frac{\mathrm{d}a}{a\dot{a}} = \mathrm{c}\int \frac{a}{\dot{a}}\,\mathrm{d}z_c\,, \tag{5.7}$$

$$T(z_c) = \int \mathrm{d}t = \int \frac{\mathrm{d}a}{\dot{a}} = \int \frac{a}{\dot{a}}(1 + z_c)\,\mathrm{d}z_c\,, \tag{5.8}$$

[132] which is entirely independent of velocities and Doppler shift [131], apart from having a similar effect *on light*; but cosmology redshifts *everything*: energies, densities, temperatures... space and time themselves

[133] provided that $a(t)$ is monotonic; but we will not concern ourselves with the contrary Big Bounce scenarios

[134] Notable exceptions are the relative radial distance between source and lens used in gravitational lensing and the relative transverse distance, i.e. projected size, of standard rulers [521].

where the limits of integration are implied. Thus, an important quantity — which fully encodes the history of the Universe — has emerged: the relative time derivative of the scale factor, also called the *Hubble parameter*: $H(z_c) \equiv \dot{a}/a$. Its present-day value, the *Hubble constant* $H_0 \equiv H(0)$, sets convenient length/time scales (units[127]) for measuring the Universe:

Hubble parameter & constant

- the Hubble distance: $D_H \equiv c/H_0$, is not the size of the Universe; for all we know, it is infinite, but the (current) radius of the *observable* Universe is $z_c \to \infty$ in eq. (5.7);
- the Hubble time: $T_H \equiv 1/H_0$, is not the age of the Universe ($z_c \to \infty$ in eq. (5.8));
- the critical density: $\rho_c \equiv \frac{3}{8\pi} H_0^2/G$, which defines the notion of mass in cosmology.

composition of the Universe
cosmological constant
density parameter
EOS
redshift law

2. The Universe consists of a mixture of perfect — collisionless and non-interacting — fluids: a concept that can be stretched to include also geometric factors like curvature and the *cosmological constant* (Λ) [141],[135] which exert unusual — but not unphysical — *negative* pressure. Each of them is parametrised by a (dimensionless present-day) *density parameter* $\Omega_{i0} \equiv \rho_{i0}/\rho_c$ and (possibly time-dependent) *equation of state (EOS)* $\mathfrak{w}_i(z_c)$. The latter determines its *redshift*[132] *law*, i.e. the evolution of its mass/energy density:

$$f_i(z_c) \equiv \rho_i(z_c)/\rho_{i0} = \exp\left[3 \int (1 + \mathfrak{w}_i(z_c)) \frac{dz_c}{1 + z_c}\right]. \tag{5.9}$$

Particularly, common "fluids" have a constant $\mathfrak{w}_i$, which implies

$$f_i \to (1 + z_c)^{3(1+\mathfrak{w}_i)} = \begin{cases} (1 + z_c)^4 & \text{radiation}^{136} \quad (\mathfrak{w}_r = 1/3), & (5.10a) \\ (1 + z_c)^3 & \text{matter}^{136} \quad (\mathfrak{w}_m = 0), & (5.10b) \\ (1 + z_c)^2 & \text{curvature} \quad (\mathfrak{w}_k = -1/3), & (5.10c) \\ \text{const} & \Lambda \quad (\mathfrak{w}_\Lambda = -1). & (5.10d) \end{cases}$$

[135] Einstein's attempt to restitute Aristotle, which he later called his "biggest blunder", has since been re-branded and resurrected — with an instrumental contribution from SNæ Ia — as (constant) "dark energy".

[136] In a cosmological context, "matter"/"radiation" refer to any non-/relativistic particles, respectively: i.e. baryonic and dark matter, on one hand, and primordial photons on the other. It is also conceivable to build models where a species like sufficiently — but finitely — light neutrinos *transition* from relativistic to "classical" as their density/temperature redshifts.

A cosmological model specifies the natures ($\{\mathfrak{w}_i\}$) and amounts ($\{\Omega_{i0}\}$) of a universe's constituents, collectively comprising the *parameters* C which are of interest in cosmological inference. Together with the Hubble constant (the "initial" condition), they determine the full expansion history through the Hubble parameter, whose square equals the total mass/energy density of the Universe by the first *Friedman(n) equation*[137] [160, 161]:

cosmological parameters

Friedman(n) equation

$$E^2(z_c, C) \equiv [H(z_c, C)/\mathrm{H}_0]^2 = \rho(z_c)/\rho_c = \sum_i \Omega_{i0} f_i(z_c), \tag{5.11}$$

and so C can be constrained from simultaneous measurement of (cosmological) redshift and distance (cf. eqs. (5.7) and (5.8)). In fact, one "data point", at $z_c = 0$, is immediately available and allows one component—usually curvature as the least fluid-like—to be eliminated in favour (expressed in terms) of the others:

$$E(0, C) = \sum_i \Omega_{i0} = 1 \quad \Longrightarrow \quad \Omega_{k0} = 1 - \sum_{i \neq k} \Omega_{i0}. \tag{5.12}$$

Owing to their distinct scalings, each of the remaining components comes to dominate the dynamics at different times (redshifts), and therefore, inquiries into each must consider data from different epochs. For example, analyses of the cosmic microwave background (CMB), an early-Universe probe, have determined $\Omega_{r0} \approx 10^{-5}$ and $\Omega_{m0} \approx 0.3$ (matter–radiation equality around $z_c \approx 3000$) [101]. In contrast, measuring effectively curvature requires observations from lower redshifts, e.g. of baryon acoustic oscillations (BAO) [e.g. 110] or SNæ Ia, which have hitherto been consistent with a spatially *flat Universe* ($\Omega_{k0} \approx 0$) [101]. For their analysis, one can safely disregard the cosmological influence of radiation and instead assume either the *Λ–cold dark matter (ΛCDM)* model, which supposes only non-relativistic matter and a cosmological-constant: $C_{\Lambda\mathrm{CDM}} \equiv [\Omega_{m0}, \Omega_{\Lambda 0}]$, or any exotic *dark-energy (DE)* model [see e.g. 77]; a popular parametrisation for a non-trivial EOS is $\mathfrak{w}(z_c) = w_0 + w_a\, z_c/(1 + z_c)$ [96, 324], which reduces to a Λ for $w_0 = -1$ and $w_a = 0$.

flat Universe ($\Omega_{k0} = 0$)

ΛCDM

DE

[137] which explicitates the time–time component of the Einstein field equations [140] for the FLRW metric

5.2.1 Confrontational theses

The Hubble "constant" is widely regarded as a cosmological parameter alongside $\{\Omega_{i0}\}$ and $\{\mathfrak{w}_i\}$. In fact, it was the first—and only—one, initially appearing in the Hubble–Lemaître [235, 314, 315, see also 328] "law", which expresses the local limit of eq. (5.7) as a linear relation between distance and cosmological redshift[138]:

$$D_\mathrm{c} = \int \frac{\mathrm{c}\,\mathrm{d}z_\mathrm{c}}{H} \overset{D_\mathrm{c} \ll \mathrm{c}/\mathrm{H}_0}{\Longrightarrow} \mathrm{H}_0 D_\mathrm{c} \approx c z_\mathrm{c}. \tag{5.13}$$

model-independent distance

This Taylor-series approximation can be continued to derive a *model-independent distance* expression in terms of *universal* cosmological parameters [e.g. 541, section 1.4]:

$$c_n \equiv \frac{1}{\mathrm{H}_0^n} \frac{\mathrm{d}^n a}{\mathrm{d}t^n}\bigg|_{t=0} \Longrightarrow \begin{cases} n = 1 \to 1 & \text{the linear Hubble law,} \quad (5.14a) \\ n = 2 \to -q_0 & \text{cosmic } deceleration\text{,} \quad (5.14b) \\ n = 3 \to j_0 & \text{cosmic jerk,} \quad (5.14c) \\ \dots & \text{etc.} \end{cases}$$

However, we do not regard H_0 as a parameter *of interest* ($\mathrm{H}_0 \notin C$) but instead as a fundamental constant that sets the system of units and whose value—commonly expressed[127] in km/s/Mpc—cannot be measured without reference to other scales and so is immaterial in purely cosmological analyses, like the ones from this thesis. Indeed, in all of what follows, the Hubble constant will be exactly degenerate with a parameter that sets the absolute brightness scale, which is customarily defined for a distance of 10 pc but in cosmology *must* be related to $\mathrm{D_H}$ (see fig. 6.1).

Still, at present, many researches—among them supernova cosmologists—are preoccupied with humanity's evident inability to ~~measure~~ come to an agreement[139] as to the numerical value of H_0. Naturally, this *Hubble tension* [see e.g. 127] arises only as a miscalibration between *multiple* (g)astrophysical phenomena *assumed* to be well understood: e.g. the CMB, BAO, and SNæ Ia *juxtaposed* with the local distance ladder [188].

Hubble tension

luminosity distance

The luminosity distance $D_\mathrm{L}(z, z_\mathrm{c})^2 \equiv S_k(D_\mathrm{c}(z_\mathrm{c})) \times (1+z)^2$ is widely used to express the *combined* influence on light of distance and redshift, which would not be harmful to science if distance and redshift had a perfect one-to-one correspondence, **but** they do not (cf. footnote 132 and eq. (5.13)). Suppose photons emitted with *rest-frame* wavelength λ_r and *observed* with $\lambda_\mathrm{o} = \lambda_\mathrm{r}(1+z)$, so:

rest vs. observer frame

---

[138] In the initial—and still widespread—formulation $\mathrm{H}_0 r = v_r$, this is interpreted as a recession velocity, **but** as remarked, it is instead the space between objects that is expanding *while* light is travelling.

[139] I move (not really) to replace the cæsium hyperfine transition with the current age of the Universe in the SI definition of a second and be done with it.

- their energy is lower: $hc/\lambda \propto (1+z)^{-1}$;
- their *rate* of arrival is reduced as $1/\Delta t \propto (1+z)^{-1}$;
- spectral intervals are dilated as $1/\Delta\lambda \propto (1+z)^{-1}$ (i.e. the Jacobian of $\lambda_r \to \lambda_o$).

Therefore, measurements of *integrated* flux scale overall as $(1+z)^{-2}$, whereas spectral quantities as $(1+z)^{-3}$: refer to table 7.1. On the other hand, flux *densities* are affected by distance—more specifically, by the (inverse of the) prefactor to $d\Omega^2$ in eq. (5.4): $S_k(D_c(z_c)) \equiv D_M(z_c)^2$, called the (square of) the *transverse comoving distance*.

transverse comoving distance

The luminosity distance combines the two effects for the convenience of broadband photometry, **but** its common presentation in terms of a single redshift is misleading: it is only valid if $z = z_c$, i.e. in the absence of significant peculiar velocities. Of course, if $z_{pec} \ll z$—a general rule of thumb is $z \gtrsim 0.03 \approx 10\,000\,\mathrm{km/s}/c$—, identifying the two and calculating the (transverse) comoving distance using the total redshift is admissible.

Lastly, when modelling light received at λ_o, one needs to consider the emission at the shorter $\lambda_r = \lambda_o/(1+z)$. Instead, for historical—read, incomprehensible, lost in time—reasons, astronomers prefer to work in *the same band*, say X, in both the rest and observer frames. This approach—typical of methods of (reverse) data *analysis*, rather than (forward) *modelling*—then requires calculating the hypothetical apparent brightness in $(1+z)X$ or the hypothetical absolute emission in $X/(1+z)$ using a presupposed *spectral energy distribution (SED)* of the source and the notorious *$K_X(z)$-correction* [546, 236, see also 285], just so as to preserve an appearance of the distance modulus in the presence of redshift:

SED

K-correction

$$\mu \equiv m_X - M_X = 5\log_{10}(D_L(z, z_c)/10\,\mathrm{pc}) + K_X(z). \tag{5.15}$$

This whole discussion can be evaded by explicit construction of a forward model, as we describe in chapter 11.

Chapter 6

Supernova cosmology for Nobel laureates

Less than ideal standard candles in practice, SNæ Ia need to be massaged— much more gently than other similar objects and concepts [188, 521, 228]—in order for their brightness to be usable for *precise* cosmological distance measurements.

Spherical SNæ in vacuo Even if not (yet) derivable from first principles (i.e. through physical modelling/simulation), the intrinsic brightness of *any* cosmic explosion can be derived with a simple observational procedure—and a slew of assumptions (i.e. "first principles"). Considering an expanding glowing sphere (e.g. the photosphere of δ Cephei or the ejecta of *any* SN explosion), Baade [29, later followed by 65, 292, 529] reasoned that its *bolometric luminosity*[140] L is determined by *bolometric luminosity*

- the sphere's radius R, "measurable" through the radial[141] velocity $v_R \equiv \dot{R}$, which can be extracted from the object's spectrum, and
- its temperature T, "measurable", or at least inferrable with standard methods from photometry or spectroscopy, *assuming* thermal equilibrium and Planck's law [412]:

$$L = 4\pi R^2 \sigma T^4 \quad \Longrightarrow \quad \dot{L}/L = 2\, v_R/R + 4\, \dot{T}/T\,, \tag{6.1}$$

[140] This is the astronomical term for the total flux (integrated across the full spectrum). The present argument *can* be modified to consider the emission in a number of given bandpasses (>1 necessary just for the temperature determination) since it already assumes a given spectral flux distribution.
À propos, an absolute magnitude system for bolometric luminosity and irradiance (flux density), which is anyway not realisable in practice, can be relatively painlessly introduced by stipulating e.g. 3.0128×10^{28} W and $2.518\,021\,002 \times 10^{-8}$ W/m^2 as the respective standards [342].

[141] Under spherical symmetry, *every* motion is radial, but in principle, v_R here does not necessarily stand for motion along the line-of-sight to the observer.

© The Author(s), under exclusive license to Springer Nature Switzerland AG 2026

K. Karchev, *Supernova Cosmology for the 21st Century*,
https://doi.org/10.1007/978-3-032-15072-1_6

where $\sigma \approx 5.67 \times 10^8\,\mathrm{W/m^2/K^4}$ is the Stefan–Boltzmann constant. Substituting R from the former and expressing the relative rate of change in luminosity ($\dot{T}/L$) as an (absolute) rate of change in magnitude ($\dot{M} = \dot{m}$)—which is independent of distance (and any other constant relative modification to the brightness like extinction) because of the logarithmic nature of the magnitude scale:

$$L = \pi\sigma T^4 v_R \left[\ln 100^{1/5} \times \dot{m} + 4\,\dot{T}/T\right]^{-2}. \tag{6.2}$$

This allows *calculation* of L from relatively scarce observations: a couple of photometric measurements in a couple of bands to estimate the temperature and its derivative and a spectrum to extract the ejecta velocity.

Elegant as it is, Baade's method has two glaring flaws. Firstly, it is unrealistic: SNæ are not (all and always) isotropic and opaque blackbodies,[142] and while the latter can be explicitly verified with spectral information, the former assumption is never *robust*: i.e. there is no easy way to identify whether it holds and correct the result or discard the observation so as to not propagate incorrect conclusions otherwise. And secondly, it is still very data-intensive: on one hand, it *necessitates* spectroscopic data and careful line modelling to extract a precise and accurate v_R; on the other, it calls for a *differential* measurement of temperature—an abstraction that is *expected* to be ill-defined in non-idealised cases—, which is prone to large estimation errors. As an alternative to the latter, one can trace the expansion through time and derive the radius absolutely, but that requires even more spectra or assumptions (e.g. a free-"fall" expansion dictated purely by gravity and the initial velocity imparted by the explosion). All in all, physical standardisation applies mainly in Plato's universe of universals.

Empirical standardisation[143] shuns explicit physical modelling—in accordance with the stated epistemological objective of supernova cosmology—and Platonic ideals in favour of an analysis of, by and for real observations. By effectively utilising the entire pool of data even for inference of individual-object properties, an empirical procedure can achieve greater statistical power (certainty/precision) than the *self*-standardisation described above.[144]

[142] According to Colgate [100], the blackbody assumption *may* apply only to the early phases of a SN Ia before the effects of radioactive heating and emission take over and more broadly to CC SNæ.

[142] I apologise to the poor Bayesian reader for this discussion. Standardisation is an unfortunate and inseparable part of the history of SN Ia cosmology, and sometimes it is necessary to see the depths of hell in order to appreciate heaven.

[144] While the references to BHMing here should be clear, I note why it is *not* (usually) employed in processes like self-standardisation: there is no noise, scatter, or uncertainty within Plato's *static* reality of the Forms. That is, physical modelling is conditional on *perfect* knowledge of the universals—the top hierarchical level—instead of on parameteres with varying degrees of posterior probability inferred from i.i.d. observations.

The essence of empirical standardisation, therefore, is cohesion: the approach relies on identifying a transformation of the observed data that leads to the least possible variance (across a population of objects identified to be of the same type, e.g. SNæ Ia) while preserving — better: *distilling* — the contained information about a problem of interest (e.g. absolute brightness / distance). The SN Ia class[145] already presents a fortuitous *intrinsic* homogeneity (see Minkowski's remark above), which afforded Perlmutter [404, 405], Schmidt and Riess [446] the 2011 Nobel Prize in Physics. Beyond relying on fortune,[146] the simplest approach to constructing and strengthening cohesion is by identifying and "explaining" some of the observed variation in a quantity of interest through its *correlation* with other *observable* quantities (rather than in terms of *causal* connections with other *intrinsic* properties). In statistical terms, this corresponds to the act of *conditioning*: the variance of a conditional distribution $\mathrm{p}(y \mid x)$ is never greater than that of the corresponding marginal $\mathrm{p}(y)$, which has integrated within it the *additional* variability of a $\mathrm{p}(x)$.

correlation vs. causation

The most primitive procedure for standardisation of SNæ Ia was pioneered by Phillips [410] and Tripp [515, 516] and remains, tragically, in wide use until today. In it, the "absolute magnitude"[147] M of a SN Ia is estimated by a linear combination of *covariates* $\hat{\mathbf{x}}(\mathbf{d})$ deterministically extracted from data:

covariates

$$M^s = M_0 + \boldsymbol{\alpha} \cdot \hat{\mathbf{x}}(\mathbf{d}^s) + \epsilon^s, \tag{6.3}$$

where s labels individual SNæ Ia (N_{SN} in total considered). Empirical standardisation strives to minimise the average *residual scatter*, $\sigma_0 \equiv \sum_{s=1}^{N_{\mathrm{SN}}} (\epsilon^s)^2 / N_{\mathrm{SN}}$, across the analysed sample, i.e. the mean squared error (MSE) of the estimator from M^s. Given pre-determined covariates $\{\hat{\mathbf{x}}(\mathbf{d}^s)\} \to X$ and $\{M^s\} \to Y$ derived from the observed brightnesses[148] $\widehat{m}(\mathbf{d}^s)$ of SNæ at known distances, standardisation corresponds to performing a *linear regression* to determine the *standard* SN Ia absolute magnitude M_0 and *correlation coefficients* $\boldsymbol{\alpha}$. In cosmological inference, on the other hand, each distance modulus is calculated under

residual scatter

linear regression

standard SN Ia

correlation coefficients

[145] Sample selection is, of course, one way of ensuring homogeneity, but it sacrifices constraining power (if the rejected examples are still informative but the inference procedure is not sufficiently sophisticated to process them) and is prone to inadvertent biases: see subsection 8.3.6.

[146] In the interest of fairness, both teams extended the simplistic one-parameter model (standard absolute brightness at maximum + Gaussian noise) with corrections related to colour and light curve shape, as we describe next, and even considered Malmquist bias, but given the quality and quantity (42 + 16 = 58 SN Ia) of the data they analysed, the modifications hardly had any effect on the results [405, fig. 5].

[147] This is usually taken at peak in a specified band: traditionally B, but since this choice influences the power of standardisation (the resulting cohesion), recently Avelino et al. [27, see also earlier references therein] have advocated for standardisation in the near infrared (NIR). While this detail is immaterial to the general argument presented here, it will influence later analyses (including chapter 12).

[148] These should correspond in nature to the modelled absolute M chosen as per footnote 147.

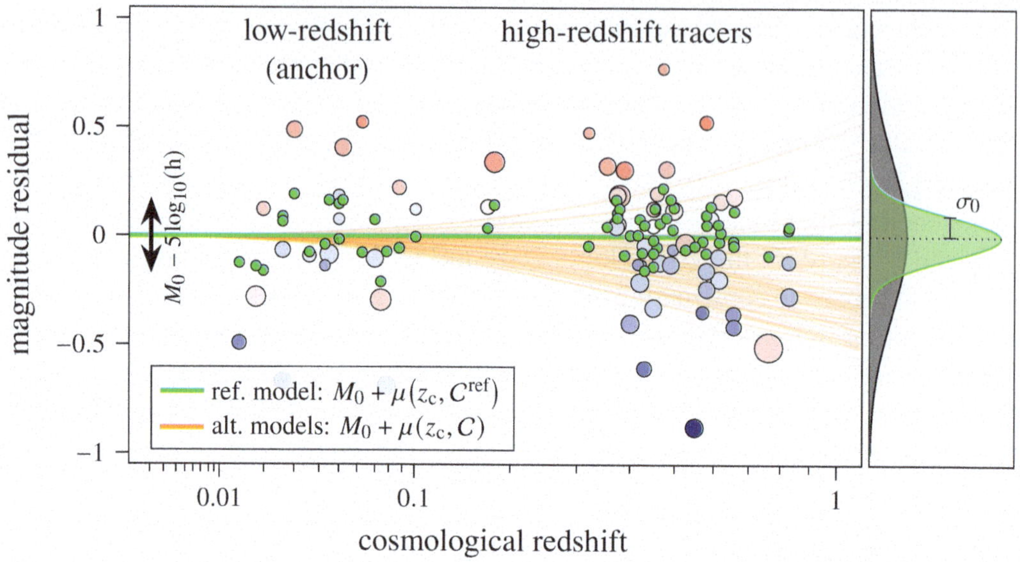

Figure 6.1: Proverbial SN standardisation using two covariates (represented as the colour and size of the markers). Correcting for the bluer–brighter and broader–brighter correlations results in the green points, which are much more concentrated (with a residual/internal scatter σ_0) around the true underlying model (green line, C^{ref}), which allows it to be picked from among alternatives (orange lines, random C). Note that the effects of M_0 and $H_0 = h \cdot 100\,\text{km/s/Mpc}$ — to shift the Hubble diagram rigidly in magnitudes — are completely degenerate, so one of them can be arbitrarily fixed.
*The redshifts in this visualisation are taken from [405], but the remaining values are fictitious.

assumption of the C and a cosmological redshift (cf. eq. (5.15)), and a non-linear fit for C (possibly in conjunction with M_0 and $\boldsymbol{\alpha}$) can be performed, as depicted in fig. 6.1.

witchcraft
bluer/broader–brighter

The difficult part of standardisation is devising informative $\hat{\mathbf{x}}$ in the first place.[149] In keeping with the spirit of empiricism, early proposals relied on meticulous examination of observational data and domain expertise to derive natural-intelligence optimised summaries that reflect two *intuitions* (illustrated in fig. 6.1): *bluer–brighter* and *broader–brighter*. The former, expressed through the observed colours, is related to extinction, which simultaneously dims and reddens light (but never brightens and bluens it); we ex-

PCA

[149] An experienced data scientist might immediately resolve to *principal component analysis (PCA)* [467], but that was not a viable option in the data-poor days; and besides, PCA requires strictly structured data, i.e. equal-dimensional for each s, which SN Ia observations are not: see section 7.1. Still, the empirical summaries discussed in this paragraph (colours and shapes) *are* linear combinations of (regularised) data indeed derivable with PCA.

pound on this in subsection 8.3.3. The latter is a largely empirical observation that can be parametrised through a measure of the light-curve shape: e.g. $\Delta m_{15}(B)$, the increase in B-band magnitude 15 days—found to "provide the greatest discrimination"—after maximum light [410, 197, see also 428, 429, 430].

The final option, preferred since [445, 404], is based on fitting a parametrised *template* (discussed in section 8.1) to each light curve $\mathbf{d}^s$ and using the recovered (e.g. MAP / MLE) values as covariates. Note that this still corresponds to a *deterministic*—yet hardly tractable—transformation (compression/summarisation) of the original data into estimators of a model's parameters: $\hat{\mathbf{x}}(\mathbf{d}^s)$, through the fitting procedure. It also challenges the interpretation of standardisation: it is more reasonable to believe—and more powerful in practice (see section 8.2) to assume—that the absolute magnitude, an intrinsic property, is related to the *latent* (true/intrinsic) values of the model's (other) parameters rather than to *measurements* thereof.

Accounting for measurement uncertainty is, in general, a major issue for empirical standardisation, which ultimately reduces to a comparison of two deterministic estimators:

$$\widehat{M^s} \equiv \widehat{m}(\mathbf{d}^s) - \mu(\widehat{z_c}(\mathbf{d}^s), C) \quad \leftrightarrow \quad \widehat{M}(\mathbf{d}^s) \equiv M_0 + \boldsymbol{\alpha} \cdot \hat{\mathbf{x}}(\mathbf{d}^s). \tag{6.4}$$

The natural interpretation in this case is frequentist: uncertainties from the observables ($\widehat{m}(\mathbf{d}^s)$, $\hat{\mathbf{x}}(\mathbf{d}^s)$, and—very importantly—$\widehat{z_c}(\mathbf{d}^s)$, which estimates z_c from data) are propagated *linearly* and combined *in quadrature* to form the variance of the estimator of the residual:

$$\widehat{\epsilon^s} \equiv \widehat{M^s} - \widehat{M}(\mathbf{d}^s) = (\widehat{m}(\mathbf{d}^s) - \mu(\widehat{z_c}(\mathbf{d}^s), C)) - (M_0 + \boldsymbol{\alpha} \cdot \hat{\mathbf{x}}(\mathbf{d}^s)) \tag{6.5}$$

$$(\sigma^s)^2 \equiv \sum_{\substack{x_i, x_j \\ \in \{\widehat{m}, \hat{\mathbf{x}}, \widehat{z_c}\}}} \Sigma_{x_i x_j} \frac{\partial^2 \widehat{\epsilon^s}}{\partial x_i \partial x_j} = \left(\sigma^s_{\widehat{m}}\right)^2 + \left(\sigma^s_{\widehat{z_c}} \left.\frac{\partial \mu(z, C)}{\partial z}\right|_{z = \widehat{z_c}(\mathbf{d}^s)}\right)^2 + \sum_{\substack{\hat{x} \in \hat{\mathbf{x}} \\ \alpha \in \boldsymbol{\alpha}}} \left(\alpha \sigma^s_{\hat{x}}\right)^2 + \ldots \tag{6.6}$$

where $\boldsymbol{\Sigma}$ is the observational covariance matrix, and we have omitted the cross-terms in the expansion for clarity. This leads finally to the quintessential *fitting objective / likelihood* (see section 1.3) of orthodox SN cosmology [e.g. 192, and most followers thereof]: *χ^2 fit*

$$-2L_{\{\mathbf{d}^s\}}(M_0, \boldsymbol{\alpha}, C) = \chi^2 \equiv \sum_{s=1}^{N_{\mathrm{SN}}} \frac{\left[\widehat{\epsilon^s}(\mathbf{d}^s, M_0, \boldsymbol{\alpha}, C)\right]^2}{[\sigma^s(\mathbf{d}^s, M_0, \boldsymbol{\alpha}, C)]^2 + \sigma_0^2}, \tag{6.7}$$

where the addition of σ_0^2 accounts for the *scatter* of the estimator*s* in eq. (6.5) within the population (in addition to their individual observational variances from eq. (6.6)). This still falls short of proper probabilistic modelling on two counts:

1. Since σ_0 is not known *a priori*, it also needs to be inferred, but a pure eq. (6.7) would trivially set it to infinity. Instead, it is determined *separately* either
   - *prior* to the fit, e.g. from external observations with diverse but *known* distances or from sub-samples with unknown but *similar* distances, i.e. binned in redshift [see e.g. 44, subsection 5.5]
   - or—contentiously, yet commonly—by setting it so as to ensure that the *best-fit* χ^2 per degree of freedom is unity. Naturally, this requires an iterative procedure of re-fitting with different constant σ_0 until an anticipated convergence.
2. The $\boldsymbol{\alpha}$ are present in the denominator of eq. (6.7) (they control the contribution of the covariate uncertainties ($\sigma^s_{\hat{x}}$) to the total σ^s), and this can "bias" the fit towards larger values than suggested purely from the numerator. In other words, eq. (6.7) allows standardisation to *increase* the residual scatter without penalty. To counteract this, Astier et al. [21, and followers thereof] *fix* $\boldsymbol{\alpha}$ only in the denominator, i.e. perform inference with fully fixed uncertainties and again re-iterate using the new best-fit values.

These problems are not exclusive to standardisation: in fact, they will reappear in the traditional procedure for correcting selection effects (cf. subsection 8.3.6); rather, they are a general shortcoming of the frequentist approach to uncertainty in a *hierarchical* setting.

Chapter 7

Supernova cosmology for data scientists

7.1 Digital photometry: how raw can you go

A datum of photometry, d, is a measurement, taken at time[150] t and in band/filter f, of a source's *photon*[152] *flux density*: *photon flux density*

$$R_f(t) \equiv \int \frac{F(t,\lambda)}{hc/\lambda} T_f(\lambda)\,\mathrm{d}\lambda\,, \tag{7.1}$$

where $\mathrm{h\,c}/\lambda$ is a photon's energy, and all quantities are, understandably, in the observer's frame. The *spectral flux density* $F(t,\lambda)$ is the prime object of source modelling/inference. *spectral flux density* Beyond the intrinsic brightness (spectral flux) $\Phi(t_\mathrm{r},\lambda_\mathrm{r})$ of the source itself (for SNæ Ia described in section 8.1), naturally described in its rest frame, a model needs to account for *propagation effects*, namely extinction—in the host and in the Milky Way (MW)—, *propagation effects* redshift, and distance, as we elaborate in chapter 11.

On the other hand, the *transmission function* $T_f(\lambda)$ encodes all *"instrumental" effects* *transmission function* and specifications, which include the filter and camera wavelength responses and atmospheric absorption (if the observation is ground-based). The former two are precisely mea- *instrumental effects* surable in laboratory conditions and usually stable throughout the operation of an instru-

[150] Naturally, this is given in more convenient coordinates than those used in cosmology (e.g. for the metric in eq. (5.4)); namely, (as I said, for historical reasons) in (modified) Julian days (MJD) with origin exactly 166 years and one day before the deadline for submission of this thesis—arguably much less arbitrary than the last coincidence of the solar (28 yr), lunar (19 yr), and *Roman-taxation* (15 yr) calendar cycles[151] in 4713 BC.

[151] I am sure these were *designed* to be co-prime with least common multiple (ΛCM) $3\times4\times5\times7\times19 = 7980$.

[152] This is due to the use of *charge-coupled devices (CCDs)* in modern observatories—miniature photovoltaic *CCD* cells that produce electricity from starlight via Einstein's Nobel-worthy photoelectric effect (and more than a century of engineering advances). The earlier photographic plates, on the other hand, measure total deposited energy and typically have a highly non-linear response (in a variety of aspects) [276].

© The Author(s), under exclusive license to Springer Nature Switzerland AG 2026

K. Karchev, *Supernova Cosmology for the 21st Century*,
https://doi.org/10.1007/978-3-032-15072-1_7

Table 7.1: Radiometric terminology (co-opted from Wikipedia:Radiometry).

	quantity	symbol	redshift	distance
luminosity	(energy) flux	$L \equiv \Delta E/\Delta t$	$\propto (1+z)^{-2}$	(absolute)
	spectral flux	$\Phi \equiv \Delta L/\Delta\lambda$	$\propto (1+z)^{-3}$	(absolute)
intensity	flux density	$E \equiv L/\text{area}$	$\propto (1+z)^{-2}$	$D_\mathrm{M}(z_\mathrm{c})^{-2}$
	spectral flux density	$F \equiv \Delta E/\Delta\lambda$	$\propto (1+z)^{-3}$	$D_\mathrm{M}(z_\mathrm{c})^{-2}$
	→ photon flux density	$R_f \equiv \int F/E_\gamma(\lambda)\, T_f(\lambda)\, \mathrm{d}\lambda$		

air mass · *calibration*

ment; in contrast, the latter varies based on the altitude of the source above the horizon—published "filter" transmissions usually include the contribution of a unit of *air mass* (i.e. apply for observations near the zenith)—and atmospheric conditions, i.e. cloudiness. This and the residual possibility of variation in the calibration of the optics and electronics—e.g. across the focal plane / camera sensor—, makes determining the absolute *normalisation*[153] of T_f a major challenge for astronomical measurements.

photometric standard · *zero point* · *ADU*

Instead, for every "pointing", simultaneous auxiliary measurements are made of *photometric standard* sources, whose F^0 is known (or prescribed).[154] Ultimately, this results in a *noisy measurement* of the experimental *zero point*: ZP, the magnitude of an (hypothetical) emitter that produces—under the specific conditions of each observation—an instrumental readout of 1 ADU (an *analog-to-digital unit*[127]). The uncertainty in its determination is usually expressed as a normal posterior, with mean and (usually small) standard deviation released alongside the source measurements.

signal $[\mathrm{e}^-]$

In combination with the photometric standard, for which a photon flux density R_f^0 can be defined in analogy with eq. (7.1), the expected *signal* (number of photoelectrons, e^-) from the source can be written as

$$\langle d \rangle_\mathrm{src} = \frac{R_f(t^{s,i})}{R_f^0} \frac{10^{0.4\times\mathrm{ZP}}}{\mathrm{gain}}, \tag{7.2}$$

gain $[\mathrm{ADU}/\mathrm{e}^-]$

where the *gain* is an instrumental setting[155] that dictates the conversion between ADU

[153] Alas, astronomy is not an absolute science: life would be too easy if it were. In fact, the magnitude system[125] was devised exactly to measure brightnesses *in practice* through *comparisons* (ordering/sorting).

magnitude system

[154] This attempt for astronomical absolution[153] has created two[citation needed] major photometric standards defining separate *magnitude systems*: Vega and AB. The former is tied to the eponymous star (for historical reasons [35]) but not really [257]: its agnosticist definition ("$m_\mathrm{Vega} \equiv 0$ in any filter, now let's move on"), has later been restated in terms of experimental phenomena [203] and theoretical models (of white dwarf as the simplest astrophysical object!) [53]. On the other hand, the AB(solute) system stipulates a standard (null-AB-magnitude) $F^\mathrm{AB} \equiv \lambda^2 \times 10^{3.56}\,\mathrm{Jy}/c$ [392, with a "minus-sign" typo in the defining (unnumbered) equation], leaving its practical realisation to the observers.

(readout) and photoelectrons (signal/data). Finally, the data is a Poisson realisation:

$$d \sim \text{Pois}(\langle d\rangle_{\text{src}} + \langle d\rangle_{\text{bg}}), \tag{7.3}$$

that contains a "noise" contribution from the electronics, scattering in the atmosphere ("sky signal", depending e.g. on the Moon's phase and the time of night), and the continuous host light. Calculating/measuring this *background* term is a science of itself [45, see eg], as it depends on a wide variety of factors: on the atmospheric *seeing*, *diffraction* in the optical system, instrumental effects in the electronics (*dark & readout current*), the exposure setting (integration time) and the telescopes's light-collecting area, and on the procedure for measuring fluxes, both of the transient as it is transpiring, and of its host.

background

seeing & diffraction

dark & readout current

A common simplification of the above (forward, in the style of our inference framework) instrument description is calibration, which *reduces* the data to a Gaussian approximation of the Poisson distribution in the well-known limit of a large rate $\langle d\rangle_{\text{src}} + \langle d\rangle_{\text{bg}} \gg 1$:

data reduction

$$\begin{aligned} d &\rightsquigarrow \mathcal{N}(\langle d\rangle_{\text{src}} + \langle d\rangle_{\text{bg}}, \langle d\rangle_{\text{src}} + \langle d\rangle_{\text{bg}}) \\ \implies d - \langle d\rangle_{\text{bg}} &\rightsquigarrow \mathcal{N}(\langle d\rangle_{\text{src}}, \langle d\rangle_{\text{src}} + \langle d\rangle_{\text{bg}}). \end{aligned} \tag{7.4}$$

Now, the background level $\langle d\rangle_{\text{bg}}$, whose *a priori* calculation is involved beyond feasibility, as discussed above, can be *measured* from *auxiliary data* (b) of e.g. a sourceless sky and the transientless host (before/after the transpiration) to produce an estimate/or

calibration data (auxiliary)

$$b \sim \text{Pois}(\langle d\rangle_{\text{bg}}) \approx \langle d\rangle_{\text{bg}}, \tag{7.5}$$

which is then subtracted from the observation to produce a *calibrated "flux"*[156]:

calibrated flux

$$\texttt{FLUXCAL} \equiv d - b, \tag{7.6}$$

treated (and released) as the data. Its (Gaussian) uncertainty is calculated according to eq. (7.4), which requires—circularly—knowledge of the "true signal" $\langle d\rangle_{\text{src}}$. When forward modelling, this is readily available through eqs. (7.1) and (7.2); for calibration purposes, one can apply the same strategy as for the background:

$$\texttt{FLUXCALERR}^2 \begin{cases} \approx \langle d\rangle_{\text{src}}(\ldots) + b & \text{(semi-forward modelling)}, \\ \equiv d \approx \langle d\rangle_{\text{src}} + \langle d\rangle_{\text{bg}} & \text{(calibration)}. \end{cases} \tag{7.7}$$

[155] It concerns digitalisation with limited precision (bits) and so may differ e.g. when determining the zero point and the various backgrounds, which have different intensities.

[156] Calibrated *flux* is a misnomer—beyond the fact that it measures, nominally, (photon) flux *density*—: e.g., since the Poisson distribution does not satisfy the linearity implied for the normal in eq. (7.4), and the background level is anyway estimated, its subtraction from a particular observation can result in *negative* calibrated "flux"; this is a known problem, especially when expressing FLUXCAL in—logarithmic—*magnitudes*, usually dealt with by *discarding* problematic measurements: they evidently contain more noise than signal.

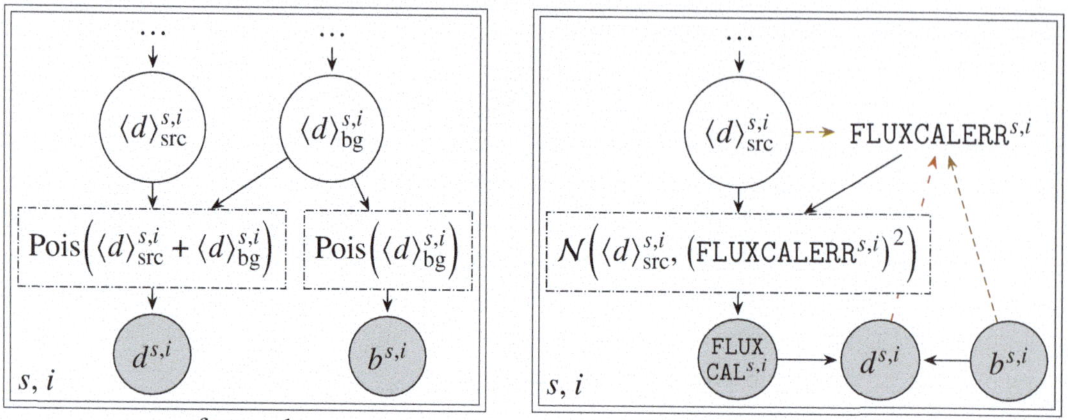

forward semi-forward / calibration

Figure 7.1: Two approaches to instrumental effects (both apply independently to each pointing, labelled s, i as per subsection 7.1.1). A fully forward model considers the generating process for all data: primary (d) and auxiliary (b). In contrast, calibration splits the data into contributions from "signal" (assumed Gaussian) and "noise"/background (directly estimated from b). The contribution of the *signal* to its own uncertainty (alternative dashed lines) is either modelled directly (semi-forward model) or estimated directly from d. Notice that this latter case creates a loop in the graph.

The two alternatives are depicted as dashed connections in the right-hand graphical model in fig. 7.1. Notice that "calibration" results in a loop and so is, technically, improper: it does not capture the dependence of the noise on the signal but rather fixes it to an estimate based on the observed d and b.

Finally, the assumed sampling distribution/likelihood under this approximation is

$$\texttt{FLUXCAL} \rightsquigarrow \mathcal{N}\left(\langle d \rangle_{\text{src}}, (\texttt{FLUXCALERR})^2\right). \tag{7.8}$$

In this case, linear re-scaling does not modify inference—in contrast to the original Poisson sampling—, so the ZP and gain can be *set* arbitrarily in a data release: common values are ZP $\to$ 25 or 27.5 or 30 mag and gain $\to$ 1.

7.1.1 Transient photometry: light curves and surveys

LC

A *light curve (LC)* comprises the N^s_{obs} photometric measurements $\mathbf{d}^s \equiv \left[d^{s,i}\right]_{i=1}^{N^s_{\text{obs}}}$ of a particular transient s. Essential for its interpretation is the associated metadata (cf. sections 1.1 and 4.2): the times, bands, gains, and zero points $\left[(t, f, \text{gain}, \text{ZP})^{s,i}\right]_{i=1}^{N^s_{\text{obs}}}$ pertinent to each data point. In case of calibrated data, this also includes the background measurements and the combined uncertainties $\left[(b, \texttt{FLUXCALERR})^{s,i}\right]_{i=1}^{N^s_{\text{obs}}}$.

A collection of light curves (and metadata), in turn, is a photometric *survey* (or compilation): the data set $\{\mathbf{d}^s\}_{s=1}^{N_{\text{SN}}}$ analysed in SN cosmology. It is thus a catalogue, in the sense of section 4.2, on two levels. Firstly, it is truly a data *set* $\mathbf{D}$, for its size, the number of observed (detected/selected) transients, N_{SN}, is *a priori* undetermined. And secondly, each light curve $\mathbf{d}^s$ is a catalogue in itself: the number of observations and associated metadata, even when pre-determined by e.g. a particular observing strategy, are in general different among the transients. *survey*

Certainly, as with any catalogue, one can argue that the *raw* data has a much better defined format: e.g. a collection of images (pixel readouts) taken at pre-determined times and in pre-determined filters (this includes the auxiliary observations used for calibration). It is the *identification*, i.e. detection, of objects that introduces the stochasticity of data structure. Therefore, one may regard the catalogue size and metadata as entirely fixed (on both levels) if one is certain that this will not affect inference: e.g. the assumed model does not suppose a dependence of the observed counts on parameters of interest and/or a procedure for sample selection is not considered. In such a scenario, the data set of sets reduces to a simple *array* of size $\sum_{s=1}^{N_{\text{SN}}} N^s_{\text{obs}}$, whose elements are identifiable by the combined label s, i, and the metadata reduces to simple object-specific settings, which control the sampling distribution. We will adopt this streamlined approach for our initial forays into SN data analysis in chapters 12 and 13. *data array*

7.2 The data deluge

Before the end of the 19th century, supernovæ were being discovered "by chance" and "by eye"—i.e. not very efficiently. S Andromedæ (SN 1985) was the first to be observed through a telescope and located in a galaxy other than our own[157] [see 30]. After the identification of the Ia type and the recognition of its excellent properties as a distance indicator, targeted campaigns were instituted to discover SNæ Ia at cosmological distances. While Norgaard-Nielsen et al. [388] discovered only one in two years, the Calán/Tololo survey observed ~30 "low-z" SNæ Ia in the following years [196, 197], and the Nobel-worthy work of Perlmutter et al. [404, 405] and Riess et al. [446] added 58 objects at redshifts reaching $z = 1$. Similarly to the Universe, the field then entered a period of accelerated expansion, with discoveries streaming in from dedicated SN campaigns, general transient searches, and extensive cosmological surveys alike: see table 7.2. By the moment of this

[157] Of course, these did not exist—as universally recognised and accepted *concepts*—until after the Great Debate of 1920 [477].

writing, thousands of SNæ Ia have been observed both at low ($z \lesssim 0.1$) and high redshifts (up to and slightly beyond $z = 1$).

Despite its abundance, current SN Ia data remains scattered across redshift and wavelength (see fig. 7.2) due to the variety of observational strategies employed through the years. Moreover, each listed data set in table 7.2 has been produced with unique instrumentation, usually with a somehow non-standard set of filters, different magnitude systems and zero points, and sometimes even on different telescopes within a single survey. Moreover, spectra are usually obtained, processed, and published separately and notoriously hard to calibrate.[158] On the contrary, deriving stringent cosmological constraints relies on tracing consistently the absolute luminosities of SNæ Ia through as wide a redshift range as possible. This has motivated monumental efforts to compile, reconcile, and cross-calibrate different low- and high-redshift surveys: notable examples include

- 463 SNæ Ia in Jones et al. [261] ≡ Foundation ∩ PS1MD ,
- 740 SNæ Ia in JLA [44] ≡ SDSS ∩ SNLS,
- 1701 SNæ Ia in Pantheon(+) [474, 475] ≡ JLA ∩ PS1MD ∩ various low-z,
- 2087 SNæ Ia in Union3 [462], a compilation of 24 sources.

The preferred approach [e.g. Supercal(-fragilistic): 473, 72] is to calibrate onto the magnitude system of the instrument/survey with greatest overlap with the rest of the data sets: e.g. Pan-STARRS. This highlights the importance of large observational efforts: other such examples are the Dark Energy Survey (DES) and the Zwicky Transient Facility (ZTF).[159]

Large compilations naturally present a slew of systematic uncertainties: e.g. related to the magnitude offsets between separate surveys (or telescopes). Even within monolithic data sets calibration parameters abound: every single pointing includes supplementary measurements to determine a zero point, sky and instrumental background, and the host contribution (see section 7.1). A traditional Bayesian analysis quickly escapes the realm of computational feasibility; instead, one of two simplistic approaches is currently adopted for each of those nuisances:

measurement noise

- it is either assumed to be normally distributed and added in quadrature to a cumulative "*measurement noise*" (`FLUXCALERR`, see section 7.1);
- or it is fixed throughout an analysis, and its influence is determined by re-running the whole inference pipeline for several (usually two) representative values in order to "propagate" its "uncertainty".

Needless to say, one can do better—and has to, in order to fulfill the statistical promise of large observational data sets without introducing systematic biases.

[158] see e.g. the procedure in Betoule et al. [44] and Kenworthy et al. [277]: spoiler alert, they give up

[159] The calibration of the ZTF itself has been revealed to suffer a peculiar "pocket effect" biasing measurements by ~0.02 mag, and as a result Rigault et al. [451, cf. fig. 2] warn against using its output for cosmology.

Table 7.2: Summary (far from complete) of sources of SN Ia data in the 21st century, their wavelength and redshift coverage. The hitherto active campaigns have incorporated extensive spectroscopic followup and confirmation as a core part of their observing procedure, unlike those expected in the near future.

	survey	#^{a} of SNæ Ia	z	bands^b	references
	CfA $^{1–4}$	346	<0.1	*UBVRI*	[447, 253, 215, 216]
	Loss $^{1\&2}$	351	<0.05	*BVRI*	[165, 495]
SN Ia	SNLS	252	0.15–1.1	*griz*	[194]
targeted	ESSENCE	213	0.1–0.81	*RI*	[377]
	CSP $^{\text{I\&II}}$	259	<0.137	*ugri* *BV YJH*	[298, 411]
	SweetSpot	74	<0.1	*JHK*	[543]
	DEH$_0$VILS	83	<0.1	*YJH*	[406]
	RAISIN	37	0.2–0.6	*JH*	[262]
all-sky	Foundation	225	<0.1	*griz*	[157]
transient	PS1MD	279	0.03–0.68	*griz*	[474]
	ZTF ^{ongoing}	3628	<0.3	*gri*	[451]
all-sky^c	SDSS	1443(+677†)	<0.5	*griz*	[464]
cosmology	DES	1635†	0.1–1.13	*ugrizy*	[466]
low-blueshift	LSST ^{2025} 🤞	$\sim 10^5$/yr†	$\lesssim 1$	*ugrizy*	[240, 303]
Universe	WFIRS ^{2027} 🤞	$\sim 10^{3\dagger}$	<1.7–3	*RZYJHF*	[231]

^{a} Numbers quoted here should not be treated as exact but indicative: e.g. as *reported* in abstracts and not as *used* in analyses or contained in data releases…

^{b} The actual transmission of a “given” filter may differ significantly between instruments and pointings due to atmospheric conditions.

^{c} Transients were still observed in narrow “drilling” fields.

† supernovæ classified based on their photometry (but possibly with precise host redshift information)

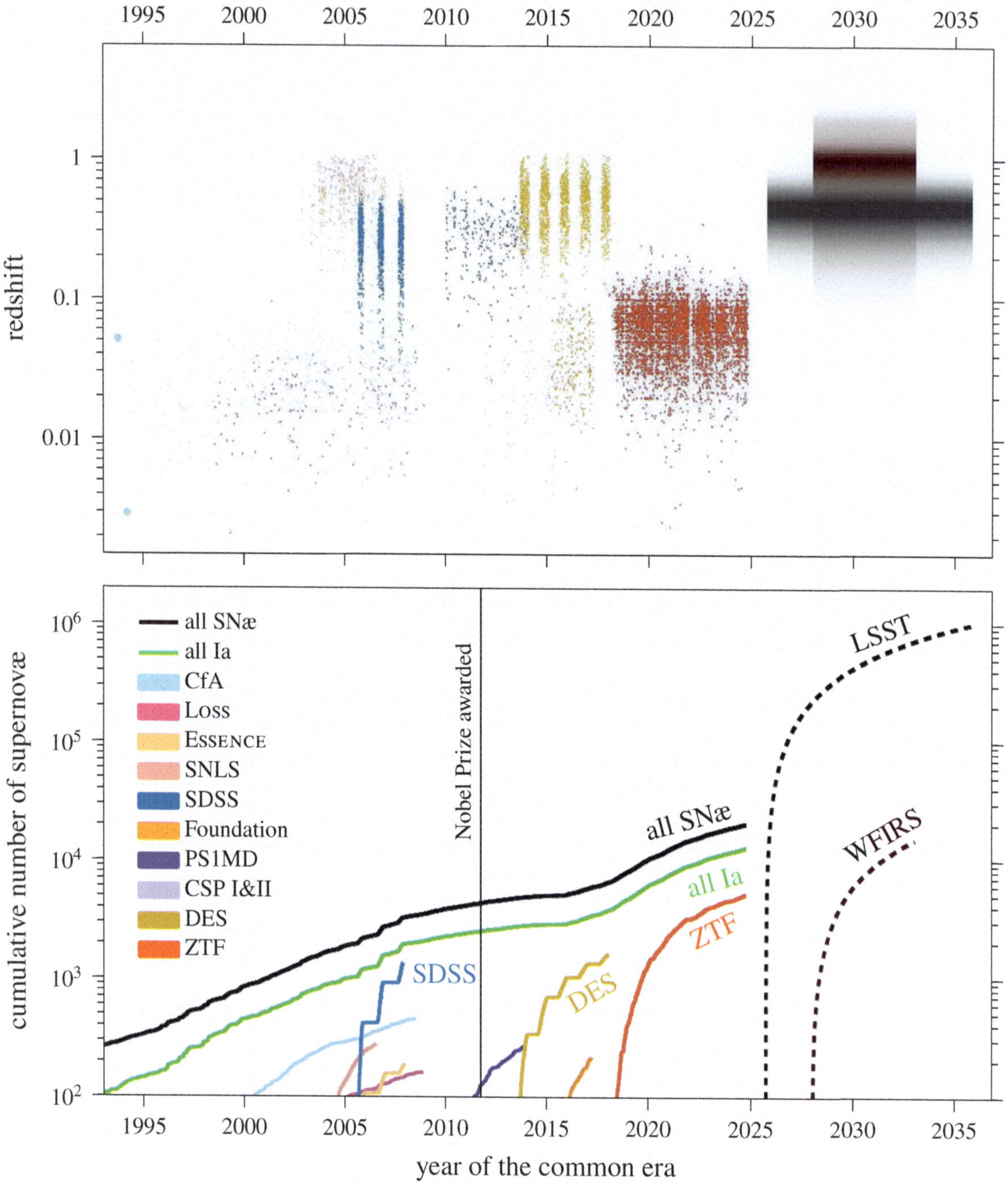

Figure 7.2: SNæ Ia discovered in the last two decades and projections for the near future. *Top*: redshifts of confirmed SNæ Ia with publicly available photometry from major surveys, including estimates for future instruments (see table 7.2 for sources). *Bottom*: cumulative number of observed SNæ (all types) and type Ia specifically with time, as well as counts per survey.

The *low-blueshift™ Universe* promises to completely disrupt the *status quo* in two respects. On one side, the Vera Rubin Observatory's *Legacy Survey of Space and Time (LSST)* [240], scheduled to begin by the end of 2025, is forecast to discover *millions* of supernovæ over the following ten years [303, fig. 11.2]: many orders of magnitude more than currently available (see fig. 7.2). In the NIR, it will be backed by the Rubin Space Telescope (RST) [492] and its Wide-Field Infrared Survey (WFIRS) [231], while (unlensed) detections with bigger telescopes like the European Extremely Large Telescope (ELT) are predicted to reach $z_c = 3$ [60]. *near future* *LSST*

Future data will be almost exclusively photometric due to the excessive resource demands of spectroscopy. This has drastic implications to transient typing[124] and redshift estimation (see subsection 8.3.7 and subsection 8.3.2, respectively, for further details), which are most reliably achieved with spectral information. In its absence, SNæ Ia (will) have to be classified solely based on light curves—or analyses (will) need to be re-formulated to rigorously handle contaminants (non-Ia transients) in the object catalogue. Similarly, redshift will need to be either extracted from the light curve or from *auxiliary observations*[102]: *host data* e.g. targeted spectroscopic followup of host galaxies *after*[160] transients have been identified or from archival data. This implies an additional analysis step: host identification, which is not a trivial task in crowded and blended fields (see subsection 8.3.5). Instead, the most ubiquitous source of redshift information will be the host photometry recorded by the survey before/after the transient.

---

[160] Measuring flux split over wavelength bins (*spectroscopy*) requires proportionally longer integration time with *spectroscopy* respect to broadband (wavelength-integrated) photometry. This is not, in principle, a constraint that cannot be overcome by technology: colossal multi-fiber spectroscopic galaxy surveys like Sloan Digital Sky Survey (SDSS) and Dark Energy Spectroscopic Instrument (DESI) have achieved miracles for cosmology, given enough time. Transients are a different beast, however: with them, time is a scarce commodity (evidenced by the name), and their occurrence is random (cf. the "revolt against determinism"), which requires rapid-response instead of massively coordinated strategies, as the ones currently employed. Robotic observatories like Pan-STARRS, GOTO, and ZTF (which has a low-resolution ($R \sim 100$) spectrograph) are a step in the right direction, but spectroscopy cannot, almost by definition, keep up with photometry. Consider, simply, that LSST light curves are expected to comprise approximately 50 broadband measurements each; this means that spectroscopic followup of *all* discovered transients can in principle achieve only a similar $R \sim 50$ on average.

Chapter 8

Supernova cosmology for statisticians

The exponential influx of SN data in the last two decades and the potential it holds for uncovering some of the deepest mysteries of the Universe has motivated the development of statistical models for its analysis that quickly surpass the sophistication of intuition-driven standardisation. On the other hand, the epoch of LSST will bring about a step-like change in both data quantity (much greater) and quality (somewhat lower due to the impossibility of spectroscopic follow-up), which will correspondingly introduce new challenges in terms of computational and modelling requirements. In this chapter, therefore, we review the existing methodologies for SN Ia cosmology and then discuss potential pitfalls in their application to future data (i.e. identify areas that require improvement) before finally presenting our vision for scalable and principled application of neural SBI in the field.

8.1 SN Ia templates

While SN Ia cosmology has remained largely empirical (with only few developments motivated by physical/mechanistic insight) since the practical breakthroughs of the 1990s, the ensuing accumulation of high-*quality* observations has enabled a truly data-driven approach centered around the creation and utilisation of a parametrised *template* for the intrinsic spectral flux timeseries $\Phi(t_r, \lambda_r)$ of SNæ Ia in their rest frame, which allows modelling measurements beyond those at peak in a given band, i.e. extracting information from / standardising the full light curve. *template*

A template is built from a large collection of photometric and/or spectroscopic data compiled to span the times (phases)[161] and wavelengths in which Φ is defined. Since observations measure spectral flux *density* in the *observer* frame, they first need to be cor-

[161] The template epoch, $t = 0$, can be freely chosen and usually represents the peak (in e.g. the B band).

© The Author(s), under exclusive license to Springer Nature Switzerland AG 2026

K. Karchev, *Supernova Cosmology for the 21st Century*,
https://doi.org/10.1007/978-3-032-15072-1_8

rected for distance, time/energy dilation, and redshift (straightforward with spectroscopy but more involved for photometry); and since the template is meant to represent the SN Ia-intrinsic brightness, extinction along the line of sight needs to be taken into account. In short, creating a template means *inferring* $\Phi(t_r, \lambda_r)$ from data.

To do this, one first needs to introduce a parametrisation. The simplest conceivable template has no parameters and thus represents the standard (average) $\Phi(t_r, \lambda_r)$ of SNæ Ia in (rest-frame) time and wavelength bins; such were built by Leibundgut et al. [311] (from 75 light curves) and Hsiao et al. [233] (from ~600 spectra from ~100 SNæ Ia). Simple single-parameter extensions, heavily influenced by the standardisation framework, were developed by Perlmutter et al. [404] (who allowed only for a uniform time-rescaling) and Riess et al. [445] (who allowed for a time-dependent flux-rescaling): see fig. 8.1. The latter—more flexible—approach led to the development of the MLCS (Multicolor Light-Curve Shape) [445, 446, 252, 254] templates defined in a discrete set of K-corrected and de-extincted broad bands.

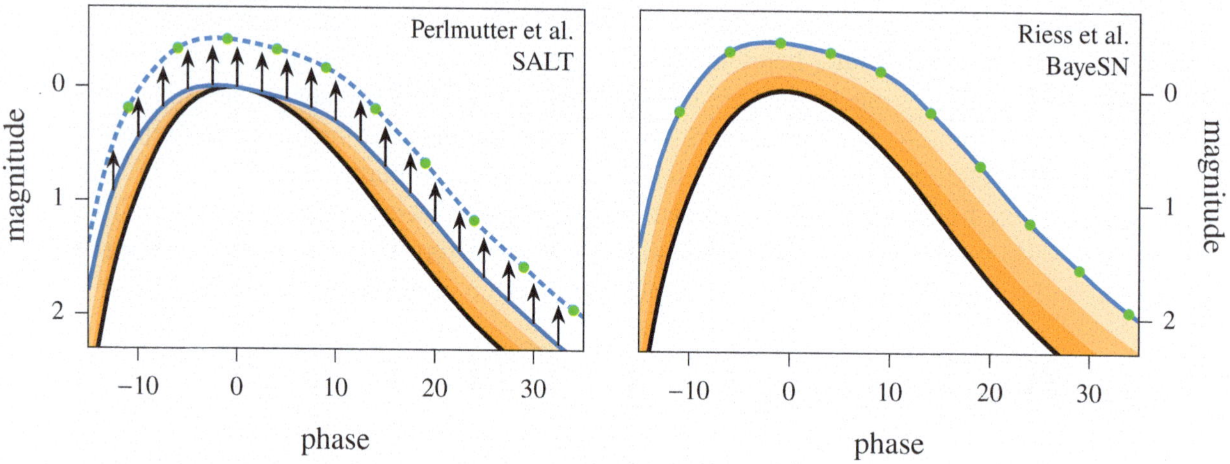

Figure 8.1: Two ways to standardise a light curve (green points) in terms of a standard template (black line). This example uses SALT with $x_1 = 3$ (and $c = 0$) for the reference and "observation" in both panels. *Left*: standardise first only the stretch in time (either directly by a parameter s as in SALT 1 or by adding x_1 times a flux component constrained to zero at the peak as in SALT 2+); then shift (Tripp-correct) coherently in brightness by an amount that *ends up* correlated with the stretch (αx_1, black arrows). *Right*: directly model the light curve with a "principal component" that is added to the reference in rtion to a "x_1". This approach, adopted by MLCS and BayeSN, does not require *post factum* correction / brightness standardisation. See also the similar fig. 18 in Perlmutter's Nobel lecture [403] and note the misrepresentation therein of MLCS [cf. 254, fig. 7].

8.1.1 The *de facto* standard(isation)

SALT

SALT (*spectral Adaptive Light-curve Template*) [192, 193, 277] is a simple extension of magnitude standardisation to a range of times and wavelengths that has emerged as the preferred method for extracting covariates and deriving standardised distance estimates in all major SN Ia cosmological studies [44, 474, 73, 109]. It falls within the larger framework of *(functional) principal component analysis (PCA)*-based templates, which includes also the more recent SNEMO (superNova EMpirical MOdels) [469] and SUGAR (SUpernova Generator And Reconstructor) [309]. They consist of a Φ_0 shared among all supernovæ (corresponding to the standard template) and N_{PCA} additional $\{\Phi_i\}$ whose contributions to a particular object's brightness are controlled by parameters $\{x_i^s\}$ (cf. eq. (6.3)):

functional PCA

$$\text{SALT}: \quad \Phi^s(t_r, \lambda_r) \equiv \Phi_0(t_r, \lambda_r) + \sum_{i=1}^{N_{\text{PCA}}} x_i^s \Phi_i(t_r, \lambda_r) + \epsilon^s(t_r, \lambda_r). \tag{8.1}$$

As in ordinary PCA,[162] the number of components can be "objectively" determined by examining the amount of residual variations $\epsilon^s(t_r, \lambda_r)$ for models with increasing N_{PCA}.[163] While SNEMO finds in it reason to employ up to 14 principal components, SALT only uses a single pair x_1, Φ_1, corresponding[164] to early light-curve shape corrections. In addition, both SALT[165] uses a fixed *colour law* with intensity (but not form) controlled again per-supernova by a parameter c^s:

colour law

$$\text{SALT}: \quad \Phi^s(t_r, \lambda_r) \to \Phi^s(t_r, \lambda_r) \times \exp(c^s \, \text{CL}(\lambda_r)), \tag{8.2}$$

meant to represent *extrinsic* reddening due to the environment. The unfortunate contradiction with the stated goal of the template as SN-intrinsic and the inability of the fixed CL to anyway account for the observed extrinsic variations beyond the amount of extinction has led to much controversy (see subsection 8.3.3) but little substantial modification to the SALT framework from the above description.

---

162 Template training procedures overcome the difficulties[149] of inconsistent temporal and spectral sampling (observations) among the different SNæ Ia through interpolation either of the template (spline-based in SALT) or of the observations (through Gaussian process regression in SNEMO and SUGAR).

163 Given a limited (finite) training set, eq. (8.1) can incorporate *all* of the ϵ into as "few" as N_{SN} components (devolving $\{x_i\}$ into indicator variables), so an alternative stopping criterion is needed: e.g. comparing ϵ with the observational noise [see e.g. 309, fig. 2]. Via an independent method, the counting statistics of "twin" SNæ Ia [144], Rubin [460] found an *intrinsic dimension* of the SN Ia spectral flux of "3–5", but this does not in general correspond to the number of *linear* parameters required to span it.

intrinsic dimension

164 In fact, in the first version of SALT [192], x_1 was forced to be Perlmutter et al.'s stretch factor s, making Φ_1 a not-exactly-principal component.

165 and, at the end of the day, also SNEMO and SUGAR after convoluted discussions

Comparing the SALT template eqs. (8.1) and (8.2) to eq. (6.3), we can assimilate the template components and the colour law as extended correlation coefficients, and the per-object parameters x_i^s and c^s as the covariates. To "train" the template, they can all be simultaneously fit[166] to a collection of observations (photometric and spectroscopic) of a training set of SNæ Ia by minimising the amount of residual variations $\epsilon^s(t_r, \lambda_r)$; i.e. by a χ^2 fit extended from eq. (6.7), which results in estimates $\widehat{x_i^s}(\{\mathbf{d}^s\})$ and $\widehat{c^s}(\{\mathbf{d}^s\})$—notice, these depend on the *full* training set, which has determined the templates and, hence, the meaning/nature of the per-object parameters. In practice, one also needs to resolve several scaling degeneracies, most importantly[167]:

- between each Φ_i and its corresponding x_i by imposing on the latter unit estimator scatter $\sum_{s=1}^{N_{SN}} \left(\widehat{x_i^s}\right)^2 \equiv 1$ across the training SNæ;
- similarly, between c and the colour law, by fixing the latter's normalisation so as to interpret the former as a colour parameter for historical reasons;
- the overall normalisation of the *observed* light of each SN Ia: $\Phi^s \rightarrow x_0^s \Phi^s$—i.e. its spectral flux *density*—by fixing a correspondence $x_0 = 1 \implies m_B = 10.5$ at peak for historical reasons.

The latter point has important cosmological implications: since the template represents a latent (not directly observable) quantity, x_0 needs to account simultaneously for the intrinsic brightness scale (per SN, i.e. M^s) and the effect of distance (but without K-corrections, which are taken into account explicitly by redshifting...). This means that *additional* (cosmological) information can be exploited in the fit by requiring that the normalisations of SNæ at the same distance (i.e. redshift / in the same galaxy) be similar, up to the intrinsic scatter σ_0. However, this pushes the boundary of non-forward non-Bayesian hierarchical modelling, and instead, this information is extracted *post factum* from the estimators $\widehat{x_0^s}$ in a *separate* cosmological fit.

Lastly, SALT needs to model the significant residuals in order to faithfully represent observations. As per the discussion around σ_0 in eq. (6.7), SALT incorporates "diagonal", i.e. uncorrelated across phase and wavelength, uncertainty of Φ_0 and Φ_1 (but not in CL, see below), propagated linearly and combined in quadrature. Since the traditional χ^2 adjustment would combine the residuals across all phases and wavelengths and leave a $\sigma_0(t_r, \lambda_r)$ largely undetermined, Guy et al. [193, subsection 6.1] assume its functional dependence matches that of the propagated epistemic uncertainty (estimation variance of the template) and re-scale it within bins. Kenworthy et al. [277, eqs. (3) and (15)] introduce a slight improvement and model it as a spline-based surface, similarly to the principal components,

---

[166] In a slight departure from the present description, SUGAR pre-defines the covariates by extracting informative features from the spectrum "at" maximum light and then determines the non-principal components that correlate with them.

[167] See Kenworthy et al. [277, subsection 2.1]), from where we take the concrete examples here.

and fit for it with the full Gaussian likelihood, which includes an uncertainty "regularisation" in the normalisation/entropy term.[168] Still, the diagonal description $\sigma_0 \to \sigma_0(t_r, \lambda_r)$ cannot capture brboadband variation caused by combination of the "scatter" at a range of wavelengths, and so SALT resolves to an *ad-hoc* K-correction uncertainty, which resembles a learnt uncertainty in CL). As in canonical standardisation, these involved procedures employed to determine and apply template variance [see also 113] are symptomatic of the general unsuitability of reverse (model-fitting) methodologies for representing uncertainties in a hierarchical setting.

Despite all the correspondence pointed out above, which imply that template modelling can be a *full replacement* for standardisation at peak, it has been repeatedly found [e.g. 194, 44, 474, 277] that linear correlations between the derived SALT $\widehat{x_1^s}$, $\widehat{c^s}$ and $\widehat{x_0^s}$ still remain[169] and need to be corrected for by Tripp standardisation (eq. (6.3)) of the familiar form $M = M_0 - \alpha x_1 + \beta c + \epsilon$ as a result of two further modelling choices: the already mentioned restriction of the colour law and the stipulation of a zero average flux in the B band from the Φ_1 component, which artificially splits the principal variation into a stretch (Φ_1) and a magnitude shift for historical reasons: see fig. 8.1.

8.1.2 The Bayesian SN Ia template

The appropriate approach to forward modelling and uncertainty quantification for the SN Ia spectral timeseries template is Bayesian modelling of all Φ^s as *random functions* (an extension of the concept of a random variable to infinite dimensions, whereby the index of a random array is replaced by the function's argument(s)). This forms one of the **founding principles of BayeSN** [343, 344, 346]: the Bayesian SN Ia template. Very similarly to SALT, etc., it represents the spectral flux as a PCA-like series expansion, albeit in magnitudes with respect to the Hsiao et al. [233] template[170] rather than linear flux: *random function*

$$\begin{aligned}\text{BayeSN:}\quad & -2.5\log_{10}[\Phi^s(t_r,\lambda_r)/\Phi_{\text{Hsiao}}(t_r,\lambda_r)] \\ &= -19.5 + \delta M^s + W_0(t_r,\lambda_r) + \theta_1^s W_1(t_r,\lambda_r) + \epsilon^s(t_r,\lambda_r), \quad (8.3)\end{aligned}$$

where δM^s and θ_1^s—respectively, the residual coherent scatter (ϵ in orthodox standardisation (eq. (6.3))) and stretch (SALT's x_1)—are now true *random variables* (i.i.d. across the SN Ia population, i.e. object-specific parameters in a BHM) assigned population (hierarchical) priors (as detailed in fig. 8.2) in lieu of SALT's degeneracy-resolving ("*a posteriori*")

168 instead of pure χ^2 adjustment—a win for science!... They still iterate fitting separately the template and the uncertainty, of course...

169 Incidentally, Kenworthy et al. [277] improve upon previous SALT models by explicitly decorrelating the $\widehat{x_1^s}$ and $\widehat{c^s}$ from each other (across the training set)... but not from $\widehat{x_0^s}$.

170 which assumes the role of a photometric standard[154] and is normalised to a $M_B = 0$ in the Vega system

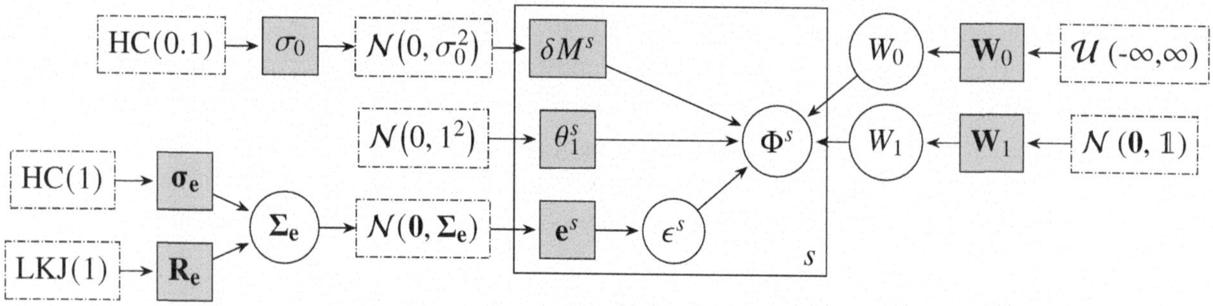

Figure 8.2: (Intrinsic) BayeSN as a hierarchical functional model for SN Ia spectral flux timeseries $\Phi^s(t_r, \lambda_r)$ (note that $W_{0,1}$ and ϵ also are random functions parametrised by random variables). Refer to Mandel et al. [346, subsection 2.5] for a full description.

constraints on estimators. Importantly, they are assumed (*a priori*) uncorrelated,[171] and the famous broader–brighter effect is instead encoded in the W_1 component, à la MLCS (cf. fig. 8.1).

processes

The template components $W_{0,1}(t_r, \lambda_r)$ and the residual perturbations $\epsilon^s(t_r, \lambda_r)$ (per-SN Ia) are also assigned prior *processes* ("distributions" of random functions) through discrete parametrisation on a fixed grid $\left[\mathbf{t}_g, \boldsymbol{\lambda}_g\right]$ in (rest-frame) phase and wavelength and 2-dimensional spline interpolation:

$$W_0(t, \lambda) = \mathrm{Spline2d}\big(t, \lambda;\ \mathbf{t}_g, \boldsymbol{\lambda}_g, \mathbf{W}_0\big), \tag{8.4}$$
$$W_1(t, \lambda) = \mathrm{Spline2d}\big(t, \lambda;\ \mathbf{t}_g, \boldsymbol{\lambda}_g, \mathbf{W}_1\big), \tag{8.5}$$
$$\epsilon^s(t, \lambda) = \mathrm{Spline2d}\big(t, \lambda;\ \mathbf{t}_g, \boldsymbol{\lambda}_g, \mathbf{e}^s\big). \tag{8.6}$$

GP

The arrays of spline knots $\mathbf{W}_{0,1}$ (shared among all SNæ Ia) are assigned weakly constraining priors (cf. fig. 8.2 and Mandel et al. [346, subsection 2.5]), allowing the template to be learnt purely from data. On the other hand, $\{\mathbf{e}^s\}_{s=1}^{N_{\mathrm{SN}}}$ explicitly model each SN Ia's residual variations. They are random *arrays* (each of size equal to the number of grid points) and are *a priori* normally distributed with covariance split into a diagonal contribution $\boldsymbol{\sigma}_\mathbf{e}$ (standard deviation at each grid point, akin to $\sigma_0(t_r, \lambda_r)$) and a correlation matrix $\mathbf{R}_\mathbf{e}$ (no equivalent in SALT) among the different spectro-temporal regions, which are also given (a final layer of hierarchical) priors. This makes eq. (8.3) a *Gaussian process (GP)*[172] [see e.g. 440] with

[171] and *a posteriori* for real data as well: see Mandel et al. [346, footnote 11]

[172] Interestingly, this formulation for the 2-dimensional distribution of a SN Ia's light is reminiscent of Karchev et al.'s [267] description of source light in a gravitational lens, down to the nonlinear transformation (redshift/lensing) and intergation (within a filter/pixel) that relate the model to observations.

mean function $-19.5 + \delta M^s + W_0(t_r, \lambda_r) + \theta_1^s W_1(t_r, \lambda_r)$ (itself hierarchically modelled with a (non-Gaussian) process) and a non-explicit albeit interpretable covariance arising from spline interpolation—which promotes smoothness—inbetween the grid points.[173]

The second tenet of BayeSN is proper and physically motivated colour modelling, specifically with respect to the variety of dust *extinction laws* that supernovæ are subjected to in their variety of hosts. In glaring contrast to SALT, BayeSN uses a parametrised colour law—namely, the one derived by Fitzpatrick [153, hereafter F99]: *extinction law*

$$\Phi^s(t_r, \lambda_r) \to \Phi^s(t_r, \lambda_r) \times \left[\mathrm{F99}(\lambda_r; R_V^s)\right]^{A_V^s}. \tag{8.7}$$

This introduces *two* new parameters per SN:

- A_V^s is the extinction *optical depth* (strength), which is proportional to the logarithm of the line-of-sight within the host and so is given a prior $\mathrm{Expon}(1/\tau)$, controlled by an *a priori* unconstrained average extinction $\tau \sim \mathrm{HC}(1)$; *optical depth*
- R_V^s is the F99 colour-law parameter, related to the properties of dust grains in the host galaxy [e.g. 132] and identified with the *total-to-selective extinction* ratio for the B and V bands: *colour-law parameter*

$$R_V = A_V/(A_B - A_V) \equiv A_V/E_{B-V}\,. \tag{8.8}$$

The distribution of dust properties—read, the hierarchical priors for the i.i.d. $\{A_V^s\}$ and $\{R_V^s\}$—has been at the heart of a high-stakes methodological standoff in the literature, whose resolution could afford a significant increase in the precision of SN Ia distance estimates through an explanation of a large fraction of the unmodelled (residual) scatter. We comment further on the topic in subsection 8.3.3 and present two SBI contributions to the subject in chapters 12 and 13.

Training BayeSN is very similar to SALT—on the surface: the model is confronted with redshift- and distance-corrected broadband-photometric and spectroscopic data, which for the purposes of dust modelling, extends into the NIR, and the *joint posterior* of all parameters (filled squares in fig. 8.2 and $\{(A_V, R_V)^s\}$) is inferred. However, BayeSN's particularly high-dimensional[174] middle (object-specific) layer—the plate in fig. 8.2—poses a computational challenge to likelihood-based techniques (even if only the global parameters are of interest for constructing the template), and this has necessitated the use of specialised high-dimensional algorithms: initially Gibbs sampling [343, 344], and then HMC (NUTS) with automatic differentiation in `Stan` and `NumPyro` [346, 184].

[173] An alternative GP-based SN Ia model was developed by Kim et al. [286], who did not expand the mean function beyond the Hsiao et al. template and used a more standard kernel-based covariance function.

[174] The grid's size—and that of $\mathbf{e}^s$—varies between implementations: 6 × 9 [346], 6 × 6 [511], 6 × 11 [537].

Cosmological inference with BayeSN is yet to be conclusively demonstrated in a way that makes full use of the model's proper hierarchical structure. Certainly, it can be— and has been [510, 262,[175] 126, 519[176]]—used to infer distances, given a fixed Φ model, but this has required breaking the degeneracy with each SN's average magnitude, i.e. with the $\{\delta M^s\}$ parameters. Perplexingly, the cited studies all consider *a priori* uncorrelated distances, relinquish the hierarchical prior $\delta M^s \sim \mathcal{N}(0, \sigma_0^2)$ in favour of an unconstrained $\mu^s + \delta M^s \sim \mathcal{U}(-\infty, \infty)$, derive per-SN independent distance moduli, and *fit* them (*vs.* external redshift constraints) with a cosmological model; in fact, Thorp & Mandel [510] *subtract* a fiducial cosmology and analyse the residuals in terms of various dust extinction configurations and *intrinsic* brightness differences in the SN Ia population rather than distance. Instead, the natural approach is to extend the BHM to include (cosmology-)parametrised distance moduli and model redshift constraints: thus, the cosmology (of interest) can be derived directly while distances are a by-product.

fitting ?!

8.2 Bayesian SN Ia cosmology

BHM with LC summaries

Proper—*Bayesian hierarchical*—SN Ia cosmology has been realised in a number of studies, but only on the level of *light-curve summary parameters*, i.e. "covariates" as extracted for Tripp standardisation, mainly using (or representing) SALT. Such analyses— with cool names like BAHAMAS (BAyesian HierArchical Modeling for the Analysis of supernova cosmology) [347, 479], UNITY (Unified Nonlinear Inference for Type-Ia cosmologY) [461, 462], Steve [221], and BIRD-SNACK (Bayesian Inference of dust law R_V Distributions using SN Ia Apparent Colours at peaK) [538]; names like Simple-BayeSN [345]; and tremendous missed opportunities like ABYSS+ (Application of BaYesian graphs to SN Ia data analysis and compression) [336]—gracefully handle the interplay between population scatter and observational uncertainties by leaning on the concept of latent (true) values $\mathbf{x}^s$ of the covariates, i.e. i.i.d. realisations of a random variable $\mathbf{x}$, of which the estimators $\widehat{\mathbf{x}^s} \equiv \hat{\mathbf{x}}(\mathbf{d}^s)$ are noisy measurements for each SN Ia. In a BHM, the latter are *also* treated as realisations of a random variable, $\hat{\mathbf{x}}$, which represents the observables: a small change of notation ($\widehat{\mathbf{x}^s} \to \hat{\mathbf{x}}^s$) but a giant leap in interpretation.

As previously alluded, the Bayesian worldview shifts standardisation (where needed, e.g. for SALT: cf. fig. 8.1) onto the latent layer, while also explicitating the "residual scatter"

[175] In Jones et al.'s cosmological analysis of the RAISIN survey, BayeSN is an awkward passenger taken along for a ride by the main driver: the SN(oo)Py fitting code [74] (a NIR extension of MLCS.)

[176] While most BayeSN analyses use HMC (NUTS), Uzsoy et al. [519] demonstrated the application of (non-amortised) VI, with a simple multivariate normal proposal, for inference from *individual* SN Ia light curves.

as resulting from a population distribution (usually normal with standard deviation σ_0):

$$M^s \sim \mathcal{N}\left(M_0 + \boldsymbol{\alpha} \cdot \mathbf{x}^s, \sigma_0^2\right), \tag{8.9}$$

The distance modulus: $\mu\left(z_c^s, C\right)$, a deterministic function of (true cosmological) redshift, z_c^s, then converts M^s to (latent) apparent magnitude, m^s. To complete the data vector, noisy measurements $\widehat{z_c}^s$[177] and $\widehat{m}^s$ are made of each SN Ia, so that $\mathbf{d}^s \rightarrow \left(\widehat{z_c}^s, \hat{\mathbf{x}}^s, \widehat{m}^s\right)$. These are related to the object-specific parameters: $\boldsymbol{\lambda}^s \equiv \left(z_c^s, \mathbf{x}^s, M^s\right)$ via a sampling distribution that is, in general, not constrained to be diagonal in any way, i.e. any element of $\mathbf{d}^s$ may depend on any other quantity of possibly a different SN.[178]

The (Bayesian) hierarchy in SN cosmology is established by the *population distribution* *population distribution*

$$\mathrm{p}\left(z_c^s, \mathbf{x}^s, M^s \,\middle|\, \boldsymbol{\gamma}\right) = \underbrace{\mathrm{p}\left(M^s \,\middle|\, \mathbf{x}^s, \not{z}_c^s, \boldsymbol{\gamma}\right)}_{\text{“standardisation”}} \underbrace{\mathrm{p}\left(\mathbf{x}^s \,\middle|\, z_c^s, \boldsymbol{\gamma}\right)}_{\substack{\text{SN Ia properties} \\ \text{\& evolution}}} \underbrace{\mathrm{p}\left(z_c^s \,\middle|\, \boldsymbol{\gamma}\right)}_{\text{rate}}, \tag{8.10}$$

where $\boldsymbol{\gamma}$ are the global parameters[179] of the model. Among them are C and M_0, $\boldsymbol{\alpha}$, σ_0—in the Bayesian setting, Tripp standardisation is just a particular parametrisation of one of the necessary conditional distributions. The remainder is more difficult to model. The last term in eq. (8.10) is the cosmological *rate of occurrence* of SNæ Ia, which is very often overlooked in SN Ia cosmology with spectroscopic (i.e. precise) redshift estimates since it contributes little to the cosmological constraints. Discussion on redshift modelling,[102] especially with regards to future photometric-only surveys, is presented in subsection 8.3.2; we use a common SN Ia rate parametrisation as a power law of $\left(1 + z_c^s\right)$ [128] in chapter 15 and improve upon it with a physically motivated forward model in chapter 16. *SN Ia rate*

On the other hand, $\mathrm{p}\left(\mathbf{x}^s \,\middle|\, z_c^s, \boldsymbol{\gamma}\right)$ represents the distribution of properties within the SN Ia population—and its redshift evolution (discussed shortly)—and thus requires more care.

177 This is an obvious simplification since the cosmological redshift is not directly measurable (cf. eq. (5.6)). We defer the subtleties to subsection 8.3.2 and here simply remark that $\widehat{z_c}^s$ *always* contains noise, e.g. related to peculiar-velocity “subtraction”.

178 Of course, by de Finetti’s theorem [563] (see also footnote 119), such cases can only be artificially constructed by “marginalising” some global parameter (if the SNæ are exchangeable, i.e. all relevant (meta)data is considered). Nevertheless, they are extremely common in SN Ia cosmology, in which the likelihood of summarised data is represented as a Gaussian with non-trivial (dense) *covariance matrix* that attempts to combine statistical and systematic uncertainties [see 103, and followers thereof], and so follows the “marginalisation” even of the correlation coefficients $\boldsymbol{\alpha}$ on top of the “usual” calibration parameters. *covariance (stat. + sys.)*

179 They, of course, require hierarchical priors, but these are largely a matter of convenience rather than physical validity, and so usually uninformative or Jeffreys-inspired distributions are chosen (see references in this section and section 14.1 for particular examples).

For covariates that are parameters of a light-curve model (e.g. SALT), it may be reasonable to assume the *empirical* distribution of the estimated values for the relevant training set[180] (e.g. [277, fig. 4]), but this is usually very different from high-redshift cosmological samples (although typically it is more "complete" in terms of sample selection (although the requirement for spectroscopy might counteract magnitude completeness...)).

More commonly, however, the choice of $\mathrm{p}(\mathbf{x}^s \mid z_c^s, \boldsymbol{\gamma})$ is made in the interest of some computational tractability or other—cf. the discussion after eq. (1.8)—motivated also by the fact that template training is usually implemented so as to "fix" it to a given distribution, e.g. a unit normal for x_1 (see above). For instance, the first SN Ia-cosmological BHM, March et al. [347], is almost entirely solvable analytically, on account of being composed primarily of normal distributions.[181]

Beyond light-curve summaries / template parameters, e.g. $x_1, c \in \mathbf{x}$, Bayesian standardisation has been extended with other properties like the mass of the host galaxy [336, 479, 345, 221, 461, 462], the location of the SN within it [217], and the "re-brightening time" [478, 121]. In general, *any* imaginable quantity x_i (see subsection 8.3.4) of which a measurement $\hat{x}_i$ can be made (and described through a sampling distribution) can be included within $\mathbf{x}$ (and $\hat{\mathbf{x}}$) and its utility for standardisation (at least to linear order) evaluated through the posterior of its corresponding correlation coefficient α_i.[182]

Beyond population modelling and rigorous uncertainty quantification, BHMing also allows seamless accounting for probabilistic effects like complicated uncertainties (in lieu of linear error propagation), e.g. in the template,[183] calibration,[184] and photometric redshift estimation: *z*BEAMS [453] (subsection 8.3.2); for sample selection: UNITY, Steve (subsection 8.3.6); and for non-Ia contamination: BEAMS (Bayesian Estimation Applied to Multiple species) [300, 453] (subsection 8.3.7).

---

[180] But beware that those may be affected by selection or estimation biases. As an extension, $\mathrm{p}(\mathbf{x}^s \mid z_c^s, \boldsymbol{\gamma})$ is often flexibly parametrised (e.g. as a skew normal in UNITY 1), selection effects are not modelled, and inference is to be understood as representing the sample of *selected* objects: $\mathrm{p}(\mathbf{d} \mid \mathfrak{s}, \boldsymbol{\gamma})$, which of course, may not be the same as the total population. The danger of doing this is to solidify the results from one study (e.g. the template training) and use them in another setting for a differently constructed sample.

[181] Crucially, this requires linearisation of the distance modulus for the purposes of redshift uncertainty propagation (see Karchev et al. [269], appendix B for more details), which is justified only when the latter is small, i.e. with spectroscopic information.

[182] Estimators can also be correlated with one another in the orthodox "inference" / standardisation framework, but this is far more prone to improper interpretation of uncertainties; indeed, it is highly dependent on the noise / estimator variance. Moreover, Bayesian modelling presents a means for "hypothesis testing" (model selection) that is easily spun out of parameter inference—and according to some, is more interpretable.

[183] See e.g. Grayling & Popovic's [184] mass-dependent spectral-timeseries modelling with BayeSN (even though their analysis is not exactly in a cosmological context).

[184] See e.g. UNITY [462, subsection 4.4]... for an example of room for improvement.

8.3 Pitfalls

In this section, we discuss the major outstanding challenges in the field of SN cosmology, how they have been approached in the literature—often in sub-optimal ways constrained by tractability even in Bayesian settings—, some of the detrimental effects this has on inference results in certain cases (hence the term "pitfalls"), and how a comprehensive SN cosmological model/simulator *should* be formulated and solved (with SBI) in order to extract maximum information from future large observational campaigns.

8.3.1 Scalability

explicit modelling

However powerful, the Bayesian approach needs to be *explicit*[29]: it requires parametrisations of all considered—physical and statistical—effects in the hierarchy: for example, to account for classification and sample selection, one introduces latent labels for the class and selection status, and for variability in the spectral timeseries, an explicit vector $\mathbf{e}^s$ for each SN Ia. These need to be assigned prior probabilities and subsequently sampled simultaneously with the (usually very few) (global) parameters of interest in any likelihood-based analysis, which presents a computational challenge even with current "modestly" sized SN Ia collections, let alone the possibility of scaling to the 10^5–10^6 SNæ from LSST.

But alas, recall that parameters are auxiliary constructs (when they are not of interest) and so can, at times, be avoided: e.g. by re-formulating the template as an implicit Gaussian process (BayeSN $\longrightarrow$ Kim et al.), through laborious analytical calculation of a modified likelihood for selected samples (Steve, UNITY), or by enumeration of the models for different classes, fitting with each (essentially), and averaging (BEAMS).

At most of those times, the computational convenience of an analytical description (over numerical marginalisation) is the sole motivation for adopting a particular convenient—but otherwise arbitrary—probabilistic description for phenomena/effects that are known to not exactly follow it. The prime example are linearly propagated Gaussian redshift uncertainties (discussed shortly), instrumental effects (Pois $\longrightarrow \mathcal{N}$, discussed in section 7.1), and systematic effects expressed through a numerically estimated covariance[178] [see 103] for highly compressed—mostly hand-crafted (or intuition-inspired), hence possibly sub-optimal—summary representations of the data.

Lastly, current analyses are ~~often~~ exclusively split into stages: first, a template is trained (usually with different, more detailed, and more robustly observed and selected data); then, distances are derived with a fixed template (with variability from the first stage incorporated in a more or less principled manner); finally, cosmological constraints are derived from the distances[185] (and usually separately inferred redshifts), after modelling of (or "correcting"

[185] The last step may differ based on the scientific goal of the analysis; e.g. Thorp & Mandel [510] investigate

for) selection biases (subsection 8.3.6).

On one hand, this approach ignores available information and hierarchical constraining power: e.g. the assumption of a particular cosmology and a set of true redshifts—if both are sampled in a MCMC—"standardises" the fluxes and allows more stringent constraints to be derived even from observations with uncertain redshift and unknown distance. And on the other hand, partitioning an analysis can lead to a hard-to-trace bias[180] from an unquestioned re-utilisation of results derived in a different setting than that of a subsequent application.[186]

SBI counters all of the above arguments by definition:

- it is implicit: likelihood evaluation or explicit probabilities are not needed;
- it is marginal: only the parameters of interest are inferred;
- it is scalable: simulation and NN evaluation are linear[188] in the size of the data set;
- it can implement end-to-end inference simply by compiling an appropriate simulator.

8.3.2 Redshifts (and velocities)

While the majority of the modelling effort devoted to SNæ Ia (and standard candles in general) is expended on inferring distances (i.e. absolute and apparent brightnesses), it is just as important to the final cosmological constraints to have access to cosmological(!) redshift estimates that are as accurate as they are precise—e.g. Wojtak et al. [547] show that a systematic offset $\Delta z_c \sim 10^{-5}$ can have a disproportionately larger impact (~1 %) on dark-energy inference [see also 325, 237]—and to employ a statistical procedure able to

---

magnitude offsets (in other cases interpreted as distances) in relation to model selection of dust laws and a magnitude step (subsection 8.3.3).

Amplifications of training-set biases

[186] A curious point in that regard was made by Hogg & Villar [227, subsection 4] and applies to any data-driven modelling (including non-simulation-based[187] ML). Every result of a training/fitting/optimisation procedure (e.g. a template) carries a sampling *bias* associated to the particular realisation of its training set (consider estimating the mean of a distribution from a finite number of samples: the result will differ from the true mean by an amount that reduces with the size of the "training set"). When the "optimal" result is employed in a separate subsequent analysis of a larger sample (hence, with smaller statistical uncertainty than the estimator variance of the first stage), the results *will* be biased. However, estimating the sample variance in the first place requires a model for how the training set has been sampled—even in the frequentist interpretation—i.e., a hierarchical prior/likelihood. In Bayesian terms, the problem can be neatly expressed and resolved by considering—as per eq. (1.4)—a *posterior* derived from the first analysis as the prior for the second; or by training the template while *simultaneously* performing "downstream" inference.

[187] The training set in end-to-end SBI is much larger than the analysed datum, which is a singleton example. This is in contrast to piecewise SBI approaches like learning a per-object likelihood, which *also suffer* from amplification of training-set biases.

[188] While some NN architectures (prominently, attention-based / transformers), have superficially quadratic complexity, they can be accelerated without significant loss of performance [e.g. 306, 293, 98] or replaced by a deep set with explicitly linear scaling, as we already discussed.

properly account for their uncertainty. Thus, the redshift pitfall of SN Ia cosmology has two entrances.

On one side, one needs to rigorously model all *other* sources of redshift besides cosmological expansion: eq. (5.6), since it is only the total redshift z that can be inferred from the wavelengths of observed spectral features. The additional contributions come from large-scale gravitational redshift in the cosmic neighbourhood of the Earth[189] and the total (line-of-sight) peculiar velocity of the light emitter[190] with respect to the Earth. While the motion of the Earth, Sun, and Milky Way can be rigorously accounted for [413, 414], the *bulk flow* of distant galaxies is known only on average [e.g. 89, 58, 59],[191] and the *motion* of a SN *within* its host is essentially[192] unconstrainable, i.e. it can only be incorporated as epistemic uncertainty.

bulk flow

proper motion

Peculiar velocities and the local gravitational redshift do not, obviously, scale with z_c, so they represent a relatively important effect only for small z_c. Early studies [e.g. 280], therefore, *discarded* SNæ Ia with $z \lesssim 0.02$. More recently, the standard approach to the conversion between total and cosmological redshift has been essentially the same across all cosmological (or not) SN Ia analyses: a mean bulk-flow velocity $\widehat{v_{\mathrm{pec}}}$ (converted to redshift) is "subtracted" from the measured total $\hat{z}$ (as per eq. (5.6)):

$$1 + \widehat{z_c} = (1 + z) / \left(1 + \widehat{v_{\mathrm{pec}}}/c\right) , \tag{8.11}$$

and a largely arbitrary scatter, e.g. $\sigma_{\widehat{v_{\mathrm{pec}}}} = 150\,\mathrm{km/s}$, is assumed to account for all remaining redshift sources, in (quadratic) addition to the measurement uncertainty $\sigma_{\hat{z}}$ of $\hat{z}$:

$$\sigma^2_{\widehat{z_c}} = \left(\sigma_{\widehat{v_{\mathrm{pec}}}}/c\right)^2 + \sigma^2_{\hat{z}} . \tag{8.12}$$

This brings us to the other opening of the pit: propagating uncertainties (as always). After noticing that the Gaussian description implied in eq. (8.12) is inappropriate when $\sigma_{\widehat{z_c}} \approx \widehat{z_c}$ since, unlike z_{pec}, the cosmological redshift is strictly non-negative (and immediately closing our eyes to this fact), we can proceed as in eq. (6.6) to linearise the distance

linear propagation of redshift uncertainty to magnitudes

[189] which, if constrained, can be used to infer a *local* value of the Hubble constant [79, 118], distinct from that at high redshifts.

[190] It has been suggested by Foley [155] that the velocity of the SN Ia ejecta can be a significant source of redshift, which furthermore depends on the phase since the luminous material decelerates with time after the initial explosion, making $z \to z(t_r)$.

[191] Recently, SNæ Ia have been *used* to constrain the nearby galactic bulk flow [129, 329]

[192] but not quite: it may either be directly estimated from a comparison of the host spectrum (redshift) and that of the SN (background subtracted), but this requires careful modelling of line profiles resulting from an unknown explosion geometry; or since it is modellable through the dynamics of the host [47], the SN's proper motion may be constrainable through its host-centric position (see subsection 8.3.4)

modulus (usually only in its low-z_c already-linear Hubble-law limit eq. (5.13)...) and derive the magnitude / distance modulus "uncertainty" that corresponds to $\sigma_{\widehat{z_c}}$:

$$\sigma_\mu = \sigma_{\widehat{z_c}} \left.\frac{\partial\mu(z_c, C)}{\partial z_c}\right|_{\widehat{z_c}} \to \frac{\sigma_{\widehat{z_c}}/\widehat{z_c}}{\ln 10^{1/5}} \quad \text{as} \quad \widehat{z_c} \to 0. \tag{8.13}$$

Linear propagation of redshift uncertainty is harmful to cosmological inference. It *always* leads to a form of Eddington bias[193] (introduced shortly): we demonstrate this in fig. 14.5. Regardless of the redshift-estimation precision, there is always a data set size large enough so that the bias is evident. Our results from chapter 14 show that, for a reasonable *photometric* redshift uncertainty ($\sigma_{\widehat{z_c}} = 0.04 \times (1 + \widehat{z_c})$) and residual magnitude scatter of $\sigma_0 = 0.1$, cosmological parameter inference is biased by 2 sigma already with 2000 SNæ Ia.

volumetric rate $[1/\text{Mpc}^3/\text{yr}]$

The solution (as always) is Bayesian hierarchical modelling. This requires, in the first place, a prior for the cosmological redshift, usually expressed through the (comoving) *volumetric rate* R of SN Ia explosions (in the rest frame, hence the time-dilation factor) and the evolution of the cosmological volume element [see e.g. 226, eq. (28)]:

$$\mathrm{p}(z_c \mid \boldsymbol{\gamma}) \propto \frac{R(z_c)}{1 + z_c} \frac{\partial V_c(z_c, C)}{\partial z_c} = \frac{R(z_c)}{1 + z_c} \frac{D_c^3(z_c, C)}{E(z_c, C)} \left[\mathrm{sinc}\left(\sqrt{kD_c^2}\right)\right]^2. \tag{8.14}$$

The distribution of cosmological redshifts (of any given "probe"), therefore, depends on the cosmological model, and so inference can be performed solely from redshift measurements,[194] without reference to brightness, standardisation, or the Hubble diagram — provided a faithful model for R and the observational selection efficiency. In parallel with the above discussion on the distribution of SN Ia parameters/covariates (especially footnote 180), R is usually given a flexible parametrisation and inferred from data: in chapter 15, we will do precisely that adopting a power law $R_0(1 + z_c)^\beta$ [128].

SFR
DTD

The rate of SNæ Ia, however, *is* susceptible to physical modelling (at least much more so than $\mathrm{p}(\mathbf{x} \mid z_c, \boldsymbol{\gamma})$) since their formation scenario is at least partially understood [426]. Combined with modelling of the *star-formation rate (SFR)* history [see e.g. 338] of their hosts, the SN Ia *delay-time distribution (DTD)* allows principled forward representation of $\mathrm{p}(z_c \mid \boldsymbol{\gamma})$ and reverse inference of its parameters, which can identify and constrain the physics of SN Ia formation, as we demonstrate in chapter 16.

[193] The case considered here corresponds to the illustration in fig. 8.5 with $\mu \to x$ and $z_c \to y$, instead of the inverse setting we will discuss in relation with selection effects. The two conditions for Eddington bias to arise: namely, a scatter in x, i.e. μ, and non-constancy of its distribution, are ensured, respectively, by eq. (8.13) and the nonlinearity of μ, which implies that $\mathrm{p}(\mu) = \mathrm{p}(z_c)/\partial_{z_c}\mu$.

[194] To first approximation, even the total number of detected objects is cosmologically informative.

Photometric redshift estimation Apart from a prior, Bayesian inference (of cosmological redshifts) requires a likelihood / sampling distribution to represent the measurement process. Presently, large analyses are transitioning from fully *spectroscopic* (i.e. relying *spec-z* on typing and, by extension, redshift estimation from a spectrum of the SN Ia itself) to "photometric" (e.g. Amalgame [422] and DES [109]), but only with regards to classification[195]; instead, redshifts are still being derived from follow-up (or archival) spectroscopy of the host.[196] Above, in eqs. (8.11) and (8.12), we presented the usual simplified Gaussian description for both peculiar-velocity ($\widehat{v_{\text{pec}}} \pm \sigma_{\widehat{v_{\text{pec}}}}$) and total-redshift ($\hat{z} \pm \sigma_{\hat{z}}$) estimation appropriate in this setting.[197] Typically, both uncertainties are of the same magnitude, with the pure $\sigma_{\hat{z}}^{\text{spec}} \approx 10^{-4}$, which means that their proper combination is paramount to avoiding biases as cautioned by Wojtak et al.; this calls for the construction and utilisation of a peculiar-velocity+redshift BHM, like the one presented by Rahman et al. [437],[198] in which to incorporate the external constraints from bulk-flow measurements.

Unfortunately, most supernovæ detected in future surveys like LSST will not be followed up spectroscopically.[160] While the spectrum of the host might be (or eventually become) available in sky regions that overlap with *galaxy* (i.e. non-transient) surveys like SDSS or DESI, the primary source of redshift information for future SN cosmology will be the transient survey's own photometric observations. One possibility to extract it is to focus solely on the SN's light curve and either fit it with a freely redshiftable template like SALT [112], or to use a bespoke method like Photo-*z*SNthesis[199] [431]. However, the more popular—and more constraining [see e.g. 365]—approach examines the broadband measurements of the host galaxy, taken before and/or after the transient [381]. While convenient and much less demanding observationally and computationally, a host *photo-z* is still *photo-z* highly uncertain, model dependent, and expected to deviate significantly from a Gaussian approximation, even exhibiting multimodality in certain cases [312, see also 25, fig. 1].

These complications, in view of the general requirements of redshift estimates for unbiased and precise SN Ia cosmology [325, see also 431, fig. 12], highlight the superiority of proper—Bayesian hierarchical—redshift inference over simplistic error propagation. Nevertheless, they also introduce a levels of intractability (i.e. photo-*z* constraints that may

---

[195] The one prominent instance of *purely* photometric SN Ia cosmology is Ruhlmann-Kleider et al. [463], but their methodology leans heavily on BBC, which we ~~calumniate~~ ~~revile~~ ~~berate~~ denounce in subsection 8.3.6.

[196] In this case, the derived redshift does not include the contribution due to ejecta velocity and the SN's proper motion, but they are usually just lumped into $\sigma_{\widehat{v_{\text{pec}}}}$ anyway…

[197] Note, this refers to the *calculation* of the redshift uncertainty, not to its propagation onto magnitudes!

[198] Incidentally, Rahman et al.'s goal is not (dark-energy) parameter inference *per se* but detecting (or rejecting) an anisotropy in the expansion [see also 490]: a hypothesis we shun as a matter of principle

[199] Curiously, and without ~~realising~~ explicitly advertising it, Qu & Sako combine a number of concepts from the present thesis to present redshift inference through simulation-based NPE implemented as classification over a discretisation of the parameter space.

not be represented as analytical distributions) to a BHM that is already computationally challenged by $O(N_{SN})$ latent parameters for the true cosmological redshifts (and true peculiar velocities).

The solution (as always) is SBI: it allows

- implicit marginalisation of (redshift) constraints represented as posterior samples from external analyses;
- simultaneous transparent modelling of all relevant sources of redshift, as well the correlations they induce (e.g. in SNæ from nearby regions of the bulk flow);
- unified inference of SN Ia properties (individual and population-level) *and* redshifts from both the light curves and associated host photometry using a single inference network, as we demonstrate in chapter 16.

8.3.3 "Interaction" with the host: dust extinction

intrinsic bluer–brighter
extrinsic redder–dimmer (dust)

Standardisation of the intrinsic brightnesses of SNæ Ia was first presented by Phillips [410] in terms solely of a stretch covariate. Later, it was improved by Tripp[200] [516] via the crucial addition[201] of a correction related to the observed $B-V$ colour, which was understood as an *intrinsic* correlation due to the explosion physics and clearly separated from the similar (both in nature (brightness–colour) and in sense (bluer–brighter / redder–dimmer)) *extrinsic* effect of interstellar dust.

The superposition of the two phenomena means that observed properties of SNæ Ia reflect neither accurately: see Mandel et al. [345, fig. 3]. As a consequence, studies which interpret the colours and magnitudes of SNæ Ia solely in terms of dust extinction & reddening [see references in 345] find optical depths (A_V) and dust laws (R_V) inconsistent with other estimates, e.g. of the Milky Way; conversely, standardisation methods (like SALT) that consider the full brightness scatter of SNæ Ia (after stretch correction) as arising from a correlation with colours infer a larger coefficient of proportionality than that appropriate for the pure intrinsic relation due to the additional effect of dust. Moreover, standardisation via a single colour–brightness relation results in a poorer fit than using an appropriate combination of the two relevant effects, as demonstrated by Mandel et al. [345], Brout & Scolnic [71], Popovic et al. [420], who perform *post factum* corrections of SALT parameters. Grayling et al. [185] compare quantitatively this "dust2dust" variant of the SALT framework[202] to the principled modelling of BayeSN, whereas Brout & Riess [70] give a

[200] to the extent that 25 years later a referee would insist that a manuscript refer to eq. (6.3), realised with the SALT x_1 and c parameters, as the "Tripp formula" rather than "Phillips relationship"…

[201] also with respect to Tripp's earlier work [515]

[202] Note that the SALT template is till trained in the original way (for historical reasons), and so the inferred colour law $\mathrm{CL}(\lambda_r)$ still incorporates a residual mixture of the dust and instinsic bluer–brighter effects.

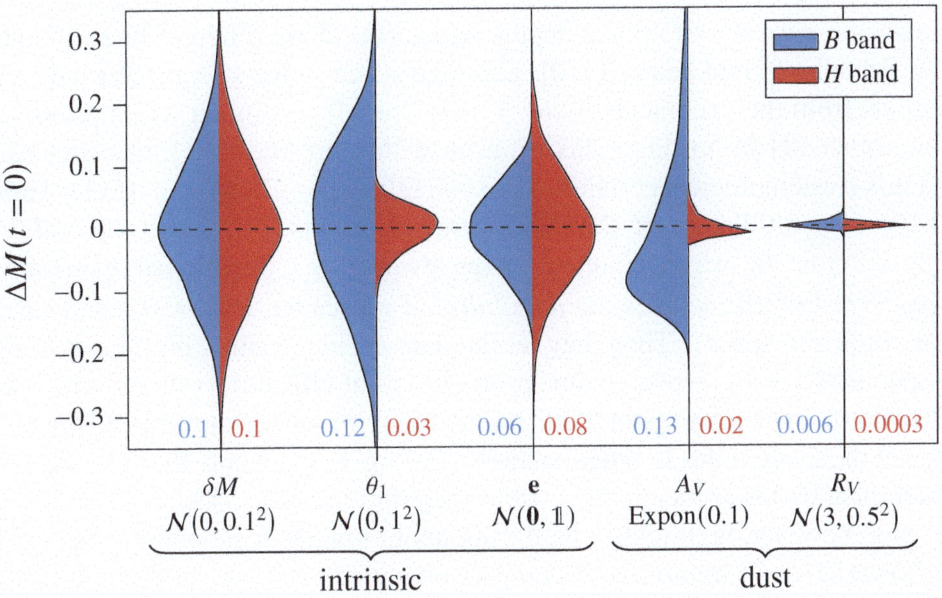

Figure 8.3: (SIDE-real) Variations in rest-frame B (visible) and H (NIR) absolute magnitudes at phase 0 (around maximum), as simulated by BayeSN, induced by varying each of the free local parameters according to its fiducial hierarchical prior, with respect to a reference value with $\delta M = 0$, $\theta_1 = 0$, $\mathbf{e} = \mathbf{0}$, $A_V = 0.1$, $R_V = 3$. Numbers along the bottom specify the standard deviation of magnitude variations in the two bands.

general overview of the modelling issues around dust that appear in SN Ia analyses.

Dust is an aspect of the physical (forward) model of SN — and not only — observations, whereas the issues mentioned above stem from an observation-centric (i.e. reverse) perspective. They are non-existent in proper forward models like BayeSN: built, as elaborated, specifically to encapsulate the physical properties and possible aleatoric variability of dust extinction. Its modularity, i.e. clear separation of the SN-intrinsic brightness distribution[203] and the external effects (dust extinction in the host and beyond, as we describe in chapter 11), makes it easy to control the exact form of the dust law, so as to e.g. match the model used in a simultaneous analysis of a SN's host galaxies, for which the usually assumed extinction curve is that of Calzetti et al. [80] rather than Fitzpatrick [153] (see chapters 11 and 16). When a simpler representation through summary parameters suffices, Simple-BayeSN or BIRD-SNACK can also be used for principled Bayesian dust modelling.

The population (hierarchical) properties of the dust in SN Ia host galaxies ($\mathrm{p}(R_V^s, A_V^s \mid \gamma)$) are more contentious. The Milky Way, for example, is most commonly assumed to have

[203] Incidentally, BayeSN does model the intrinsic bluer–brighter correlation implicitly in the first principal component W_1 and so does not need explicit accounting for it, much like the stretch correction (cf. fig. 8.1).

a single dust law and a variety of optical depths $A_{V,\mathrm{MW}}(\Omega)$ along different lines of sight, as exposed by Schlafly & Finkbeiner [470], although recent efforts [e.g. 559] have uncovered variations from the canonical $R_{V,\mathrm{MW}} = 3.1$. Initially, a similar assumption was made regarding other SN Ia-hosting galaxies, even within the BayeSN framework [e.g. 345, 346], but this epistemological certainty was soon relaxed by Thorp et al. [511], Thorp & Mandel [510] — in fulfillment of BayeSN's original promise —, and the *possibility* of significantly different R_V^s was accounted for (as always) by a flexible parametrisation $R_V^s \mid \gamma \sim \mathcal{N}(\mu_R, \sigma_R^2)$ (which includes the *possibility* of no scatter, $\sigma_R \to 0$, i.e. a single dust law across time and space). Then they let the data decide, leaning on the constraining power — specifically with respect to dust properties — of NIR observations, which are minimally affected by interstellar extinction and therefore provide a convenient "*anchor*" to which to relate the magnitudes in bluer bands[204]: see fig. 8.3 and note that variations in the NIR are significantly lower than in the visible spectrum in most cases.

NIR anchor

The same plot, however, highlights a hard truth about ~~life~~ SN Ia modelling: the effect of dust-law (R_V) variations is minuscule in comparison with the intrinsic brightness scatter of SNæ Ia: the coherent scatter δM, the stretch-related brightening (measurable through θ_1), and the residuals ϵ. This means that independent dust inference for individual objects is all but impossible: instead, constraints *need* to be pooled across the population with a hierarchical model [511, fig. 5]. Nevertheless, the results of such an analysis, as we will also demonstrate in chapter 12, exhibit a heavy case of *shrinkage*: a statistical effect whereby all marginal R_V^s posteriors concentrate toward the population mean (μ_R), are highly correlated (i.e. a "systematic" shift in μ_R induced coherent offset of all R_V^s), and attain — marginal! — *uncertainties* similar to the population *scatter* (σ_R). Shrinkage is *not* a problem *per se* but requires careful interpretation, especially of terms like "scatter", "uncertainty", and "constraint".

shrinkage

Lastly, dust-population modelling must satisfy two "physicality" constraints: $A_V > 0$ and $R_V \gtrsim 1.2$, which represent, respectively, the obvious fact that dust only absorbs and does not emit light[205] and the physics of Rayleigh scattering [see 132]. In the context of a dusty BHM, the former requirement is naturally incorporated in the support of the exponential distribution $A_V^s \sim \mathrm{Expon}(1/\tau)$ introduced above, whereas the latter needs to be enforced on the assume $\mathcal{N}(\mu_R, \sigma_R^2)$ via... truncation [511]. The goal here not computational optimisation as when we introduced truncation to SBI, but the mathematical

[204] This is partially a manifestation of the general sentiment that differential/slope measurements, e.g. colour $X - Y$, are more precise if the two individual estimates are further apart.

[205] Or does it?... See e.g. subsection π of Leja et al. [313] (and references therein): dust *does* emit due to being heated by the various astrophysical processes that transpire throughout galaxies, but its light is re-processed and emitted at very different wavelengths that that of the original radiation, so its effect on SN observations is indeed purely extinctory (and reddening).

formalism (eq. (2.24)) is exactly the same:

$$\mathrm{p}\big(R_V^s \,\big|\, \mu_R, \sigma_R\big) = \begin{cases} \mathcal{N}(\mu_R, \sigma_R^2)\,/\,c(\mu_R, \sigma_R) & \text{if } R_V^s \in [1.2; \infty], \\ 0 & \text{otherwise;} \end{cases} \tag{8.15}$$

$$c(\mu_R, \sigma_R) \equiv \int_{1.2}^{\infty} \mathcal{N}\Big(\mu_R, \sigma_R^2\Big)\, \mathrm{d}R_V^s, \tag{8.16}$$

but now the normalisation c is not a simple rescaling but crucially depends on the population parameters, modifying their marginal likelihood in a way the favours broad R_V distributions (i.e. a high σ_R), which would otherwise predict a number of objects with very low—or even negative—R_V^s in contradiction with the data. Further discussion and explanation of the two effects can be found in Grayling & Popovic [184]. As a bottom line, μ_R and σ_R no longer represent the mean and standard deviation, respectively, of $\mathrm{p}\big(R_V^s \,\big|\, \boldsymbol{\gamma}\big)$ but are simply a... convenient flexible parametrisation.

8.3.4 Interaction with the host: additional standardisation

Beyond dust extinction, the astrophysics of galaxies can also influence the population of SNæ they host. That is, it is possible to augment the model from eq. (8.10) with a set of (latent) *host parameters* $\mathbf{g}$:

host parameters

$$\mathrm{p}(z_\mathrm{c}^s, \mathbf{g}^s, \mathbf{x}^s, M^s \,|\, \boldsymbol{\gamma}) = \underbrace{\mathrm{p}(M^s \,|\, \mathbf{x}^s, \mathbf{g}^s, \cancel{z_\mathrm{c}^s}, \boldsymbol{\gamma})}_{\substack{\text{modified} \\ \text{“standardisation”}}} \underbrace{\mathrm{p}(\mathbf{x}^s \,|\, \mathbf{g}^s, \cancel{z_\mathrm{c}^s}, \boldsymbol{\gamma})}_{\substack{\text{environment} \\ \text{dependence}}} \underbrace{\mathrm{p}(z_\mathrm{c}^s, \mathbf{g}^s \,|\, \boldsymbol{\gamma})}_{\substack{\text{galaxy} \\ \text{evolution}}}, \tag{8.17}$$

which act as conditioning for all the distributions previously considered (except that the SN Ia rate is more conveniently expressed in terms of galaxy evolution referred to the DTD, as we explained above and will demonstrate in chapter 16). To motivate their explicitation in the hierarchical model, the host properties must be measured, possibly with noise: $\hat{\mathbf{g}}^s \sim \mathrm{p}(\hat{\mathbf{g}}^s \,|\, \mathbf{g}^s, \ldots)$, whence the usual game proceeds as with any SN-specific parameter: by uncovering empirical connections among estimators.

Stellar mass M_*,[206] commonly expressed in logarithmic terms: $\log_{10}(M_*/\mathrm{M}_\odot)$, is correlated with mostly any galactic property[207] and has therefore been extensively used as a proxy to approximately account for the influence of “galactic astrophysics” in general (for an excellent literature review, consult Thorp & Mandel [510]). Due to (previously) limited statistics and an already low residual scatter (~0.1 mag), the standard approach has

[206] The asterisk stands for “star” / stellar, so as avoid confusion with an absolute (stellar...) magnitude.

[207] See e.g. Heavens et al. [208], who express the full galaxy formation history through the observed stellar mass.

been to split the sample along the median mass (typically around $10^{10.5}\,\mathrm{M}_\odot$, although the location of the *mass split* can be treated as a free parameter and optimised in terms of resulting standardisation) and allow different standard brightnesses M_0^{low} and M_0^{high} in the sub-samples with low- and high-mass galaxies, respectively, while keeping the rest of the global parameters shared.[208] This is equivalent to fitting as usual after defining a mass-related standardising covariate that takes on binary values: $\hat{x}_{M_*} \in \{\text{low: } 0, \text{high: } 1\}$, and a corresponding "correlation coefficient", i.e. the so-called *mass step*[209] ΔM:

mass split

mass step

$$M^s = M_0 + \Delta M \cdot \hat{x}^s_{M_*} + \ldots, \tag{8.18}$$

Accounting for significant measurement uncertainties in the stellar mass—which most analyses do not, thus risking a trivial instance of Eddington bias—is problematic in this formulation due to the discrete nature of $\hat{x}^s_{M_*}$: Shariff et al. [479, see especially subsection 3.2], for example, achieve this by marginalising the class-assignment probabilities. In any case, studies like Shariff et al. [479], Thorp & Mandel [510] generally attempt to answer the question of whether additional standardisation *is at all needed* (after careful consideration of all *physical* effects known to correlate with the stellar mass: e.g. a trend in the dust population), or the scatter that remains is truly random by examining the posterior probability of ΔM, which is not rigorous in the Bayesian sense.[210] Of course, this approach is undertaken because of the computational burden of proper model selection. We overcome this with SBI in chapter 13 to conclusively resolve the question of the probability of existence of a mass step when confronted with the possibilities of different dust-law populations.

Local environment The (integrated) stellar mass of a galaxy is a global property and can thus only statistically (i.e. on average) have an influence on the brightnesses (and other qualities like stretch) of a SN Ia. If physical connections are to be sought, this must be done with reference to *local*[211] *characteristics* of the SN's environment, i.e. the *stellar population* in

progenitor population

[208] UNITY, on the other hand, goes all in and allows all correlation coefficients to differ. Similarly, Grayling & Popovic [184] experiment with training separate BayeSN templates for the two sub-samples.

[209] more accurately, magnitude step across the split in host stellar mass

[210] Consider, simply, that the space can be re-parametrised as $\Delta M \to \Delta M^2$, which will drastically alter the quantitative results.

host-centric location

[211] To this end, it must first be possible to resolve the *position of the SN* within its host: a task that may seem daunting at first but is alleviated by a serendipitous quirk of the Universe's hyperspherical (curved) geometry, due to which the relation between projected size and distance—i.e. the linear size of an object that attains a given angular scale on the sky—grows sub-linearly with redshift and even attains a maximum (around $z_c \approx 1.6$ in standard ΛCDM): see Hogg [226, fig. 2]. This means that a given telescope has a *guaranteed* minimal resolving power regardless of distance: for example, the Rubin Observatory's Simonyi Survey Telescope (that will execute the LSST…), with its $\Delta\Omega = 0\farcs2$ angular pixel size, will be able to resolve details/separations at least as small as $\approx$3 kpc *at all redshifts* [303].

which it is embedded. Thus motivated, numerous studies have found evidence for such correlations [448, 449, 450, 258, 260, 371, 372, 287, 288, 456, 458, 274, 275, 175, 174],[212] but definitive explanations of the causal channels — which will allow us to build a reasonable forward model with which to perform *principled* inference — are still outstanding. Nevertheless, the avenue is open for applications of rigorous simulation-based model selection (in the style of chapter 13) to assess the influence of local host properties and/or *a posteriori* inference of a multitude of correlation/standardisation parameters from abundant high-quality data like the low-redshift SNæ Ia collected by the ZTF.

Population evolution Since galaxies (and their (local) properties) evolve [366], the populations of the SNæ Ia that occur in them do too:

$$\mathrm{p}(\mathbf{x}^s \,|\, z_\mathrm{c}^s, \boldsymbol{\gamma}) = \int \mathrm{p}(\mathbf{x}^s \,|\, \mathbf{g}^s, \cancel{z_\mathrm{c}^s}, \boldsymbol{\gamma})\, \mathrm{p}(\mathbf{g}^s \,|\, z_\mathrm{c}^s, \boldsymbol{\gamma})\, \mathrm{d}\mathbf{g}^s\,, \tag{8.19}$$

where **g** acts as a nuisance parameter that can explain the tantalising clues [307, 386, 535] of a redshift-dependence in the distributions of SN Ia properties and of the dust that affects them [421, 512]. Therefore, this is a matter more of galactic astrophysics than of SN Ia cosmology, and so, grounded in the fact that Rubin et al. [462] found no evidence for time variation of the SN model *independently* of the galactic properties, we will not pursue this matter further after simply remarking that SBI can be trivially used to simplify the fully Bayesian time-evolving hierarchical UNITY analysis.

8.3.5 Host (mis-)identification

To extract (measure) any property — e.g. a photo-z, the dust law, or stellar mass — of a SN Ia host, it first needs to be identified, which is not a trivial task in crowded and blended fields revealed by deep observations. The standard "methodology", developed by [191, see especially fig. 1], is based on the (projected) distances between the SN and the centres of nearby galaxies, normalised by the (projected) sizes of the latter: the so-called *directional light radius (DLR)*.[213] The candidate with lowest DLR is picked, and ~~life~~ SN Ia modelling can go on as usual. *DLR*

Roberts et al. [453] realised this can be a prominent source of redshift *error* and included it into their comprehensive zBEAMS BHM by introducing a latent variable to represent the assignment of each SN to its host from among a number of identified nearby candidates, weighted *a priori* inversely to their separation (DLR). The redshift is then assigned a *mixture distribution* of the posterior inferences from each candidate host. In the *mixture distribution*

[212] This reference list was compiled, almost in its entirety, by Matthew Grayling 🙏

[213] The DLR, incidentally, can be used as a proxy for the local environment and hence, for standardisation [217].

limiting example of host spec-zs, this collapses onto a choice of the redshift which best matches the light curve (i.e. a comparison between separate redshift estimates from the SN and from nearby galaxies). When, on the other hand, the host redshifts are not perfectly constrained, their average represents a data-based *prior* on the SN's total redshift z.

Exact modelling of these effects quickly escapes analytical description (e.g. zBEAMS is replete with Gaussian approximations), and so more recent studies [418, 433] rely on forward simulations to better represent the host identification/association and redshift measurement (or constraint-combination) process employed in real analyses and — simply — elucidate its impact on the final results of inference...

Given the difficulty of designing an end-to-end *explicit* probabilistic model for (reverse) host-galaxy identification, the best approach is clearly to directly use the faithful forward simulations of the survey data acquisition and handling (mentioned above) for SBI.

- If, for example, a survey data release provides only a "best-guess" host, SBI will learn to *implicitly* marginalise the uncertainty this introduces in the most efficient way (given the data, e.g. by confronting the SN and host observations).
- If, alternatively, the final analysis has access to several potential hosts and their DLRs and measured properties, SBI trained on simulated realisations of the process of their compilation will learn to use the implied mixture distribution, provided the NN used (e.g. a multi-modal transformer) is able to ingest such unconventional data.

8.3.6 Selection effects

Selection effects, introduced in section 4.2, are the bane of SN Ia cosmology.[214] Two distinct types are prominent in SN Ia and can significantly bias parameter inference through an offset in the Hubble diagram (as we illustrate in fig. 8.4), yet only one is routinely discussed and accounted for.

They arise, in general, as a consequence of the preference for detecting brighter, bluer, and longer lasting SNæ. Since these properties are correlated (among each other and) with redshift and distance, this results in an offset of the observed sample from the total SN Ia population, with only the latter tracing accurately the expansion history and following the hierarchical distributions (and SN Ia template models) described in previous sections.

Malmquist bias [340, 341] hardly needs an introduction — it is the multi-faceted modification to the distribution of observed properties induced by a preference for selecting brighter, bluer, and longer lasting SNæ Ia — or an illustration [see e.g. 348, fig. 1]. As a simple consequence, the average brightness of the selected sample at a given redshift is

[214] They are also the reason why BayeSN still has not yet been fully applied to cosmological inference.

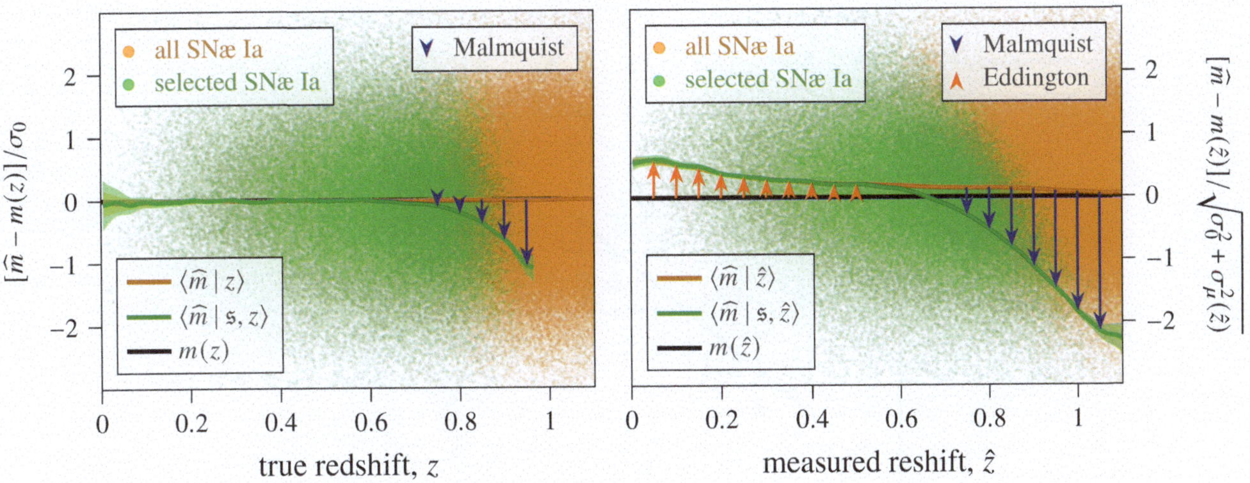

Figure 8.4: (RESSET) Two kinds of bias / offset of the Hubble diagram from the underlying magnitude–redshift relation (black). A preference for detecting brighter objects (Malmquist bias: blue) results in the the mean of the observed sample (green) appearing brighter than that of the total population (orange). Eddington bias, which can be interpreted as a preference towards more common objects, arises — also when no explicit selection is performed — from the combination of a non-constant redshift distribution and significant uncertainty in measuring the latter (e.g. photo-z): a simplified illustration is shown in fig. 8.5. For the particular case of SNæ Ia, the two effects have opposite signs and can, individually, and in combination, significantly bias cosmological inference.

shifted with respect to the total population:

$$\text{mean Malmquist bias:} \quad \langle \widehat{m} \mid \mathfrak{s}, z \rangle < \langle \widehat{m} \mid z \rangle \,, \tag{8.20}$$

and thus does not represent an accurate distance estimate. Moreover, since the strength of this effect varies with redshift, it can convincingly mimic an alternative cosmological expansion history, thus biasing parameter inference if unaccounted for.

Simultaneously, the *spread* of observed brightnesses is *reduced*:

$$\mathrm{Var}(\widehat{m} \mid \mathfrak{s}, z) < \mathrm{Var}(\widehat{m} \mid z), \tag{8.21}$$

since "extreme" examples remain undetected. This is usually reflected in χ^2 fits (via a modified $\sigma_0(z_c) < \sigma_0$), which can thus purport to attain better standardisation and enjoy a false sense of security that amplifies the effects of any mis-modelling of the mean effect (and those abound).

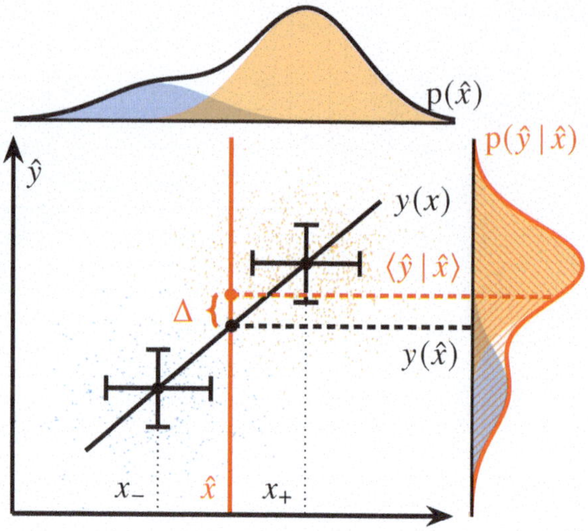

Figure 8.5: (RESSET) Eddington bias: $(\Delta(\hat{x}) \equiv \langle \hat{y} \mid \hat{x} \rangle - y(\hat{x}))$ in the response variable $(\hat{y} \equiv y(x) + \text{noise})$ caused by scatter in the independent variable $(\hat{x} \equiv x + \text{noise})$ combined with a trend in its density. For simplicity, here the distribution of the latent x has only two discrete values (x_- and x_+) with higher probability at x_+. Noisy measurements (here Gaussian with constant variance) are made at a range of $\hat{x}$, and due to the prevalence of objects with true x_+, the response variable exhibits Eddington bias towards $y(x_+)$ and away from the underlying $y(\hat{x})$. Such a bias arises in qualitatively the same way for any continuous distribution of x that varies significantly on the scale of measurement uncertainty and independently of the noise in $\hat{y}$, as long as it does not depend on x.

Eddington bias [137] is a distinct and less well-known effect due to the preference for selecting *more common* objects—even in the absence of explicit selection. In the context of SN Ia cosmology, it arises when only noisy redshift estimates are available and the distribution of SNæ is not constant with redshift: we illustrate this effect in fig. 8.5 with an idealised setup. Since the SN Ia rate increases with redshift and so does their apparent magnitude, more SNæ with a given measured $\hat{z}$ have true redshift larger than that, rather than smaller, and so

$$\text{mean Eddington bias:} \quad \langle \widehat{m} \mid \hat{z} \rangle > m\left(\hat{z}, \boldsymbol{\gamma}_{\text{fid}}\right). \tag{8.22}$$

An additional, trivial, source of non-constancy in the redshift distribution is the "lack" of SNæ with $z < 0$, which, as we demonstrate in SICRET, leads to an unavoidable offset at low redshift and incorrect inference even without selection effects and with a perfectly known redshift distribution.

Traditional approaches Acceptance criteria for building cosmological SN Ia samples are defined on two levels: during identification of astronomical transients using an arduous difference imaging pipeline [see e.g. 282, 352, 465] (*detection*) and the subsequent *selection* of a "high-quality"/cosmological SN Ia sample [e.g. 525, table 4], taking into account the availability and quality of spectroscopic follow-up of the supernova and/or its host galaxy, the fidelity of laborious light-curve model fits (e.g. with SALT), and the outputs of black-box NN classifiers (see below). This makes an exact treatment of selection effects from first principles practically impossible, necessitating the development of approximate schemes for mitigating the biases induced by selection, which fall into two major categories.

On the one hand is the typical two-step analysis, implemented as *BBC (BEAMS with bias correction)* [278], which utilises a large preliminary simulation of the SN Ia population to which the full selection procedure is applied, in order to calculate an *average*[215] offset of observed from true magnitudes[216] as a function of redshift: *BBC*

$$\Delta m(z) \equiv m(\hat{z}, \boldsymbol{\gamma}_{\text{fid}}) - \langle \widehat{m} \mid \mathfrak{s}, \hat{z}, \boldsymbol{\gamma}_{\text{fid}} \rangle , \tag{8.23}$$

This negates the selection bias for the specific choice of a (fixed) *fiducial model*: correlation coefficients, SN Ia rate with redshift, and, most importantly, cosmology. Note the degree of contrivance required to calculate this "bias" correction: it essentially adjusts the data so as to match a pre-specified model![217] Naturally, BBC with incorrect fiducial parameters leads to incorrect cosmological inference (see fig. 8.6 and even fig. 4 at the source Kessler & Scolnic [278]), which will become more visible (cf. Chen et al. [95]) with larger and more constraining data sets. *fiducial model*

While bias correction is an *ad hoc* procedure, more principled—BHM—frameworks for SN Ia cosmology have also been developed. They explicitly derive the selection efficiency for each SN as a function of all model parameters: object-specific (stretch, colour, redshift, etc.) and global (cosmology, standardisation coefficients, rates, calibration, etc.), and evaluate them at every step in an MCMC chain. But since it is unavailable from first principles, current BHM frameworks resort to fast analytic approximations (e.g. a simplistic hard magnitude cut [348]) and/or potentially inaccurate assumptions for its ana-

[215] Alongside correcting the mean brightness–redshift trend, the assumed uncertainty around it can also be modified to take into account the reduction in the variety of *observed* SNæ Ia with respect to the total population: see Kessler & Scolnic [278, subsection 5.3].

[216] Since Kessler & Scolnic [278], the framework for bias corrections has been altered and augmented, notably by Popovic et al. [419], and currently encompasses several variants with different "independent" and "response" (i.e. corrected) variables beyond redshift and magnitude: namely, the SN stretch and colour and properties of the host galaxy: see table 1 in Popovic et al. [419].

[217] One can define a less preposterous correction towards the mean of the total population ($\langle \widehat{m} \mid \hat{z}, \boldsymbol{\gamma}_{\text{fid}} \rangle$), but this fails to account for Eddington bias, as Ruhlmann-Kleider et al. [463], Chen et al. [95] demonstrate.

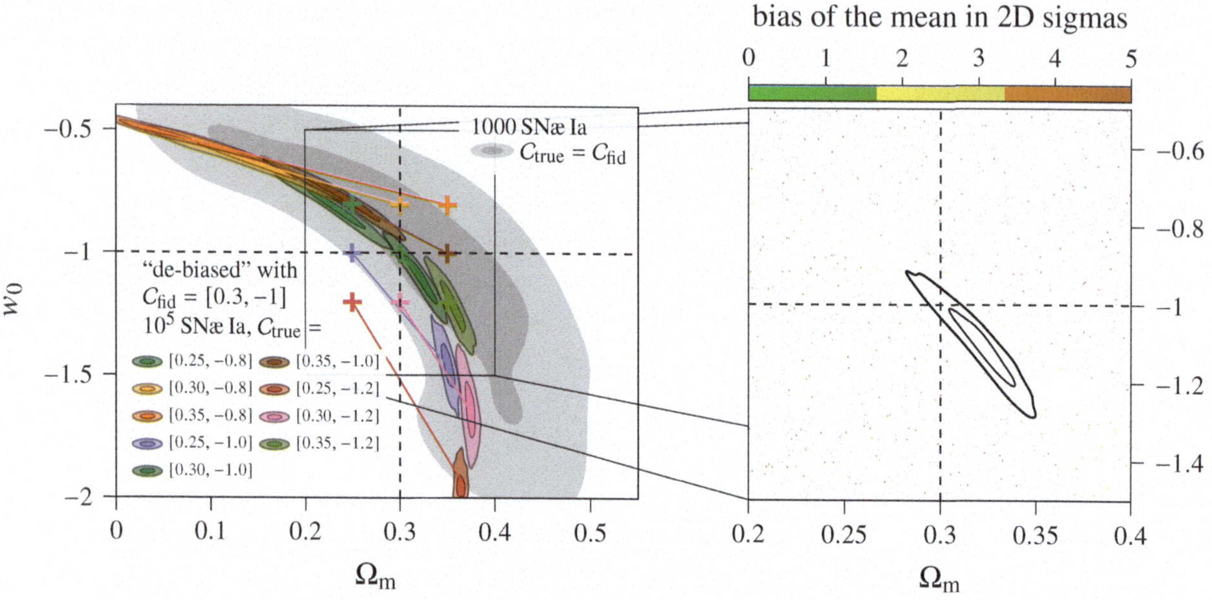

Figure 8.6: (RESSET) The inadequacy of traditional (selection-)bias correction for large data sets when the fiducial parameters are not representative of the true values. The mock data considered is described in subsection 15.3.3, and the analysis itself is detailed in RESSET, appendix C. *Left:* Posteriors (1- and 2-sigma credible regions) from 10^5 mock SNæ Ia generated with different cosmological* parameters C_{true} indicated by crosses. Instead of recovering the true parameters, "bias-corrected" posteriors trace the region in C-space most consistent with the fiducial cosmology $C_{fid} = [0.3, -1]$. This characteristic "banana"-shaped region is illustrated in pale grey through the 1- and 2-sigma credible intervals from a bias-corrected analysis of 1000 mock SNæ Ia generated from the fiducial model. *Right*: Magnitude of the systematic bias in the posterior mean in units of the statistical uncertainty (calculated from the posterior covariance matrix) as a function of the true underlying cosmology. For each point, a different mock data set of 10^5 SNæ Ia was generated and bias-corrected assuming the same C_{fid}. The black contours replicate the case $C_{true} = C_{fid}$ from the left panel. The systematic bias severely increases as one moves away from the locus of degeneracy.

*The cosmological model is a flat Universe with DE evolving with a constant (but unknown) EOS w_0.

lytic marginalisation over unobserved objects[218] [461, 462, 221]. Finally, Boyd et al. [62] present an application of a neural density estimator trained on simulations as a very flexible approximator for the intractable selection-affected probabilities of *individual objects* in a *simplified* BHM for SN Ia cosmology from summary statistics. This approach, however, still suffers from the poor scalability of sampling methods applied to large hierarchical models and the caveats of the piecewise SBI approach.[219]

The correct approach: SBI with stochastic cardinality It is an ardent claim of the present thesis that *the* correct approach to handling — rigorously — selection effects — of arbitrary complexity and in arbitrary context — is through SBI trained on examples with stochastic cardinality. We already enunciated our arguments in subsection 4.2.1 and demonstrate an application to SN Ia cosmology in chapter 15.

8.3.7 Classification

One final ingredient of cosmological analyses remains to be considered: sifting out SNæ Ia from among the many transients that are routinely observed, so as to ensure that inference is applied to data that truly represent the modelled phenomenon. In practical terms, this has been predominantly achieved through dedicated followup of the identified *candidates*, which results in very high-confidence (sub-)type determinations, taken subsequently for given and true.[220] The ability to realise this in practice, however, i.e. a survey's *spectroscopic efficiency*, is an integral part of its selection detection/selection procedure and can thus weaken its output in two respects: first, by significantly reducing the final sample size, and second, by introducing strong brightness-related selection effects. Consequently, there has been a tendency towards formulating analyses so as to handle non-Ia *contaminants* without explicit/deterministic typing [222, 259, 109], which will become indispensable when confronting future-sized data sets not susceptible to spectroscopic follow-up.

candidates

spectroscopic efficiency

contamination

[218] For instance, UNITY effectively assumes that the redshift distribution of the total population is independent of redshift and is anyway chosen out of computational convenience.

[219] In a private communication, Boyd et al. [62] have shared encouraging results that demonstrate the scalability of their method to 50 000 objects (reliant on their NUTS implementation) after arduous re-training of the neural estimator. Still, their setup is intentionally simplified in order to enable this precision: they assume the selection is independent of the global parameters varied in the secondary fit; and besides, the theoretical considerations related to MCMC sampling and finite precision (including the discussion of training-set bias amplification from footnotes 186 and 187) and the overall superiority of the end-to-end approach in terms of elegance remain valid, but possibly not to a *utilitarian*.

utilitarianism

[220] Still, the classification scheme itself is not set in stone: it is, after all, empirical[124] and judged solely based on its ability to produce samples useful for cosmology. For example, SLSNæ, which are usually excluded from standard-SN Ia analyses, have grown into a standard candle of their own, with a separate past [385], present [239], and future [238, 256].

BEAMS

Kunz et al. [300] had a first key (and surprising...) realisation in this regard: that uncertainty (as to the type of a given object) is easiest (and most rigorously) handled in a Bayesian framework: *BEAMS (Bayesian Estimation Applied to Multiple Species)* [300], even if the overall analysis is not Bayesian. As we already discussed in the context of host identification (subsection 8.3.5), the formalism represents the data on each transient by a mixture model (in exact parallel with our previous discussion on NRE, model selection, and model averaging):

$$\mathrm{p}(\mathbf{d}^s \mid \boldsymbol{\theta}) = \sum_{m=1}^{N_{\mathrm{mod}}} \mathrm{p}(\mathbf{d}^s \mid M_m, \boldsymbol{\theta})\, \mathrm{p}(M_m \mid \boldsymbol{\theta}), \tag{8.24}$$

where $\{M_m\}$ are all the transient models considered, and $\boldsymbol{\theta}$ are the parameters of interest, which are assumed to be super-modal, i.e. not associated with any M_m or, otherwise, shared among all (cosmology is of the former type). Evidently from eq. (8.24), BEAMS requires explicit models for the "contaminants" in order to define $\{\mathrm{p}(\mathbf{d}^s \mid M_m, \boldsymbol{\theta})\}$ and derive with them N_{mod} constraints on $\boldsymbol{\theta}$ from each object[221] by assuming, in turn, that its true class is M_m, before finally averaging the results in proportion to prior probabilities $\{\mathrm{p}(M_m \mid \boldsymbol{\theta})\}$, which represent the expected overall abundances of objects from the different classes.

The BEAMS procedure can, thus, be wasteful if objects are relatively unambiguously classifiable, i.e. it is unsuitable for realising minor corrections due to a small (but not negligible) amount of contamination. Consequently, in recent years, the focus has shifted back to (quasi-)deterministic classification, with the minor effects of contamination corrected by the old-fashioned method of uncertainty inflation [see 284].

PLASTICC

A principal driver for method development in this area has been the *Photometric LSST Astronomical Time-series Classification Challenge (PLASTICC)*[222] [506, 283, 223]: a large-scale simulation of the various transients expected to be observed by LSST (in reasonable proportions) meant to train and evaluate[224] classification methods that use only photometry as input. While later techniques (SuperNNova [369], SuperNNova [523], SCONE [432], and

[221] In fact, the exposition here, like the original proposition and its applications, assumes conditionally independent objects, which is admissible only if $\boldsymbol{\theta}$ represents all global parameters, but often[178] not realised in the practice of SN cosmology. In the opposite case, an exponential number $N_{\mathrm{mod}}^{N_{\mathrm{SN}}}$ of results for $\boldsymbol{\theta}$, corresponding to all the possible classifications of N_{SN} objects, must be considered.

[222] preceded by the mere supernova photometric classification challenge [281] and superceded by the daunting[223] Extended LSST Astronomical Time-series Classification Challenge (ELASTICC) [376]

[223] Francisco Förster, Anais Möller, private communication

[224] PLASTICC poses an interesting challenge to purely data- (or simulation-)driven inference: covariate shift, i.e. a mismatch between the distribution of training and evaluation/test data (see e.g. Autenrieth et al. [26] for a general method for its resolution and the recent applications to photo-z estimation [25, 374]). Curiously, covariate shift, as defined [373], does not affect posterior-density inference, *by definition* 💁‍♂️.

Gagliano et al. [162]) have relied extensively on NNs, PLASTICC's winner, avocado[225] [57], uses the ~~antique~~ time-tested boosted decision tree method.

Given that contamination is already handled in a simulation-driven manner, it is not a big stretch● to imagine simulation-based *inference with mixed-SN samples*: the classification procedure employed by the real data acquisition and processing pipeline[226] simply needs to be applied as part of the simulation framework, alongside the already incorporated detection and selection steps, and the SBI network trained on the realistically "contaminated" samples: it will then learn to implicitly consider all individual sampling distributions (in the proportion to which they are present in the "censored" data, i.e. cleaned of obvious non-Ias) and provide the final result of eq. (8.24).

SBI with contamination

Interlude: neural networks for SN science

Prior to the works presented in part IV of this thesis, cutting-edge machine-learning and neural methods were relatively rare in SN science. Neural SBI had been applied on a number of occasions[227] for cosmological inference from summary statistics [10, 8, 533, 534, 94], as a component of a BHM that corrects selection effects [62], and for inferring the properties of individual SNæ Ia from interpolated[228] light curves [522, 431]. These studies found the capabilities of fixed-size NNs like MLPs and convolutional neural networks (CNNs) and the conceptual simplicity of NPE sufficient for their idealised purposes. On the other hand, the RNN architecture emerged as the go-to choice for classification [369, 523, 162, mentioned above] due to its ability to gradually accumulate live-streaming information; and finally, in concert with the ground-breaking attention- and transformer-based advances in NLP, multi-modal processing, and towards general artificial intelligence, simpler (arguably) models found their way into un/semi-supervised learning of informative fixed-size representations (summaries) from light-curve data [130, 370, 6, 560].

We expand this diversity with two new additions:

- the Super Tuple: a middle ground between a fully-connected (single-layer) perceptron and a CNN (spiritually) but in essence, a glorified GP / PCA;
- the conditioned deep set: a simple BHM-inspired extension to the scalable and versatile stochastically-sized-set aggregator.

---

[225] which bears a striking resemblance to the contemporary STACCATO (synTHetically Augmented light-curve classificATion) [444], which also uses GP augmentation but replaces the boosted tree with a random forest...

[226] for LSST, this will be outsourced to seven external data brokers, each of which will process the observational data stream with separate—sometimes proprietary—tools and classifiers, which might pose an obstacle to faithful simulation (but not to members of each respective broker...)

[227] Honorary mentions go to the early non-neural ABC analyses [542, 251, 39] in the field.

[228] using GP regression, as popularised by STACCATO and avocado

8.4 The future: grand unified SN (Ia) cosmology

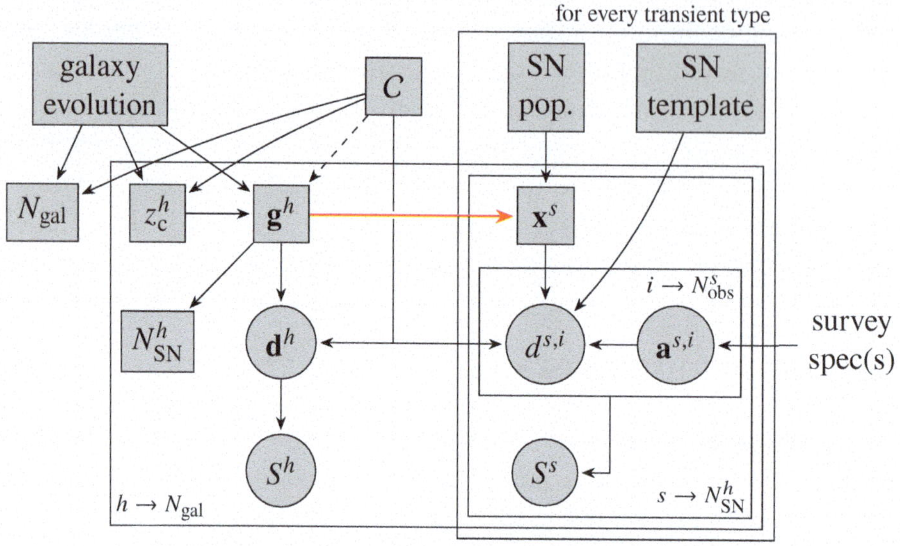

Figure 8.7: Grand unified SN (Ia) cosmology.

In the preceding sections, we repeatedly found that current analyses adopt a piecewise inference approach: e.g. deriving a template from low-redshift spectra, then fitting it to measure distances; measuring total colour and then decomposing it into intrinsic and dust-related; inferring host and SN properties separately and then correlating them *a posteriori*; calculating an average de-biasing magnitude shift and applying it as a fixed correction to the data; classifying SNæ (or identifying hosts, or binning by stellar mass) and only afterwards dealing with the uncertainties with *ad hoc* procedures.

In opposition to partial problem-solving, and inspired by rocket science, we now describe qualitatively our vision for a single-stage-to-cosmology pipeline, pieces of which are developed and presented in the remaining parts of this thesis. In the spirit of SBI (and since we already presented the required inference/methodological developments in part I), the following exposition takes the form of a forward pass through a grand unified simulator for future SN Ia (and adjacent) data, depicted in fig. 8.7.

In the beginning, we pick a cosmology C, which sets the expanding universe in motion and allows us to also sample any other global parameters to complete $\boldsymbol{\gamma}$, including: the DTD (or a marginal SN rate), the SN Ia template globals (and similar models for other transients we would like to model).

Then, we realise the galaxy model.[229] First, we determine the total count of

[229] If we feel particularly adventurous, we can run a full N-body+hydro simulation [295] or use a shortcut like

galaxies in the universe (down to a given low-mass and high-redshift cutoff that we expect to never be detectable under any circumstances: we have no proof they even exist), N_{gal}, as a stochastic (Poisson) sample from a model of structure formation. Then, by current parametrisations, we draw that many (cosmological) redshifts according to cosmic star formation ($\mathrm{p}(z^h_{\mathrm{c}} \mid \gamma)$) and stellar masses from a redshift-dependent mass function ($\mathrm{p}(M^h_* \mid z^s_{\mathrm{c}}, \gamma)$). Then, we synthesise stellar populations by sampling star-formation histories conditional on the final M^h_* and galaxy ages ($T^h_{\mathrm{age}} = T(\infty) - T(z^h_{\mathrm{c}})$). This gives us the galaxy properties $\mathbf{g}^h$, which *include* extinction parameters R^h_V and A^h_V (or other parameterisation, depending on the employed dust law (and notational convention)).

If we have data on the intra-host locations of our transients;[230] we distribute different populations and dust extinctions as a function of galactocentric radius. We collect all intragalactic populations across all galaxies in the universe, label each with p, and ask of each: what is the rate of occurrence of SNæ Ia (and all other transients we have decised to model), as calculated from the distribution of stellar ages and the DTD. Then we multiply by the duration of the survey and sample—*within each individual population*—the number of SNæ Ia, etc. that go off while we are looking. In the vast majority of populations, this will be zero, but we *will* end up with N_{SN} supernovæ correctly distributed according to local and host-averaged rates. To each SN, we can now associate a stellar population (and dust): $s \to p$ (one p may have produced multiple s).

Alternatively, we can consider whole galaxies on average—i.e. each galaxy contributes once its total stellar population and possibly a *random sample* of dust properties to emulate spacial variety—and do as above ($h \leftrightarrow p$). In the end, we have a list of transients of each type and their associated host properties.

It is time to sample SN parameters from $\mathrm{p}(\mathbf{x}^s \mid \mathbf{g}^s, \cancel{z^s_{\mathrm{c}}}, \gamma)$. Here, it is reasonable to assume that the age of the universe at the time of each SN's explosion (z^s_{c}) does not play a direct role; rather, it is the properties of the particular stellar population that determine $\mathbf{x}^s$, e.g. stretch and intrinsic colour.

Alternatively, if we are not modelling hosts, we can employ a shortcut marginal SN Ia rate $\mathrm{p}(z^s_{\mathrm{c}} \mid \gamma)$, e.g. a parametrised power law with a free normalisation, and sample redshifts from that. Depending on the level of sophistication we are eager to explore, we then simulate the SN parameters independently or given the redshift (i.e. from an evolving or

---

`Jerald`. It's just a matter of computational resources.

[230] It has been found—*a posteriori*—that SNæ Ia, for example, follow total light [427], but this is not guaranteed for other transients; this point requires domain expertise.

fixed population) $\mathrm{p}(\mathbf{x}^s \,|\, \not{z}_{\mathrm{c}}^{s}, \gamma)$. At this point, we similarly sample dust parameters. Lastly, given the templates (models) for each transient type that ~~God~~ we sampled in the beginning of the universe, we create the spectral flux timeseries of each object in its restframe and extinguish it for dust and distance by $4\pi D_{\mathrm{M}}^2$. No redshift yet?[231]

Now we model peculiar velocities. This is complicated and involved: it may suffice to use external constraints with uncertainties like Rahman et al. [437]; or we might generate plausible correlated velocity fields *a priori* assuming nonlinear dynamical theory.[229] Once we have those, we can calculate and apply the *total* redshift.

In any case, we should (have already) distribute(d) the SNæ across the sky and observational time. Now we simulate the survey. The LSST engineers and associated scientists have done this numerous times using dedicated software [553]. This provides us with the sequence of metadata (times, bands, zero points,[232] sky background, instrument noise, etc.) for the list of observations of the sky location of each of our simulated SNæ, which now allows us to simulate light curves (i.e. a list of observed noisy measurements, possibly with host light added as well (see closely below)). Passing them through the anticipated detection and selection procedure, e.g. through the full difference imaging pipeline, SNR trigger, transient classifier, template fitter, and sample cuts results in our selected transient data set $\{\mathbf{d}^s\}_{s=1}^{N_{\mathrm{SN}}}$ and meta—treated now as auxiliary—data.

> If we are modelling hosts, we subject them to the same observational ~~torture~~ procedure: generate broadband fluxes from their sampled properties and simulate the noise present in 10 yr co-add photometry: $\{\mathbf{d}^h\}$, the host side of our data set. We may call it a day here and assume perfect host identification, i.e. $\mathbf{d}^{h(s)}$; or we can apply a deterministic method like picking the galaxy with least DLR for each observed (detected) SN; or associate to each s a *set* of h with corresponding DLRs, resulting in the extraction of a mapping $\{s \to h^s\}$ or $\{s \to \{h\}^s\}$ and possibly spatially resolved observations ($\{\mathbf{d}^p\}$).

And this is how we say a universe is simulated. Others [295, 102, 302, 407, 465] say it has already been done.

Finally, we may explore combining data sources. For example, some transients *will* be followed up spectroscopically. From those data we might able to extract explicit redshift and type (notice that until now, we have only been considering photometric observables (band fluxes)) and form a multi-modal catalogue that represents *one* training example for our neural network. Then we simply ask it: was this C sampled jointly or marginally with the catalogue?

---

[231] unless we want to account for peculiar velocity with respect to the host dust?

[232] We can go as deep as we want her, as we discussed. It will probably be sufficient to have a summary zero-point measurement as a Gaussian with a small uncertainty, from which we sample

Part III

Simulators for supernova cosmology

Chapter 9

CLIPPPY: probabilistic programming

CLIPPPY is a convenience Layer for Inference and Probabilistic Programming in `Python`. Originally designed to automate many common tasks related to defining, manipulating, and exploiting forward models (while trying not to be annoying in the meantime), it now incorporates—alongside venerable utilities for VI and interfaces to other packages for more conventional Bayesian inference[233]—a ~~full~~ somewhat-fledged collection of SBI routines: in fact, it contains the implementations of all primary inference methods presented and used in this thesis, e.g. hierarchical and set-based TMNRE and neural model selection.

CLIPPPY's foundation is the `Pyro` *probabilistic programming (PP)* framework[234] [46], which implements the idea of forward modelling: i.e. of representing a model (a *joint* distribution of several random variables) through one particular decomposition into an orderable series of conditional probabilities: *PP*

$$\mathrm{p}(A, B, C) \iff \begin{Bmatrix} a \sim \mathrm{p}_A = \mathrm{p}(A); & \to (a, \mathrm{p}_A) \\ b \sim \mathrm{p}_B = \mathrm{p}(B \,|\, a); & \to (b, \mathrm{p}_B) \\ c \sim \mathrm{p}_C = \mathrm{p}(C \,|\, a, b); & \to (c, \mathrm{p}_C) \end{Bmatrix} \equiv \text{trace}, \tag{9.1}$$

i.e. a stochastic program that walks a (particular) directed acyclic graph (DAG) of the model, recording the *computed* distributions (p_A, p_B, p_C) for each random variable it en-

`https://github.com/kosiokarchev/clipppy` and *JOSS*, in prep. CLIPPPY grew out of a set of command-line utilities for orchestrating VI fits known as `pyrofit`, and so "CLI" initially stood for the more common "command-line interface". An alternative name was PYRATTT: `Pyro` AT The Terminal.

233 e.g. `emcee` [158] (affine-invariant MH MCMC) and `dynesty` [491] (nested sampling)

234 While they are usually described as "languages", `Pyro` and its `JAX`-based [63] counterpart `NumPyro` [409] (as well as the other big PP framework, `PyMC` [1]) do not define an explicit *syntax*—unlike BUGS (Bayesian inference Using Gibbs sampling) [335], `Stan` [88, 87], and `Turing.jl` [167] (to list but a few)—but provide their functionalities using existing constructs (functions, contexts, etc.) of the `Python` general-purpose programming language.

© The Author(s), under exclusive license to Springer Nature Switzerland AG 2026
K. Karchev, *Supernova Cosmology for the 21st Century*,
https://doi.org/10.1007/978-3-032-15072-1_9

counters (A, B, C) and randomly sampled values of each (a, b, c)—used for computing the downstream distributions—into a so-called *trace*. Immediately, this allows evaluation of the joint probability of the particular sampled values by simple multiplication:

execution trace

$$\mathrm{p}(a,b,c) = \mathrm{p}_A(a) \times \mathrm{p}_B(b) \times \mathrm{p}_C(c). \tag{9.2}$$

autograd

In addition, Pyro makes use of PyTorch's [400] *automatic differentiation (AD)* engine to calculate the *gradient* of $\ln \mathrm{p}(a,b,c)$ with respect to all variables via the chain rule:

$$\begin{aligned}
\frac{\partial \ln \mathrm{p}(a,b,c)}{\partial a} &= \frac{\partial \ln \mathrm{p}(a)}{\partial a} + \frac{\partial \ln \mathrm{p}(b \mid a)}{\partial a} + \frac{\partial \ln \mathrm{p}(c \mid a,b)}{\partial a}, \\
\frac{\partial \ln \mathrm{p}(a,b,c)}{\partial b} &= \frac{\partial \ln \mathrm{p}(b \mid a)}{\partial b} + \frac{\partial \ln \mathrm{p}(c \mid a,b)}{\partial b}, \\
\frac{\partial \ln \mathrm{p}(a,b,c)}{\partial c} &= \frac{\partial \ln \mathrm{p}(c \mid a,b)}{\partial c}.
\end{aligned} \tag{9.3}$$

conditioning

The true power of PP, however, lies in its ability to *manipulate* the forward pass, of which the quintessential example is *conditioning* on a given value by simply assigning it instead of sampling from the relevant calculated distribution (but still recording the latter):

$$\mathrm{p}(A, C \mid B = b_o) \iff \left\{\begin{array}{ll} a \sim \mathrm{p}_A = \mathrm{p}(A); & \to (a, \mathrm{p}_A) \\ b \leftarrow b_o;\ \mathrm{p}_B = \mathrm{p}(B \mid a) & \to (b_o, \mathrm{p}_B) \\ c \sim \mathrm{p}_C = \mathrm{p}(C \mid a, b_o); & \to (c, \mathrm{p}_C) \end{array}\right\} \equiv \text{cond. trace.} \tag{9.4}$$

The product of probabilities from this conditioned trace is proportional to the *posterior probability* of A, C at the randomly sampled a, c (given $B = b_o$) and can therefore be used as a low-effort-required "likelihood" (unnormalised posterior) evaluator for *MCMC-proposed* parameter values $a_{\text{new}}, c_{\text{new}} \to \psi_{\text{new}}$ by conditioning on them as well as on $b_o \to \mathbf{d}_o$. In combination with AD, a PP framework can thus automatically implement any method for likelihood-based (reverse) inference given only an appropriately expressed forward model.

9.1 CLIPPPY for SBI

Apologia bis After this author's conversion to the SBI ideology (and TMNRE's emergence as his inference method of choice), CLIPPPY assumed a new role as the repository of his custom implementations[235] of the methods he applies in his research (part IV). Since we already covered the techniques themselves in part I and the implementation details that we left out do not represent significant advances over other similar software[236] (in the opinion of this author)—with the exception (in the same opinion) of the elegance of their *software design* (but not necessarily *engineering*), which allows easy extension and code reuse—, we will presently discuss only two aspects of the pipeline. *software design vs. engineering*

9.1.1 Stochastification

CLIPPPY contains extensive utilities for easily defining forward models[237] interfaced with `Pyro`'s tracing mechanisms. Here I only highlight its two main features.

- The `Sample` class represents a random variable through its distribution, i.e. one line in eqs. (9.1) and (9.3); at runtime (i.e. in a forward pass), it first *computes* the distribution and them samples a value from it. `Sample` implements a sophisticated procedure for representing arrays of i.i.d. variables and handling them in parallel using batching; this extends even to batches of *sets* with unequal cardinalities, which will prove immensely useful in chapter 15
- The `Stochastic` class transforms a deterministic function's arguments into random variables that are sampled (evaluated) before every invocation; i.e. it declares a simple parametrised forward (sub-)model. Most commonly, the computed values—which can also be en-`Capsule`-ated and reused across the model without re-evaluation—are the result of a `Sample` component, although no restriction is placed by `Stochastic`: they can be fixed values of any type or come from other non-`Sample` calls, including other `Stochastic`s.

---

[235] to the negative dismay of his doctoral co-advisor, author of the prominent SWYFT (stop wasting your 🤬 time) [360, 362] library

[236] The community *seems* to be gravitating towards the aptly named `sbi` [505] package, which I too would recommend for beginners, i.e. scientists who want to solve a trivial scientific problem that has nonetheless long evaded them because of dimensionality or tractability issues; nevertheless, "practitioners" of SBI, i.e. researchers developing methods to tackle new *types* of inference challenges *seem* to prefer working within their own frameworks...

[237] In fact, CLIPPPY defines its own "language"[234] (using `YAML`) for specifying graphical models structurally rather than procedurally, which allows introspection/modification even before the forward pass.

9.1.2 Truncation

Sequential and active learning are intrusive techniques that require modification of the prior within a forward model, i.e. a higher-level conditioning on a given distribution. CLIPPPY provides utilities for implementing this in general and for constructing *truncated* univariate and low-dimensional multivariate distributions on the fly (i.e. when the relevant stochastic site is encountered). These hypotheses already cover the majority of scientific use cases, in which the priors are simple, analytic, and generally separable over the model's (global) parameters.

ConUnDis

CLIPPPY's constrained univariate distribution: $\texttt{ConUnDis}\left[\mathrm{p}_X; x_{\mathrm{min}}, x_{\mathrm{max}}\right]$, represents a truncated version of $\mathrm{p}_X \equiv \mathrm{p}(X \mid Y)$ restricted to the interval $[x_{\mathrm{min}}; x_{\mathrm{max}})$. In a forward pass, when the random variable X is encountered, CLIPPPY first computes p_X given the upstream values $Y = y$, then samples an x using the inverse of the CDF:

$$x = \mathrm{F}_X^{-1}(u), \quad \text{with} \quad u \sim \mathcal{U}(\mathrm{F}_X(x_{\mathrm{min}}), \mathrm{F}_X(x_{\mathrm{max}})). \tag{9.5}$$

Consequently, CLIPPPY can only constrain in this manner analytic distributions for which the CDF F_X and its inverse F_X^{-1} are tractable. The final record in the trace is $(x, \texttt{ConUnDis}\left[\mathrm{p}_X; x_{\mathrm{min}},$ and if its probability is requested, CLIPPPY computes it as

$$\texttt{ConUnDis}\left[\mathrm{p}_X; x_{\mathrm{min}}, x_{\mathrm{max}}\right](x) = \begin{cases} \mathrm{p}(x \mid y)/c(y) & \text{for } x_{\mathrm{min}} \leq x < x_{\mathrm{max}}, \\ 0 & \text{otherwise,} \end{cases} \tag{9.6}$$

where $c(y) \equiv \int_{x_{\mathrm{min}}}^{x_{\mathrm{max}}} \mathrm{p}(x \mid y)\,\mathrm{d}x = \mathrm{F}_X(x_{\mathrm{max}}) - \mathrm{F}_X(x_{\mathrm{min}})$ is the normalisation, which we now allow to depend on the upstream parameters for which p_X was calculated. In fact, CLIPPPY makes a separate note of all "constrained probabilities" c encountered in a trace since their product is the ratio between the original and constrained model densities:

$$\mathrm{p}(x, y) = c(y)\,\tilde{\mathrm{p}}(x, y) \tag{9.7}$$

and can, therefore, be used to correct SBI trained with $\tilde{\mathrm{p}}(x, y)$.

Chapter 10

φυtorch: physics on ~~steroids~~ GPUs

φυtorch[238] is an effort to collect high-performance utilities generally useful to physicists who deal with computations (simulation, data analysis, etc.). It is similar to the immensely popular CPU-only libraries: `NumPy` [200], `SciPy` [527], and `Astropy` [22, 23] (by which it is inspired) but is based on `PyTorch` [400] and thus offers support for massive parallelism on graphics processing units (GPUs) and seamless automatic differentiation (AD).

End-to-end differentiable GPU-based simulators are highly valuable for inference since, apart from parallelising and accelerating computations, they enable high-dimensional score-based analyses with HMC or VI (section 1.2) and can be used in the context of SBI to enhance the training of neural networks as discussed in subsection 2.2.1 and eq. (2.20). A number of such simulators (and emulators) have already been developed in cosmology: e.g. for large-scale structure [368, 367, 54, 204, 205, 242, 319, 320, 321, 496, 301], strong lensing [97, 267, 189, 164], gravitational waves [104, 138, 551], and other related fields like stellar astrophysics [334], exoplanets [4, 181, 272, 230, 557], instrument modelling and experimental design [417, 550, 322, 531, 497]. Particularly relevant to this thesis are the differentiable stellar population synthesis code of Hearin et al. [206], the NN emulator of galaxy photometry `Speculator` [11], and the `NumPyro` implementation of BayeSN [184], as well as our own supernova Light-curve simulator (SLiCsim), presented in chapter 11.

Currently, `phytorch`'s toolbox includes general-dimensional *linear interpolation*, polynomial root finding, a growing collection of special functions, and two large sub-libraries:

linear interpolation

- `units`, `quantities`, & `constants`: are the basis of physical reasoning; they are what separates it from mathematics and connects it to the real world; `phytorch` implements the concept of a *"`Quantity`"* (similar to `Astropy`'s) that combines a *value* and a *unit* and tracks the changes to both throughout a computational graph, while also providing

quantitative computation

quantity (value × unit)

[238] `https://github.com/kosiokarchev/phytorch` and *JOSS*, in prep.; the name is a play on "physics (φυσικη) in `PyTorch`", "py" → "phi/φ", and "future" and in code is (de-)stylised as `phytorch`.

© The Author(s), under exclusive license to Springer Nature Switzerland AG 2026
K. Karchev, *Supernova Cosmology for the 21st Century*,
https://doi.org/10.1007/978-3-032-15072-1_10

checks of consistency: i.e. for every (numerical) operator[239] of `PyTorch`, `phytorch` defines a set of admission criteria for the inputs and a rule for the units of the output;

- `cosmology`: modelled after `Astropy`'s eponymous module, `phytorch` provides a wide array of cosmo*graphic* functionalities implemented for models within the FLRW-metric family: ΛCDM and the most popular parametrisations of evolving dark energy; the structure is as abstract and modular as possible so as to allow flexibility, code reuse, and easy extension / addition of functionality; the key component, which ultimately compute distances in the different models, are the (currently) two drivers: a numerical integrator reliant on `torchdiffeq` [93] that is general as to the cosmological model but limited as to parallelism and speed, and a custom analytic (and analytically differentiable) driver, which we describe in ~~extensive~~ ~~some~~ minor details next.

AADcosmo

Analytic auto-differentiable ΛCDM cosmography

All of the cosmological simulators listed above require calculating cosmographic distances (see section 5.2), which are a key ingredient in the modelling and data analysis of standard candles, sirens, and rulers, volumetric rates and densities, the cosmic microwave background radiation, Ly α forests in quasar spectra, as well as in the studies of galaxy properties and evolution. In the general case, cosmographic calculations require evaluating integrals like eqs. (5.7) and (5.8) *numerically*, which is both slow and not trivially parallelisable, while requiring a *further* numerical integration for the gradient calculation [93]: this is the approach chosen by the prominent `jax-cosmo` library [81].

numerical integration
adjoint method

For certain special cosmologies, analytic results have been discussed in the literature and implemented e.g. in `Astropy`. For example, the comoving distance in a (spatially) *flat ΛCDM* universe ($\Omega_{k0} = 0$) can be expressed using the Gauss hypergeometric function [31], the Legendre elliptic integrals [357, 555, 556, 142], which are also applicable in the non-flat case [146, 508, see also references therein], and the Carlson elliptic form [327].[240] Dabrowski & Stelmach [111] present a general solution, i.e. valid also in the presence of radiation and curvature, that makes use of the Weierstrass elliptic function.[241] Finally, Valkenburg [520, appendix B] use Carlson's basis to solve for the time coordinate by considering the slightly more general Lemaître–Tolman–Bondi (LTB) metric [316, 317, 513, 56], which describes the evolution of spherically symmetric but not necessarily homogeneous "dust" (i.e. cosmological fluid, not a SN-cosmological pitfall).

flat ΛCDM

[239] Propagating units backwards with the automatically computed gradients is a work in progress.
[240] Liu et al. mention the applicability of Carlson's formulation to the non-flat case but do not elaborate.
[241] which is, in my opinion, more of theoretical than of practical importance since methods for its numerical evaluation are hard to come by

In AADcosmo, I describe the first unified analytic framework for cosmographic distance calculations applicable to the general Λ–cold dark matter cosmology with non-zero curvature and in the presence of radiation (ΛCDMr). My solution is based on the Carlson elliptic integrals [83]—*special functions* that form an alternative to Legendre's basis [308] more suitable to analytic rather than geometric problems—and boasts two key advantages to other formulations. First, the fast and rapidly converging algorithms for evaluation of the Carlson integrals[242] [84] make their massively parallel implementation for GPUs straightforward. Moreover, their derivatives can also be calculated analytically and without reference to additional special functions, enabling automatic differentiation of the computed distances with respect to the cosmological parameters. Below, I summarise the framework before presenting a basic application within a toy model of SN Ia cosmology solved with high-performance/dimensional gradient-enabled likelihood-based inference techniques.

ΛCDMr

special function

ΛCDMr cosmography

The homogeneous and isotropic ΛCDMr cosmological model, which supposes the presence of classical (baryonic and dark matter) and relativistic (photons, etc.) fluids, as well as cosmological constant-like dark energy and a general curvature—see section 5.2—, is completely described by a dimensionless Hubble parameter (eq. (5.11)) of the form[243]

$$E^2(z) = \Omega_{r0}(1+z)^4 + \Omega_{m0}(1+z)^3 + \Omega_{k0}(1+z)^2 + \Omega_{\Lambda 0}, \tag{10.1}$$

where $\Omega_{r0}, \Omega_{m0}, \Omega_{\Lambda 0} \equiv C_{\Lambda\text{CDMr}}$ are the dimensionless parameters of ΛCDMr, and eq. (5.12) sets $\Omega_{k0} = 1 - \Omega_{r0} - \Omega_{m0} - \Omega_{\Lambda 0}$. In what follows, it will be more useful (and insightful) to reformulate eq. (10.1) in terms of its polynomial roots $\{r_j\}_{j=1}^m$:

$$E^2(z) = \alpha_m \prod_{j=1}^{m} (z - r_j), \tag{10.2}$$

where $m = 4$, $\alpha_m = \Omega_{r0}$ is the leading coefficient of the polynomial.[244] This corresponds to the re-parametrisation[245] $C_{\Lambda\text{CDMr}} \to \alpha_m, \{r_j\}_{j=1}^m$, which can be calculated with well-known explicit algebraic formulæ [459] or established numerical methods, e.g. via the com-

[242] In fact, these are the basis for some numerical implementations of Legendre's integrals themselves [424].

[243] Throughout this chapter, we will only mean *cosmological* redshift and drop the subscript.

[244] One can consider equivalently the radiation-less case, for which $m = 3$ and $\alpha_m = \Omega_{m0}$.

[245] Inspecting $\{r_j\}_{j=1}^m$ provides insight into the physicality of different cosmological parameters: notably, the existence of a positive real root (only for certain $C_{\Lambda\text{CDMr}}$ with $\Omega_{m0} < 0$ or $\Omega_{\Lambda 0} > 1$: see fig. 1 in AADcosmo) implies a divergence at finite redshift, i.e. "no Big Bang" as commonly labelled on plots.

Jacobian

panion matrix[246] [424, section 9.5]. A method for the efficient calculation of the *Jacobian* of this transformation is presented in AADcosmo, appendix A.

Cosmographic distances (reprise) are integrals of expressions containing $E(z)$ along redshift (cf. eqs. (5.7) and (5.8)). Here, we will shift away from our previous egocentric perspective and allow measurements starting and ending at arbitrary z_1 and z_2 (but still constrained to the line of sight):

$$D_c(z_1, z_2) = D_H \times \int_{z_1}^{z_2} \frac{dz}{E(z)}, \tag{10.3}$$

$$T(z_1, z_2) = T_H \times \int_{z_1}^{z_2} \frac{dz}{(1+z)E(z)}. \tag{10.4}$$

The Carlson symmetric elliptic basis

elliptic integrals
elliptic basis

The integrals eqs. (10.3) and (10.4) cannot be expressed for general $C_{\Lambda\mathrm{CDMr}}$ as elementary functions. However, when $E^2(z)$ is a polynomial of degree up to four, they are instances of *elliptic integrals*, which can be reduced to a linear combination of a small set of *basis functions*. The Carlson symmetric form [83, see also 86] is the natural choice when dealing with rational functions: by preserving the original permutation symmetry in the polynomial roots, it unifies different cases that select between branches of the square root, thus simplifying the end result. The functions which form the basis for reduction are:

$$R_F(x_1, x_2, x_3) \equiv \frac{1}{2}\int_0^\infty \frac{dz}{\sqrt{(z+x_1)(z+x_2)(z+x_3)}}, \tag{10.5}$$

$$R_J(x_1, x_2, x_3, w) \equiv \frac{3}{2}\int_0^\infty \frac{dz}{(z+w)\sqrt{(z+x_1)(z+x_2)(z+x_3)}}. \tag{10.6}$$

It is useful to introduce also the degenerate versions:

$$R_C(x_1, x_2) \equiv R_F(x_1, x_2, x_2), \tag{10.7}$$

$$R_D(x_1, x_2, x_3) \equiv R_J(x_1, x_2, x_3, x_3). \tag{10.8}$$

All Carlson integrals are well-defined for all complex x_1, x_2, x_3 except the non-positive reals (for which the integrand has poles along the integration path) and for all non-zero w (the Cauchy principal value is assumed if $w \in \mathbb{R}_{<0}$). R_F and R_J are symmetric in $\{x_1, x_2, x_3\}$, while R_D is only symmetric in $\{x_1, x_2\}$, and R_C is not symmetric.

[246] Matrix formulations are advantageous since ML libraries and hardware are usually highly optimised for the operations involved. In some cases, numerical stability might even be better than when using the analytic formulæ directly.

Computing derivatives of the Carlson integrals is *closed*, i.e. does not require any other special functions [86, eq. (19.18.1); 548]: *closed framework*

$$\frac{\partial R_F}{\partial x_3} = -R_D/6, \tag{10.9}$$

$$\frac{\partial R_J}{\partial x_3} = \frac{1}{2}\frac{R_J - R_D}{w - x_3}, \tag{10.10}$$

$$\frac{\partial R_J}{\partial w} = \frac{3}{2}\left\{\frac{w^2 R_F - 2wR_D + \prod_{i=1}^{3}\sqrt{x_i}}{w\prod_{i=1}^{3}(w - x_i)} - \left(\sum_{i=1}^{n}\frac{1}{w - x_i}\right)\frac{R_J}{3}\right\}, \tag{10.11}$$

where R_F and R_D are evaluated at (x_1, x_2, x_3), and R_J at (x_1, x_2, x_3, w).[247] Derivatives with respect to x_1 and x_2 are obtained by symmetry, while those of R_C and R_D via their respective definitions and the chain rule.

Explicit formulae

The final formulæ[248] for the comoving distance and lookback time in terms of the Carlson symmetric integrals are

$$D_\mathrm{c}(z_1, z_2) = \mathrm{D_H} \times \Omega_{\mathrm{r}0}^{-\frac{1}{2}} \times 2\Delta z \times R_F\left(u_{12}^2, u_{13}^2, u_{23}^2\right), \tag{10.12}$$

$$T(z_1, z_2) = \mathrm{T_H} \times \Omega_{\mathrm{r}0}^{-\frac{1}{2}} \times \frac{2\Delta z}{r_i + 1} \times$$
$$\times\left\{R_F\left(u_{12}^2, u_{13}^2, u_{23}^2\right) - \left[\frac{(\Delta z)^2}{3}\frac{d_{ij}d_{ik}d_{il}}{d_{i5}} R_J\left(u_{12}^2, u_{13}^2, u_{23}^2, u_{i5}^2\right) + R_C\left(s_{i5}^2, q_{i5}^2\right)\right]\right\}, \tag{10.13}$$

[247] In eqs. (10.10) and (10.11) the limits have to be explicitly implemented when arguments are repeated (e.g. $w = x_i$ or $x_i = x_j$: see [548] for details). The same applies to higher-order derivatives. In general, the limits of repeated arguments either have to be evaluated numerically or hard-coded.

[248] These but scratch the surface. I have also considered (and implemented in `phytorch`) a third quantity: the *absorption distance*, used predominantly in the analysis of the Ly α forest and is related to the intersection probability of the line of sight with objects of constant comoving number density and proper cross section. For clarity and brevity, I omit it from this thesis, but the formulophilic reader can feast their eyes on the monstrosities that are eqs. (3.9) and (B.3) in AADcosmo. Moreover, AADcosmo (and `phytorch`) contains the explicit formulæ for the radiation-less case, which are less frightening to behold and arguably more applicable in practice. Finally, the general reduction schemes of Carlson [85], Gray [183], sure to please any positronic brain with their algorithmic *elegance*, are also included in `phytorch`, both numerically and symbolically. *absorption distance* *elegance*

where $\{i, j, k, l\}$ is any permutation of $\{1, 2, 3, 4\}$, $\Delta z \equiv z_2 - z_1$, $d_{ij} \equiv r_j - r_i$ (with the special cases $d_{i0} \equiv -1$ and $d_{i5} = -(1 + r_i)$), and

$$u_{ij} \equiv \sqrt{z_2 - r_i}\sqrt{z_2 - r_j}\sqrt{z_1 - r_k}\sqrt{z_1 - r_l} + \sqrt{z_1 - r_i}\sqrt{z_1 - r_j}\sqrt{z_2 - r_k}\sqrt{z_2 - r_l},$$

$$u_{i5}^2 \equiv u_{ij}^2 - (\Delta z)^2 \times d_{ik} d_{il} \frac{d_{j5}}{d_{i5}} \quad \rightarrow \quad u_{i0}^2 \equiv u_{ij}^2 - (\Delta z)^2 \times d_{ik} d_{il},$$

$$s_{i5}^2 \equiv q_{i5}^2 + (\Delta z)^2 \times d_{k5} d_{l5} \frac{d_{j5}}{d_{i5}} \quad \rightarrow \quad s_{i0}^2 \equiv q_{i0}^2 + (\Delta z)^2,$$

$$q_{i5}^2 \equiv \frac{(z_1 + 1)(z_2 + 1)}{(z_1 - r_i)(z_2 - r_i)} u_{i5}^2 \quad \rightarrow \quad q_{i0}^2 \equiv \frac{u_{i0}^2}{(z_1 - r_i)(z_2 - r_i)}.$$

beauty Admittedly, these expressions are not *pretty*, nor is analytically deriving their gradient — with respect to $C_{\Lambda\mathrm{CDMr}}$ no less — easy. But the sequence of algebraic steps they represent is trivial for a computer program, like `phytorch`'s cosmology module, to implement and trivial for an autograd engine like `PyTorch`'s to backpropagate through.

Application: basic Bayesian SN Ia cosmology

The cosmographic utilities of `phytorch` were used for all "large and involved" analyses in part IV, although differentiability is rarely needed with SBI. Still, for some comparisons with "traditional" methods, where propagating redshift uncertainty passes through the gradient of the distance modulus (eq. (8.13)), and hence of the (luminosity…) distance, we do employ the fast and parallelisable AD framework presented above.

Here, we perform a first demonstration of the scalability of high-dimensional inference afforded by a GPU-accelerated and end-to-end differentiable likelihood for SN cosmology.[249] We define a simple yet non-trivial (i.e. not analytically exactly marginalisable) model, presuming to have measured with noise the (cosmological) redshifts and derived perfectly standardised distance moduli[250] of a complete sample[251] of N_{SN} confirmed type Ia SNæ. *A priori*, we take a simple analytically parametrised gamma distribution of redshifts:

$$z_c^s \sim \gamma(2, \beta) \sim z_c^s \exp(-\beta z_c^s), \tag{10.14}$$

which resembles (qualitatively) a SN Ia population *observed* with conventional brightness selection (cf. fig. 15.3) but is in reality just the toy model used by zBEAMS. For both ob-

[249] Soon after (according to some accounts [518], simultaneously), Uzsoy et al. [519] applied VI for non-hierarchical inference from individual SN Ia light curves with an even simpler MVN proposal.

[250] i.e., ignoring uncertainties in the standardising covariates and correlation coefficients, or measurement uncertainties altogether, leaving — crucially — only the residual scatter σ_0

[251] In fact, it only needs to be independent of the global parameters, i.e. a random sub-sample.

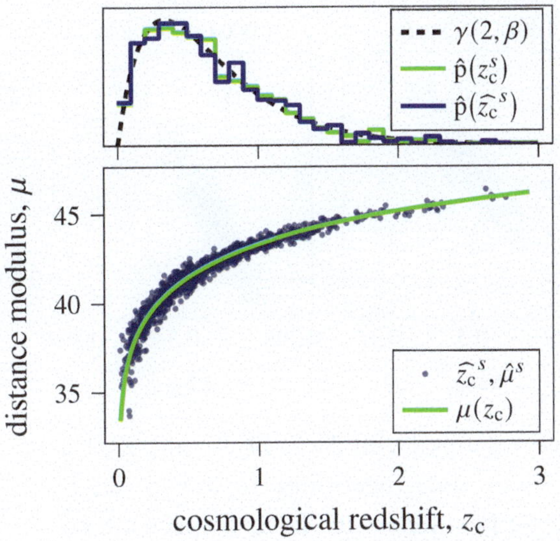

(a) Example mock data with 1000 SNæ Ia. *Top:* histograms of the true and measured redshifts, z_c^s and $\widehat{z_c}^s$, in comparison with the z_c prior, also used to draw z_c^s in the simulator. *Bottom:* observed Hubble diagram as dots and the underlying true relation as a line.

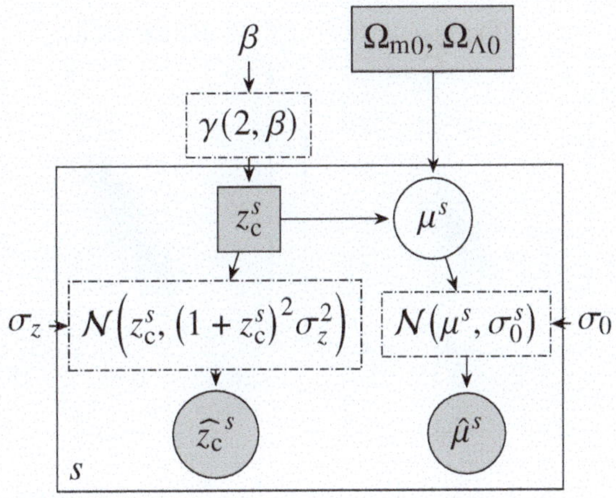

(b) Graphical representation of the toy model of SN Ia cosmology, also used to generate mock data. β, σ_0, σ_z are (fixed) model inputs in this example while $C_{\Lambda\mathrm{CDM}} \equiv \Omega_{m0}, \Omega_{\Lambda 0}$ and $\left[z_c^s\right]$ are being jointly inferred from data $\mathbf{d} \equiv \left[(\widehat{z_c}, \hat{\mu})^s\right]$.

Figure 10.1: Mock SN Ia data and the model used to generate and analyse it.

servables, we assume uncorrelated normal sampling distributions[252]:

$$\widehat{z_c}^s \sim \mathcal{N}\left(z_c^s, \left(1+z_c^s\right)^2\sigma_z^2\right), \tag{10.15}$$

$$\hat{\mu}^s \sim \mathcal{N}(\mu^s, \sigma_0^s), \tag{10.16}$$

where the (true) distance modulus is calculated deterministically from the cosmological model and the (true cosmological) redshift: $\mu^s \equiv \mu\left(z_c^s, C\right)$. For simplicity, and because radiation does not affect distances at low redshift (as discussed), we will assume the radiationless ($\Omega_{r0} = 0$) ΛCDM with free parameters $C_{\Lambda\mathrm{CDM}} \equiv [\Omega_{m0}, \Omega_{\Lambda 0}]$. Finally, because the example is already far removed from reality, we will simply fix $\beta = 3$, $\sigma_z = 0.04$, $\sigma_0 = 0.14$. This model is depicted graphically in fig. 10.1b and used to generate mock data sets[253] with number of SNæ Ia ranging from 1000 to 10^6: the smallest of them is depicted in fig. 10.1a.

[252] We make a slight meaningful distinction between the two, for in an "equivalent" likelihood, one would assume the redshift uncertainty scales with the *observed* $\widehat{z_c}^s$ rather than true z_c^s.

[253] This study is likelihood-based so the data is fixed, both in size and order, so it is in fact an array $\mathbf{d} \equiv \left[(\widehat{z_c}, \hat{\mu})^s\right]$.

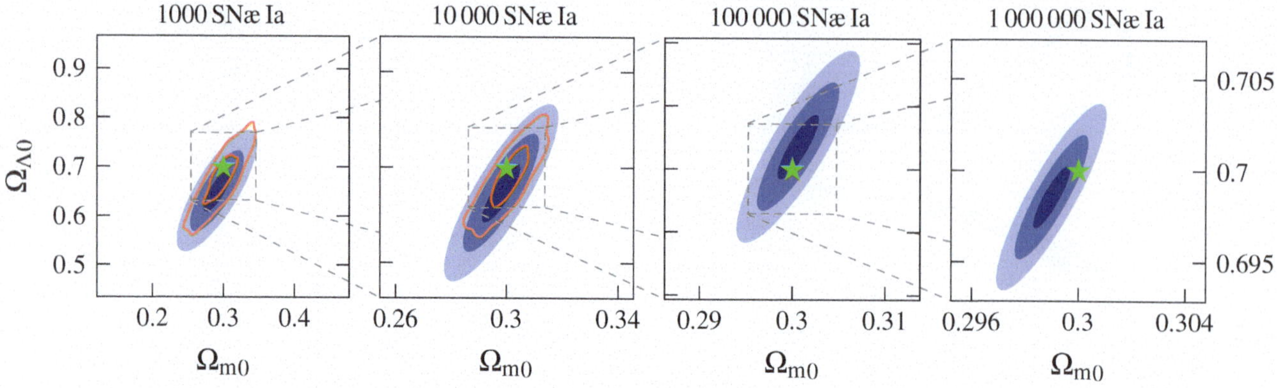

Figure 10.2: Posterior 1-, 2-, and 3-sigma credible regions (with 39.3, 86.5, and 98.9 % credibility, respectively) for Ω_{m0} and $\Omega_{\Lambda 0}$ from mock data (according to the model in fig. 10.1b) with increasing numbers of observed type Ia supernovæ. Blue ellipses are results of a VI fit as described in the text, while the red contours were derived with HMC (only performed up to 10 000 SNæ Ia and only up to 2-sigma shown). Note the different scales for each plot. The values used to produce the mock data are indicated with a star.

Results from HMC sampling and VI fits to the mock datasets are presented in fig. 10.2. Because of the computational cost of HMC (generating 1000 posterior samples with 10 000 SNæ Ia took ≈2 h on a high-end workstation), it was only applied to the datasets with 1000 and 10 000 SNæ Ia. In contrast, VI can analyse quickly (in ≈1 h on the same workstation with an NVIDIA A100 GPU) up to 10^6 SNæ Ia. To enable this, it resorts to a *partial* multivariate normal (PMVN) proposal distribution, which accounts for correlations among the two cosmological (global) parameters and between them and each individual SN's latent redshift but ignores additional posterior correlations between different SNæ: full details are given in Karchev et al. [267, subsection 4.2]. Due to the conditional structure of the model, a PMVN is sufficient in this case, especially for large observed samples when the cosmological posteriors do approach Gaussianity. (HMC makes no such assumptions, so its results can be considered a more accurate representation of the true posterior.)

While the particular results of fig. 10.2 are not a focus of this thesis, we note that VI is successful and efficient for this simple model, with the posterior size shrinking in each dimension as $1/\sqrt{N_{\mathrm{SN}}}$, as expected, and covering the parameter values used to produce the mock data. Extending the inference to more realistic models and real data, however, requires significant improvements to the guide so that correlations in high dimensions are properly accounted for (e.g. a dense covariance matrix, a normalising flow, or a score-matching proposal). Regardless, time (and section 1.3) showed that VI—and likelihood-based inference in general—is a *dead end* for SN Ia cosmology.

dead end

Chapter 11

SLiCsim: light curves for the ML era

SLiCsim[254] is the world's first generic GPU-accelerated auto-differentiable supernova Light-curve simulator. It is partly a port to the `PyTorch` ecosystem of the `sncosmo` [33] package and partly an attempt to re-build the concept of SN simulation from the ground up[255] with a focus on performance through batched parallelism and with ML/SBI applications in mind.

Simulating a SN (Ia) in SLiCsim involves three main component classes: a `Source` model subjected to propagation `Effects` results in the integrated photon flux density at a requested (observer-frame) time in a given filter (eq. (7.1)); this is then input to an `Instrument` that produces the final mock datum. All components are parametrised, and all parameters can be provided explicitly for a given forward evaluation. Thus, although SLiCsim itself only handles deterministic ("physical") aspects of a simulation, it can be

---

[254] `https://github.com/kosiokarchev/slicsim`; the code was introduced and first used in SIDE-real

[255] after more than a decade of reliance on *SNANA* [279]: the field's *de-facto* standard. It is telling as to the philosophy of SNANA that its name evokes the *analysis* of SNæ rather than their simulation; and indeed, a large fraction of its functionality concerns fitting light-curve models (MLCS, SALT, etc.) to observed data and dealing with probabilistic effects like selection corrections (BBC is an indispensable and inseparable part of SNANA) post-fitting. Still, on the side of its simulator (`snlc_sim.exe`...), SNANA has accumulated over the years an overwhelming number of bells and whistles (many related to details of the observation procedure—SNANA *is* how raw you can go—and many related to post-fit / analysis methodologies, suddenly parachuted in the middle of a forward simulation: e.g. K-corrections and parametrised efficiencies) and as a result has all but monopolised the trust of the community. While it has facilitated most of the excellent SN Ia cosmology realised since its inception, SNANA's opinionated, bespoke, and restrictive design—not to mention installation, configuration, and input/output, although those have been improving—imposes a significant barrier-to-entry on proponents of models and techniques that somehow contradict SNANA's method (pursuit of knowledge). For example, SNANA is not built for Bayesian inference; yet the developers of BayeSN have found themselves hammering its sizable latent layer into SNANA's rigid light-curve model specification so as to perform a selection-bias-corrected non-fully-Bayesian cosmological analysis to the field's satisfaction. *SNANA*

© The Author(s), under exclusive license to Springer Nature Switzerland AG 2026
K. Karchev, *Supernova Cosmology for the 21st Century*,
https://doi.org/10.1007/978-3-032-15072-1_11

transformed into a stochastic forward model through CLIPPPY's utilities. Furthermore, within the simulator, all parameters *broadcast* against each other to easily produce convenient batches of simulations — without shape-incompatibility errors if some care and thought are employed. This leads to massive performance gains when the SLiCsim is deployed (transparently with `PyTorch` and φυtorch) on a GPU. Lastly, SLiCsim makes use of φυtorch's framework for quantitative computation, i.e. assigns and manipulates physical quantities with units — from the emission in erg/s/Å to the signal in e^-.

broadcasting

`Source`

A *`Source`* defines the spectral flux distribution (total emitted energy per unit time and unit wavelength interval) of a supernova (in its rest frame): $\Phi^s_\mathrm{r}(t_\mathrm{r}, \lambda_\mathrm{r}) \equiv \Phi(t_\mathrm{r}, \lambda_\mathrm{r}; \mathbf{x}^s, \boldsymbol{\gamma})$, as a function of input SN-specific and global parameters (e.g. a stochastically sampled *template*). Currently, only SN Ia models are implemented: SALT, SNEMO, and BayeSN (and the underlying Hsiao template), with various pre-trained versions of each available. In addition, as alluded, BayeSN can be simulated *a priori* by making any/all template parameters (those outside the plate in fig. 8.2) stochastic as part of $\boldsymbol{\gamma}$.[256]

`Effect`

The SN's light is then modified by a number of parametrised *`Effects`*, each of which takes the output quantity of the previous one (starting with $\Phi^s_\mathrm{r}(t_\mathrm{r}, \lambda_\mathrm{r})$) and produces a (generally) different quantity. In the order that "light" encounters them:

1. `Extinction`(e.g. R^s_V, A^s_V) from dust in the host is modelled — as per BayeSN's second tenet and eq. (8.7) — explicitly and separately from the intrinsic Φ^s_r as a simple rest-frame wavelength-dependent multiplication. The various parametrisations from Fitzpatrick [153], Calzetti et al. [80], Fitzpatrick & Massa [154], Noll et al. [387], Kriek & Conroy [297] are available.[257]

2. `Redshift`(z^s) represents the combined effect of cosmic expansion and all peculiar velocities. It takes as input the *total redshift* z^s and has the effects of dilating wavelength and phase and suppressing luminosity three-fold as described above:

$$\Phi^s_\mathrm{o}(t_\mathrm{o}, \lambda_\mathrm{o}) = \frac{\Phi^s\left[t_\mathrm{o}/(1+z^s)\,,\,\lambda_\mathrm{o}/(1+z^s)\right]}{(1+z^s)^3}. \tag{11.1}$$

   Naturally, `Redshift` does not include distance-related dimming.

3. `Distance`(D^s_M) transforms a luminosity into an intensity (cf. table 7.1) over a sphere:

$$F^s_\mathrm{o}(t_\mathrm{o}, \lambda_\mathrm{o}) = \frac{\Phi^s_\mathrm{o}(t_\mathrm{o}, \lambda_\mathrm{o})}{4\pi\left(D^s_\mathrm{M}\right)^2}, \tag{11.2}$$

[256] Note that BayeSN's residual perturbations $\mathbf{e}^s \to \epsilon^s(t_\mathrm{r}, \lambda_\mathrm{r})$ can still be sampled even when the template's mean and principal component are fixed.

[257] These were extracted from the `extinction` package [32].

assuming three-dimensional space. Importantly, the transverse distance to the supernova, D^s_{M}, need not be defined in a cosmological context or in terms of a redshift. It is simply given as a quantity in units of distance (usually Mpc), possibly derived from calibrated measurements. Nevertheless, when a

`CosmologicalDistance`(z^s_{c}) effect is intended, the correct distance to use here is the transverse comoving—and *not* the luminosity—distance:

$$D^s_{\text{M}} \equiv D_{\text{M}}(z^s_{\text{c}}) = \sqrt{S_k(D_{\text{c}}(z^s_{\text{c}}))} = D_{\text{c}}(z^s_{\text{c}}) \operatorname{sinc}\left(\sqrt{k(D_{\text{c}}(z^s_{\text{c}}))^2}\right), \tag{11.3}$$

where by taking a square root, we have assumed we do not cross a metric divergence.[258]

4. `Extinction`$(A^s_{V,\text{MW}}, R^s_{V,\text{MW}})$ from dust in the MW is equivalent to that in the host but is applied in the observer frame;[259] and since comprehensive constraints are available from various observations, the relevant parameters are often fixed to measurements from e.g. Schlafly & Finkbeiner [470], based on the SN's sky location.

5. `Phaseshift`(Δt^s) is finally needed to account for the translation invariance of ordinary—as opposed to egocentric cosmological—time keeping[150]:

$$t_{\text{o}} = t - \Delta t^s, \tag{11.4}$$

where Δt^s is the common-clock time that corresponds to the origin of the `Source` model's phase coordinate.[260] Wavelengths are, instead, absolutely defined, so $\lambda_{\text{r}} = \lambda$.

Thus, we have reached the top of the Earth's atmosphere. Converting from the spectral flux density $F(t, \lambda)$ to a photometric measurement d is the purpose of an *`Instrument`*, as we described in detail in section 7.1. SLiCsim can calculate both "raw" photoelectron counts (eq. (7.3)) or calibrated fluxes (eq. (7.8)), provided the relevant *instrumental (pointing) metadata*:

`Instrument`

pointing metadata

$$\mathbf{a}_{\text{obs}} \equiv \begin{cases} (t, f, \text{gain}, \text{ZP}, \langle d \rangle_{\text{bg}}) & \text{for a CountsInstrument,} \quad (11.5) \\ (t, f, \text{FLUXCALERR}) & \text{for a FluxcalInstrument.}^{261} \quad (11.6) \end{cases}$$

[258] See fig. 1 in AADcosmo, which was deemed too insignificant for exposition in this thesis.

[259] This is implicit in the placement of this `Effect` after `Redshift`; a slight inconsistency might arise due to the motion of the Earth and Sun within the frame of Galactic dust, but this is hardly important for sweeping broadband photometry.

[260] This also takes care of the finite speed of light, if anyone was curious.

[261] As mentioned, calibrated fluxes are typically expressed with irrelevant gain = 1 and ZP = 27.5 mag.

Field

A survey is represented in SLiCsim through the concept of a *Field*, which may contain multiple (N_{SN}) SNæ all observed—and hence simulated—under the same conditions. It is a data structure that encapsulates the magnitude system used and an array of metadata $\left[\mathbf{a}^i\right]_{i=1}^{N_{\mathrm{obs}}}$ for each of N_{obs} pointings. The number of supernovæ and the observational metadata may be *stochastically* simulated (from a rate model for the former (cf. eq. (8.14)) and via e.g. LSST's `rubin_sim` [553] for the latter) or fixed to actual values for an already conducted survey. The latter strategy, which we adopt in chapter 12, explicitly conditions the simulator on $N_{\mathrm{SN}} \leftrightarrow N_{\mathrm{sel}}$, which is inappropriate for modelling surveys that suffer selection effects, as we explained in subsection 4.2.1; in such cases, the more involved fully stochastic setup must be implemented.[262]

photometric data vector

The final output from simulating a `Field` is an *ordered list of (ordered lists of)* mock measurements $\mathbf{d} \equiv [\mathbf{d}^s]_{s=1}^{N_{\mathrm{SN}}} \equiv \left[\left[d^{s,i}\right]_{i=1}^{N_{\mathrm{obs}}^s}\right]_{s=1}^{N_{\mathrm{SN}}}$. Combined with the observational metadata, as well as meta information $\mathbf{a}_{\mathrm{SN}}$ identifying each object (see section 12.2 for a concrete example), these can be transformed into a set of (sets of) mock measurements:

$$\mathbf{D} = \left\{\left\{\mathbf{a}_{\mathrm{obs}}^{s,i}, d^{s,i}\right\}, \mathbf{a}_{\mathrm{SN}}^{s}\right\}, \tag{11.7}$$

whose structure and size may vary from realisation to realisation. In this form, it is easy also to simulate and combine mock observations from multiple fields: indeed, in an extreme scenario of a targeted/follow-up survey (like the CSP, which we analyse in chapters 12 and 13), every SN is a field entire of itself.

ConstantSource

Simulating galaxy photometry can be achieved simply by defining a *ConstantSource* which does not vary with time. To it, too, the `Effects` listed above can be applied (although the effect of "host" dust extinction is commonly included *within intrinsic* galaxy-light models) and its photometry calculated through the same `Instrument`. Currently, only an experimental interface to a custom `PyTorch` port of the `Speculator-α` emulator [11] has been implemented for use in chapter 16, but even at this level, it already allows coordinating the simulated properties (e.g. dust extinction, progenitor population, etc.) of SNæ and their hosts, as we envisioned.

[262] Assuming that a survey is designed before it is performed, i.e. its time/filter scheduling is not influenced by ongoing observations (and hence cannot contain useful information), the survey specification can still be fixed to the observed metadata, but the SNæ and their placement within `Fields` must still be realised stochastically: see e.g. the recent `skysurvey` package.

Part IV
Science

Chapter 12

SN Ia Dust Extinction with NRE applied to real data

SIDE-real

In this chapter, we present the first fully simulation-based hierarchical analysis of the light curves of a population of low-redshift SNæ Ia. We focus, on one hand, on building a complete hardware-accelerated and parallelisable Bayesian forward model for photometric SN Ia surveys, implemented with SLiCsim and Clipppy. It includes stochastic variations in each SN's spectral flux (based on a pre-trained BayeSN model), probabilistic extinction from dust in the host and in the Milky Way, the effects of redshift and distance, and realistic instrumental noise. Secondly, we present efficient complete hierarchical inference of the SN Ia absolute magnitudes and host-galaxy dust properties, both at the population level and of the parameters of the individual objects. Using TMNRE and a bespoke network architecture we call a Super Tuple, we implicitly marginalise over 4000 latent variables (for a set of $\approx$100 SNæ Ia) while analysing directly the uncompressed collection of light curves, thus circumventing the expensive individual-object fitting stage present in all current studies. Exploiting the amortisation of our inference procedure allows us to obtain coverage guarantees for the results through Bayesian validation and frequentist calibration. Furthermore, we show a detailed comparison with joint likelihood-based inference, implemented through Hamiltonian Monte Carlo, on simulated data and then apply the trained TMNRE to the light curves of 86 SNæ Ia from the Carnegie Supernova Project, deriving marginal posteriors in excellent agreement with previous work and alternative methods. While here we focus on parameter inference, in the following chapter, we will use the same simulator to perform Bayesian model selection of the dust and brightness populations of SNæ Ia; moreover, coupling the forward model to the inference machinery demonstrated in later chapters will let us analyse high-redshift surveys and derive stringent and principled cosmological constraints.

© The Author(s), under exclusive license to Springer Nature Switzerland AG 2026

K. Karchev, *Supernova Cosmology for the 21st Century*,
https://doi.org/10.1007/978-3-032-15072-1_12

Table 12.1: SN Ia parameters, (hierarchical) priors and values used to generate mock data in SIDE-real. For local parameters, the support and size of the sampled "vector" are listed. See also fig. 12.1 for a (directed) graphical representation of the model.

parameter		prior	mock value
BayeSN template	$\mathbf{W}_{0,1}$ $\mathbf{\Sigma_e}$	fixed	M20
BayeSN "stretch" parameter	θ_1^s	$\mathcal{N}(0, 1^2)$	$\in \mathbb{R}^{\otimes N_{\rm SN}}$
BayeSN residual variations	$\mathbf{e}^s$	$\mathcal{N}(\mathbf{0}, \mathbf{\Sigma_e})$	$\in \mathbb{R}^{N_{\rm grid} \otimes N_{\rm SN}}$
covariance of $\mathbf{e}^s$	$\mathbf{\Sigma_e}$	fixed	M20
abs. magnitude offset	δM^s	$\mathcal{N}(0, \sigma_0^2)$	$\in \mathbb{R}^{\otimes N_{\rm SN}}$
abs. magnitude scatter	σ_0	HalfCauchy(0.1)	0.088
host dust-law parameter	R_V^s	$\mathcal{N}(\mu_R, \sigma_R^2)$	$\in [1.2; \infty)^{\otimes N_{\rm SN}}$
"mean" R_V^s	μ_R	$\mathcal{U}(1.2, 5)$	3.0
"st. dev." R_V^s	σ_R	HalfNormal(2^2)	0.5
host optical depth	A_V^s	Expon$(1/\tau)$	$\in \mathbb{R}_+^{\otimes N_{\rm SN}}$
mean opt. depth	τ	HalfCauchy(1)	0.329
cosmological redshift	z_c^s	fixed	$= z^s$
total redshift	z^s	fixed	$= \hat{z}^s$
measured redshift	$\hat{z}^s$	fixed	K17
MW dust-law parameter	$R_{V,\rm MW}$	fixed	3.1
MW optical depth	$A_{V,\rm MW}^s$	fixed	$= \hat{A}_{V,\rm MW}^s$
measured $A_{V,\rm MW}^s$	$\hat{A}_{V,\rm MW}^s$	fixed	SF11
time offset	Δt^s	$\mathcal{U}(\pm 7.5\,\rm d)$ [a]	$\in [\pm 5\,\rm d]^{\otimes N_{\rm SN}}$

[a] with respect to an initial estimate, `SEARCH_PKMJD`

12.1 Forward modelling probabilistic SN Ia light curves

The forward model used in SIDE-real is depicted fig. 12.1, while its random variables and the hierarchical distributions they are simulated from are detailed in table 12.1. We use the SLiCsim code, developed originally for the present application and described in detail in chapter 11.

In brief, we use a `Source` based on the BayeSN template derived by Mandel et al. [346, hereafter M20] from 79 nearby SNæ Ia with high-quality optical and NIR observations. It defines a 6×9 grid spanning the (rest-frame) ranges $t \in [-10; 40]\,\rm d$ and

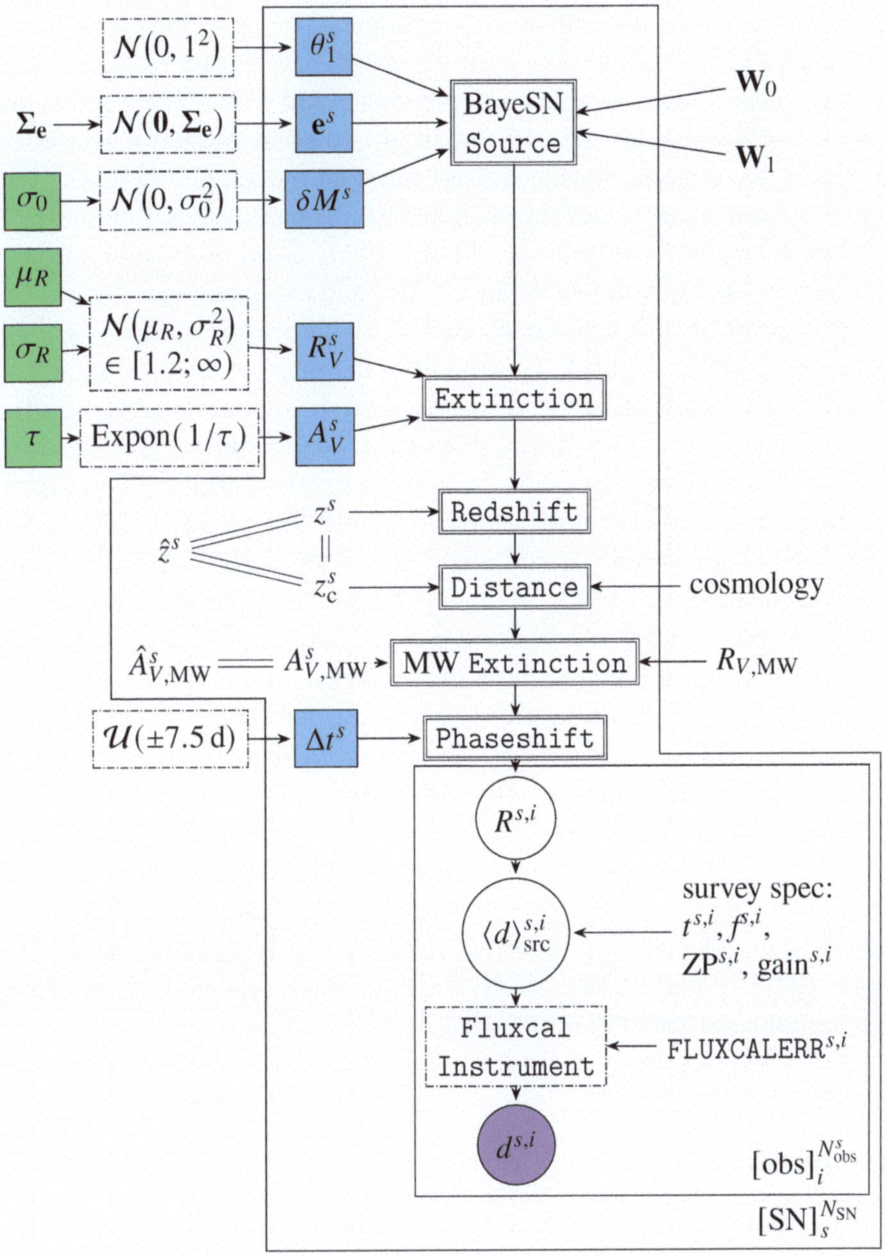

Figure 12.1: The forward model for SN Ia light curves. Green and blue shadings indicate, respectively, the global and SN-specific parameters, and the purple circle is the data (we use the simplified `FLUXCAL`-based instrument model (eq. (7.8))). Double-stroked boxes are different SLiCsim components described in chapter 11. In this particular study, the (fixed) model inputs are the BayeSN template ($\mathbf{W}_{0,1}$, $\boldsymbol{\Sigma}_\mathbf{e}$), externally measured Milky-Way dust-law and optical depths ($R_{V,\mathrm{MW}}$, $A_{V,\mathrm{MW}}^s$), redshifts ($\hat{z}^s$), the cosmological model, and the survey specification (metadata), including flux uncertainties (`FLUXCALERR`).

$\lambda \in [0.3; 1.85]\,\mu$m. We fix its principal components $\mathbf{W}_{0,1}$ and the common covariance matrix $\boldsymbol{\Sigma}_{\mathbf{e}}$ of the residual variations to their posterior means from M20, leaving only the coherent scatter σ_0 as a free SN-population parameter and 44 variables controlling the *intrinsic* brightness of each SN Ia: δM^s, θ_1^s, and the 42-component array $\mathbf{e}^s$ (the perturbations are fixed to zero at the extreme wavelengths of the grid, reducing the free $\mathbf{e}^s$ to 6×7).

We apply host-related dust `Extinction` (eq. (8.7)) parametrised by the F99 law and place population priors for the object-specific R_V^s and A_V^s as proposed by Thorp et al. [511], Thorp & Mandel [510], Grayling & Popovic [184] and explained previously. Concretely, we restrict the support of the Gaussian-shaped R_V^s hyperprior to $[1.2; \infty)$ as in eq. (8.15) (using CLIPPPY's truncation utility). This modifies the interpretation of μ_R and σ_R (which no longer represent the population mean and standard deviation) and affects, albeit mildly, their inferred values (consult Grayling & Popovic [184] for further discussion). In total, host dust is described by three population parameters and two further SN-specific variables.

Instead, we account for MW dust deterministically: following Schlafly & Finkbeiner [470, hereafter SF11], we assume an isotropic F99 dust law with $R_{V,\mathrm{MW}} = 3.1$ and perfectly measured MW optical depths $A_{V,\mathrm{MW}}^s$ at the sky locations of the SNæ, extracted from the SF11 maps. These assumptions can be easily relaxed and the MW dust properties inferred or marginalised with SBI similarly to those of the hosts.

This study does not aim to demonstrate proper redshift inference (see chapters 14 to 16). Instead, it targets a small well-observed SN Ia sample for which spectroscopy is available; therefore, we assume perfect measurements of the total redshifts: $\hat{z}^s = z^s$. When generating and analysing mock data, we will furthermore disregard non-cosmological contributions and set also $z_c^s = z^s = \hat{z}^s$. In contrast, peculiar velocities *are* present in the real data we consider and need to be accounted for. For the purposes of comparison with previous work, we will ~~extend~~ debase our pipeline by using *corrected* redshift estimates $\widehat{z_c}^s$ (as described in Thorp & Mandel [510, hereafter TM22]) and augmenting the σ_0 with an additional SN-dependent magnitude uncertainty propagated linearly from a peculiar velocity correction uncertainty of 150 km/s as lamentingly described in eqs. (8.12) and (8.13):

$$\delta M^s \leadsto \mathcal{N}\left(0, \sigma_0^2 + \left(\sigma_\mu^s\right)^2\right) \quad \text{with} \quad \sigma_\mu^s \equiv \frac{\sqrt{(150\,\mathrm{km/s}/c)^2 + \sigma_{\hat{z}}^2}}{\widehat{z_c}^s} \times \ln 10^{-1/5}. \tag{12.1}$$

Moreover, because the sample is at low redshift and offers no prospects of constraining cosmological parameters, we will adopt a *fixed*[263] flat ΛCDM model with matter density $\Omega_{\mathrm{m}0} = 0.28$ and Hubble constant $\mathrm{H}_0 = 73.24\,\mathrm{km/s/Mpc}$ and use it to calculate *fixed* `CosmologicalDistances` from the (fixed) $\widehat{z_c}^s$.

[263] Part of the motivation for making this choice was the desire for comparison with TM22 and (as we describe below) with the likelihood-based BayeSN code, which at the time did not have hierarchical-distance fitting functionalities.

Finally, we use the simplified[263] Gaussian description of calibrated fluxes (a `Fluxcal Instrument`) with fixed $\texttt{FLUXCALERR}^{s,i}$ for all observations (since detailed descriptions of the CSP observations — zero points, gains, and background fluxes — are not available in `SNANA`) and a toy model for the uncertain time of maximum — i.e. `Phaseshift` — by allowing it to fall within a 15-day time window around an initial estimate (`SEARCH_PKMJD`).

12.1.1 ~~Fake~~ Mock data

For the purposes of validating the inference procedure, we generate mock data designed to mimic the third data release of CSP, as presented in Krisciunas et al. [298, hereafter K17] and included in `SNANA` [279]. We extract the list of observation times ($t^{s,i}$), bands ($f^{s,i}$), and noise estimates[264] ($\texttt{FLUXCALERR}^{s,i}$) for each pointing for each SNæ Ia in the data release, their spectroscopic redshift ($\hat{z}^s$) and the Milky Way colour excess E_{B-V}, which we convert into $\hat{A}^s_{V,\mathrm{MW}} = 3.1 \times E_{B-V}$ (since we assume isotropic $R_{V,\mathrm{MW}} = 3.1$ for the Milky Way). These constitute the metadata, i.e. the inputs to the graphical model in fig. 12.1 (together with the M20 template).

Since the M20 model, which we use both to generate and analyse the mock data, was not trained on u-band observations and outside the rest-frame time range $[-10; 40]$ d, we exclude the corresponding entries from our CSP-like setup. For the global parameters (σ_0, μ_R, σ_R, τ), we set ground-truth values as listed in table 12.1, informed by the posterior means reported in M20, whereas the object-specific ones we sample from their priors,[265] also listed in table 12.1.

The mock data set contains $N_{\mathrm{SN}} = 134$ SNæ Ia with a total of $\sum_{s=1}^{N_{\mathrm{SN}}} N^s_{\mathrm{obs}} = 13\,202$ flux measurements.[266] Figure 12.2 depicts it four times, coloured according to the values of the different SN-local parameters. The impact of A_V and θ_1 are clearly evident as gradients (shifts) of the light curves, while that of δM and R_V less so, in accordance with the earlier discussion (fig. 8.3), and this has an impact on hierarchical inference, as we will demonstrate in section 12.3. Notice also the reduced spread[147] of SN Ia light curves in the infrared bands, where measurements allow disentangling pure-magnitude from colour-and-magnitude variations (respectively described by δM and R_V).

---

264 To facilitate the likelihood-based comparison, we increase very small reported `FLUXCALERR`s to be at least 0.01 mag, as has also been done for the real data.

265 For Δt^s, we restrict the mock values to the range $[-5; 5]$ d in order to avoid boundary effects.

266 For the mock data that we generate ourselves, we consider all SNæ from the CSP data release, instead of the restricted "clean" sample in subsection 12.1.2.

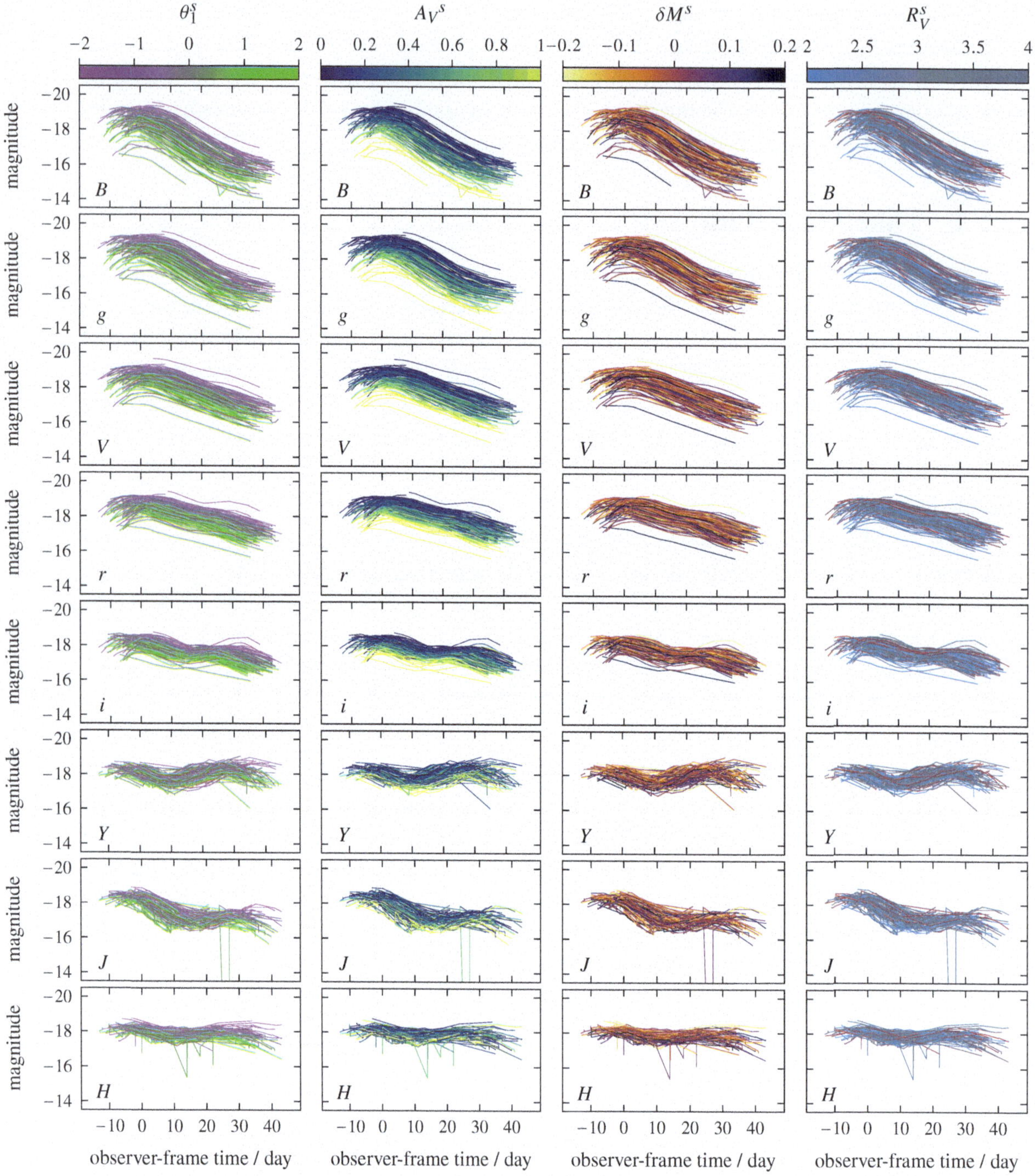

Figure 12.2: Mock light curves that we generate and analyse, corrected for cosmological distance (but not for redshift). While we work entirely in linear (flux) scale, for presentation purposes, this figure is in magnitudes. Each column shows the same light curves but coloured according to a different SN-local variable, as indicated on the top. Each row is a different CSP band: from bluest (top) to (infra)reddest (bottom). Different SNæ might have observations in different sets of the bands. All plots have the same scale and limits: notice that the diversity in redder bands is smaller, owing partly to the weaker effect of dust extinction. See also fig. 8.3.

12.1.2 Real data

We also consider a real data set comprising the light curves of 86 non-peculiar SNæ Ia from CSP identified and analysed by TM22. As mentioned, for each SN we have *two* separate (fixed) redshifts: $z^s = \hat{z}^s$ and $z_c^s = \widehat{z_c}^s$, the former acting to redshift the light curves, while the latter is only used to calculate `CosmologicalDistances`.[267] To further match TM22's setup, we also fix the standard deviation of coherent residual scatter $\sigma_0 = 0.088$ and the time offsets $\Delta t^s = 0$ instead of inferring them for this data set.

12.2 Hierarchical NRE with the Super Tuple™

We approach the inference task with joint-to-marginal neural ratio estimation as introduced in subsection 2.1.3 and expounded for the solution of a hierarchical model in section 4.1. In principle, we could use the simplest fully-connected multi-layer perceptron (MLP), inputting a data vector formed by the concatenation of each SN's light curve: $\mathbf{d}^1 \parallel \mathbf{d}^2 \parallel \ldots \parallel \mathbf{d}^{N_{\mathrm{SN}}}$, but this architecture would (mostly) unnecessarily connect individual raw observations of different SNæ, introducing excessively many network weights while providing no obvious way to extract object-specific summaries.

Instead, we introduce a network architecture that can ingest the peculiar data that a collection of light curves is and output *simultaneously* all the requested ratios as in fig. 4.2; I call it the Super Tuple,[268] and we found it is fast to train (both in terms of number of steps and time per step) and with present-day data sets achieves the best performance while requiring reasonable resources.

The Super Tuple's layout is based on the realisation that survey data $\mathbf{d} \equiv [\mathbf{d}^s]_{s=1}^{N_{\mathrm{SN}}}$ simulated as above (see also chapter 11) is a *tuple*: an *ordered* (indexable) collection of different-sized objects, whose structure — crucially — does not change between mock realisations due to the conditioning on the number of observed SNæ and the observation metadata. As

tuple

[267] As discussed in Grayling & Popovic [184], dust inference is only mildly affected by the peculiar velocity model, which mainly trades off against the residual scatter σ_0.

[268] Initially, following the reformulation of the data vector into a nested set of sets as in eq. (11.7):

$$\mathbf{d} \equiv \left[\left[d^{s,i}\right]_{i=1}^{N_{\mathrm{obs}}^s}\right]_{s=1}^{N_{\mathrm{SN}}} \to \mathbf{D} \equiv \left\{\left\{\mathbf{a}_{\mathrm{obs}}^{s,i}, d^{s,i}\right\}\right\} \tag{12.2}$$

I preferred — for its *elegance and generality* — a nested *un*conditioned, i.e. plain, deep set architecture (for the conditioned one had not been invented yet), which was proving dramatically slow to train (but still trainable, given enough time). For the application in SIDE-real, I finally succumbed and overnight before a talk I had to give at a conference implemented and trained the dumbest architecture I could think of, fully conditioned on the data structure — a usual driving force in network design — and implemented using a `for loop`; of course, it performed way better (in terms of training/validation win) and in a fraction of the time. In the name of science, we went ahead with it, but I gave it the affectionate name супер тъпа ("super dumb" in Bulgarian).

~~*elegance & generality*~~

for loop

embedding

a key first step, therefore, we use a collection of $N_{\rm SN}$ *distinct* learnable *embeddings* of the unequal-length $\mathbf{d}^s$ into a space of fixed dimension:

$$\mathbf{f}^s \equiv \texttt{SNEmbed}_s(\mathbf{d}^s). \tag{12.3}$$

We found that even a single linear layer (per SN) works well in our particular setup. The embeddings are stacked along a batch dimension and processed in parallel by a single *shared* component to derive final featurised representations of each supernova:

$$\mathbf{d}^s \equiv \texttt{SNHead}(\mathbf{f}^s). \tag{12.4}$$

These can be perceived as a standardised summary representation of each light curve and will later serves as the primary source of information for object-specific inference.

Summariser

Information across the tuple is then aggregated using a data-"set" Summariser:

$$\mathbf{S} \equiv \texttt{Summariser}\left([\mathbf{d}^s]_{s=1}^{N_{\rm SN}}\right). \tag{12.5}$$

dropout

We use a simple fully connected summariser that initially concatenates $[\mathbf{d}^s]_{s=1}^{N_{\rm SN}}$. While this approach is memory- and compute-intensive, it is simple, and in chapter 14 we will show that it does scale to the expected $\sim 10^5$ SNæ even on current (early 2020s) hardware. We prevent overfitting in this layer via stochastic *dropout* [220].

Finally, a number of ratio-estimator networks estimate ratios. For a group of global parameters of interest $\boldsymbol{\gamma}_g$:

$$\text{global:} \quad \ln \hat{\mathrm{r}}(\boldsymbol{\gamma}_g, \mathbf{d}) = \texttt{RatioEstimator}_{\boldsymbol{\gamma}_g}\left(\boldsymbol{\gamma}_g, \mathbf{S}\right), \tag{12.6}$$

and similarly for the local parameters of interest $\boldsymbol{\lambda}_g^s$ of object s:

$$\text{local:} \quad \ln \hat{\mathrm{r}}\left(\boldsymbol{\lambda}_g^s, \mathbf{d}\right) = \texttt{RatioEstimator}_{\boldsymbol{\lambda}_g}\left(\boldsymbol{\lambda}_g^s, \mathbf{S}, \mathbf{d}^s, \mathbf{a}_{\rm SN}^s\right), \tag{12.7}$$

where $\mathbf{a}_{\rm SN}^s \equiv \left[\hat{z}^s, \hat{A}_{V,\rm MW}^s\right]$ are the object-specific settings/metadata (cf. eq. (1.9), fig. 1.2, , and section 4.1), which completely identify the SN whose parameters are being inferred. As we noted previously, the presence of the global $\mathbf{S}$ accounts for *a posteriori* correlations between the parameters: i.e. we infer the posterior $\mathrm{p}(\boldsymbol{\lambda}^s \mid \{\mathbf{d}^s\})$ instead of $\mathrm{p}(\boldsymbol{\lambda}^s \mid \mathbf{d}^s)$. In a hierarchical model in which $\{\boldsymbol{\lambda}^s\}$ are *a priori* conditionally independent given global parameters $\boldsymbol{\gamma}$, i.e. $\mathrm{p}(\{\boldsymbol{\lambda}^s\} \mid \boldsymbol{\gamma}) = \prod_s \mathrm{p}(\boldsymbol{\lambda}^s \mid \boldsymbol{\gamma})$, this corresponds to the marginalisation $\int \mathrm{p}(\boldsymbol{\lambda}^s, \boldsymbol{\gamma} \mid \mathbf{d})\, \mathrm{d}\boldsymbol{\gamma}$ instead of simply $\mathrm{p}(\boldsymbol{\lambda}^s \mid \mathbf{d}, \boldsymbol{\gamma})$ as was done in previous hierarchical SBI analyses of permutation-invariant data [455, 210].

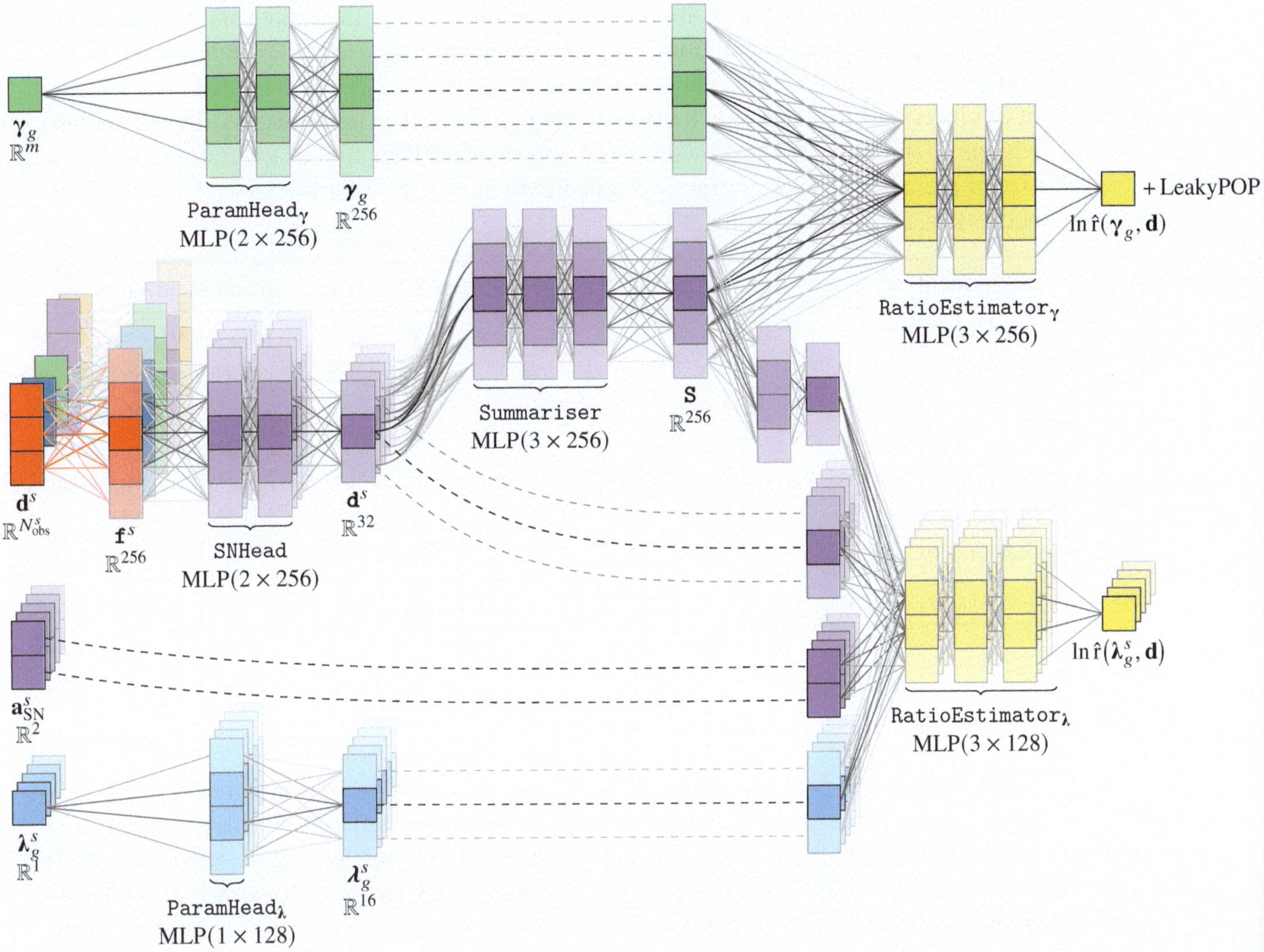

Figure 12.3: Architecture of the Super-Tuple-based complete hierarchical neural ratio estimator. Solid lines represent linear connections (followed inside the MLPs by batch normalisation and ReLU non-linearity). Dashed lines, on the other hand, connect layers that are duplicated for presentation (identity operation). When multiple parameter groups are being inferred, there are multiple parameter heads, whereas here only one is shown for clarity. Notice the similarity with fig. 14.2, the main addition here being the SN-embedding layers: N_{SN} distinct components (thus coloured diversely) with different input sizes but the same output dimension, which allow the $[\mathbf{f}^s]$ to be stacked into a single tensor along a batch dimension and processed in parallel before being flattened out for input into the summariser.

Table 12.2: Details about the components of the Super Tuple: their input and output dimensions and specific implementations. For all components, we use MLPs, indicating here the number and size of the hidden layers as $\mathrm{MLP}(n_{\mathrm{hidden}} \times d_{\mathrm{hidden}})$. Each hidden layer consists of a fully connected layer, batch normalisation, and a ReLU non-linearity. Inputs are also whitened (shifted and scaled by the mean and standard deviation of the training set). The size of global-parameter groups is denoted with $m = 1$, or 2 for the group $[\mu_R, \sigma_R]$. The network is also depicted in fig. 12.3.

component	inputs ∈ space	→	output ∈ space	implementation
$[\texttt{SNEmbed}_s]_{s=1}^{N_{\mathrm{SN}}}$	$\mathbf{d}^s \in \mathbb{R}^{N^s_{\mathrm{obs}}}$	→	$\mathbf{f}^s \in \mathbb{R}^{256}$	$\mathrm{Linear}(N^s_{\mathrm{obs}} \to 256)$
SNHead	$\mathbf{f}^s \in \mathbb{R}^{256}$	→	$\mathbf{d}^s \in \mathbb{R}^{32}$	MLP(2 × 256)
Summariser	$[\mathbf{d}^s] \in \mathbb{R}^{32 \times N_{\mathrm{SN}}}$	→	$\mathbf{S} \in \mathbb{R}^{256}$	MLP(3 × 256)
$\texttt{ParamHead}_{\gamma_g}$	$\gamma_g \in \mathbb{R}^m$	→	$\boldsymbol{\gamma}_g \in \mathbb{R}^{256}$	MLP(2 × 256, 256)
$\texttt{RatioEstimator}_{\gamma_g}$	$\boldsymbol{\gamma}_g, \mathbf{S} \in \mathbb{R}^{256+256}$	→	$\ln \hat{r}(\gamma_g, \mathbf{d}) \in \mathbb{R}^1$	MLP(3 × 256) + LeakyPOP
$\texttt{ParamHead}_{\lambda_g}$	$\lambda^s_g \in \mathbb{R}^1$	→	$\boldsymbol{\lambda}^s_g \in \mathbb{R}^{16}$	MLP(1 × 128)
$\texttt{SummaryHead}_{\lambda_g}$	$\mathbf{S} \in \mathbb{R}^{256}$	→	$\mathbf{S}_{\lambda_g} \in \mathbb{R}^{16}$	MLP(1 × 128)
$\texttt{RatioEstimator}_{\lambda_g}$	$\boldsymbol{\lambda}^s_g, \mathbf{S}_{\lambda_g}, \mathbf{d}^s, \mathbf{a}^s_{\mathrm{SN}} \in \mathbb{R}^{16+16+32+2}$	→	$\ln \hat{r}(\lambda^s_g, \mathbf{d}) \in \mathbb{R}^1$	MLP(3 × 128)

To enhance the network expressivity, we first "featurise" the raw parameters by passing them through a $\texttt{ParamHead}_\gamma$, whose output $\boldsymbol{\gamma}_g$ is concatenated to the dataset summary and input into the global ratio estimator. For latent-variable estimators, $\boldsymbol{\lambda}^s_g \equiv \texttt{ParamHead}_\lambda(\lambda^s)$ is concatenated with a processed $\mathbf{S}_{\lambda_g} \equiv \texttt{SummaryHead}_\lambda(\mathbf{S})$, which extracts the relevant summaries from $\mathbf{S}$, with the pre-processed data $\mathbf{d}^s$, and with the auxiliary inputs $\mathbf{a}^s_{\mathrm{SN}}$. Lastly, to enhance constraining power when the posterior is significantly more concentrated than the prior, we use the leaky parity-odd power (POP) activation layer [248] on the output of global-parameter ratio estimators.

All network components (SN embedders, the global summariser, and all ratio estimators) are implemented as MLPs with batch normalisation and rectified linear unit (ReLU) non-linearities. Details about layer sizes are given in table 12.2, and the network is depicted in fig. 12.3. We train using the joint-to-marginal BCE objective (eqs. (2.11), (2.14), and (2.15)) as implemented in CLIPPPY and summed over all marginal parameter groups (i.e. $\boldsymbol{\theta}$ representing, in turn, each of the global parameters and each of the local parameters, summing also over the N_{SN} SNæ).

12.2.1 NRE training

We implent and train a neural ratio estimator based on the Super Tuple architecture in CLIPPPY. Concretely, we target multiple parameter groups—global $\boldsymbol{\gamma}_g \in \{\tau, \sigma_0, [\mu_R, \sigma_R]\}$ and local $\boldsymbol{\lambda}_g^s \in \{\Delta t^s, \theta_1^s, A_V^s, \delta M^s, R_V^s\}$—simultaneously by training separate ratio estimators for each at the same time as a single data pre-processor $\mathbf{d} \to \left[[\mathbf{d}^s]_{s=1}^{N_{\text{SN}}}, \mathbf{S}\right]$.

We generate a training set of 256 000 mock survey realisations and use 6400 additional examples for validation, plotting posteriors, and calibration. To prevent overfitting, while training, we resample the instrumental noise (eq. (7.8))—this effectively augments the training set while avoiding the expensive part of the simulator—and stochastically "drop out" 50 % of the summariser input [220]. We optimize using Adam [289] with the default `PyTorch` momentum settings, a decaying learning rate schedule ($\gamma = 1/1.5$ every 10 epochs) over 100 epochs, and with mini-batch size of 128 examples. The results we present below use the checkpoint that performed best on the validation set. Training on one NVIDIA A100 GPU took ≈5 h to converge, in addition to ≈30 min needed to generate the training set. Evaluating a single set of marginal posteriors then takes on the order of milliseconds.

12.2.2 Validation with HMC

To validate our NRE results, we run a likelihood-based analysis (using the hierarchical likelihood that corresponds to the forward model in section 12.1) with HMC and consider the resulting posterior the ground truth. We use the code outlined in [184], which is based on the implementation of NUTS [225] in `NumPyro`. We run 4 chains and draw 500 samples each after 500 burn-in steps, which takes ≈30 min when run in parallel on 4 NVIDIA A100 GPUs. We verify convergence using the Gelmann-Rubin R number, the effective sample size, and other standard diagnostics as described in Grayling & Popovic [184]. We remind the reader that, as any likelihood-based method, this HMC analysis requires sampling the joint posterior of all model parameters, including in this case 4 global parameters and $N_{\text{SN}} \times (5 + 42) = 6298$ object-specific ones, most of which describe the residual light-curve variations through $\mathbf{e}^s$. In contrast, our SBI methodology implicitly marginalises $\mathbf{e}^s$ and estimates 3 global (since we group $[\mu_R, \sigma_R]$) and 5 SN-specific marginal posteriors (the latter evaluated N_{SN} times for the final results).

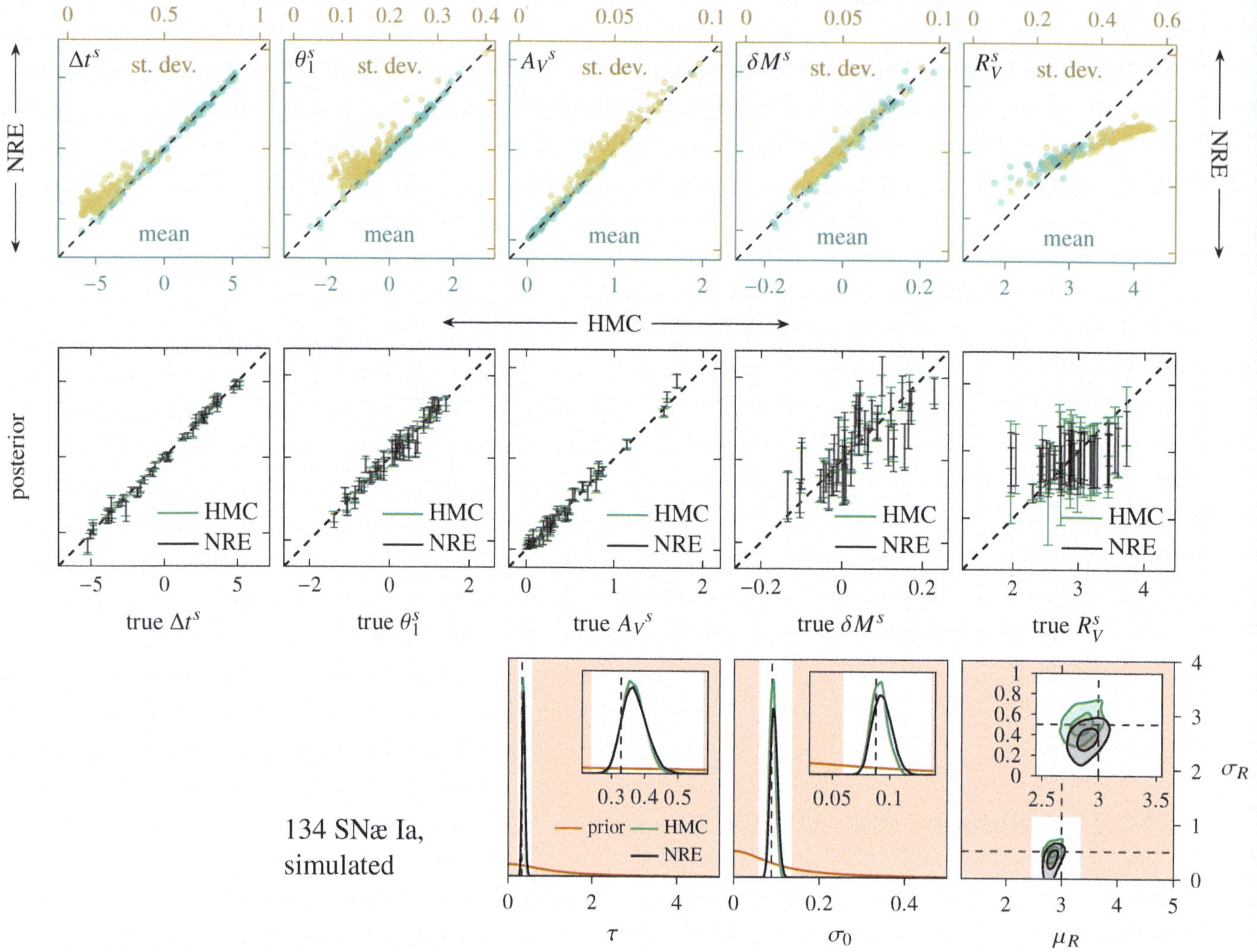

Figure 12.4: Inference results from the mock data set. *Top*: moments of the marginal posteriors of the local parameters of the N_{SN} supernovæ, as indicated in the top-left corner of each plot. Means (standard deviations) are shown in teal (ochre) with scale indicated below (above) the plot. The abscissa (ordinate) coordinate comes from the HMC (NRE) posterior, so that the diagonal indicates matching moments from the two methods. *Middle*: the same per-object marginal posteriors (mean± 1 standard deviation) plotted against the true values in the simulation. Only every third error bar is plotted for clarity. *Bottom*: posterior densities (in the 2-dimensional plot, 1- and 2-sigma credible regions) for the global parameters, as inferred by HMC and NRE, compared with the prior density and the true value used to simulate the mock data. Shaded regions indicate the truncation used for re-training the μ_R–σ_R NRE, depicted in the inset (the estimators for τ and σ_0 were not re-trained).

12.3 Results and discussion

12.3.1 Comparison of marginal posteriors for the mock data

We plot the marginal NRE posteriors, evaluated by weighting prior samples from the validation set by $\hat{r}(\boldsymbol{\theta}, \mathbf{d})$, in fig. 12.4 compared with the ground-truth marginalised HMC posteriors and the true parameter values used to generate the mock data.

We observe excellent agreement between NRE and HMC posteriors for the global parameters: τ, σ_0, $[\mu_R, \sigma_R]$, with similar uncertainties from the two methods and relative shifts of at most about 1σ. Since the ratio estimator for $[\mu_R, \sigma_R]$ is the most challenging, we re-trained it after truncating the global-parameter priors, as described by Miller et al. [361] and illustrated in the figures.

Similarly, SN-local parameters, with the exception of R_V^s, are very well recovered and in agreement with HMC. A detailed comparison of the first two moments of the 1-dimensional marginal posteriors is shown in the top row of fig. 12.4. In general, NRE exhibits slightly larger uncertainties for most parameters, as was previously observed in SICRET. It is important to note that R_V^s inference is almost entirely population-driven. Since constraints from individual objects are weak, the hierarchical structure induces shrinkage, as we anticipated. This is not an artefact of the inference procedure used but rather a feature of the hierarchical model itself, and is observed for both NRE and HMC. Thus, small changes in the μ_R–σ_R posterior shift all the R_V^s marginals coherently, leading to similar offsets from the HMC results for individual objects. We note that, while the $N_{\mathrm{SN}} + 2$-dimensional joint μ_R–σ_R–$\{R_V^s\}_{s=1}^{N_{\mathrm{SN}}}$ posterior can be studied using HMC, with NRE, we only derive marginal posteriors [however, see 14, 326].

12.3.2 Results on real data

Similarly, for the real CSP data (subset, as described in subsection 12.1.2), we plot the marginal posteriors for all free SN-specific and global parameters in fig. 12.5 in comparison with previous results from TM22, who also relied on HMC. The NRE- and HMC-derived posteriors are in good agreement with about 1-sigma offset and similar sizes, as was the case when analysing the simulated data set. In fig. 12.6, we focus specifically on population-level dust inference, depicting the Bayesian results derived with the two methods as well as regions with calibrated exact (frequentist) confidence. Since the NRE is already nearly optimal, this procedure returns almost exactly the regions of corresponding credibility, serving as further reassurance of their correctness.[269]

[269] For completeness, refer to appendix A in SIDE-real, where we perform both Bayesian validation and frequentist calibration, as described in subsection 2.3.2, of the global real-data inference network.

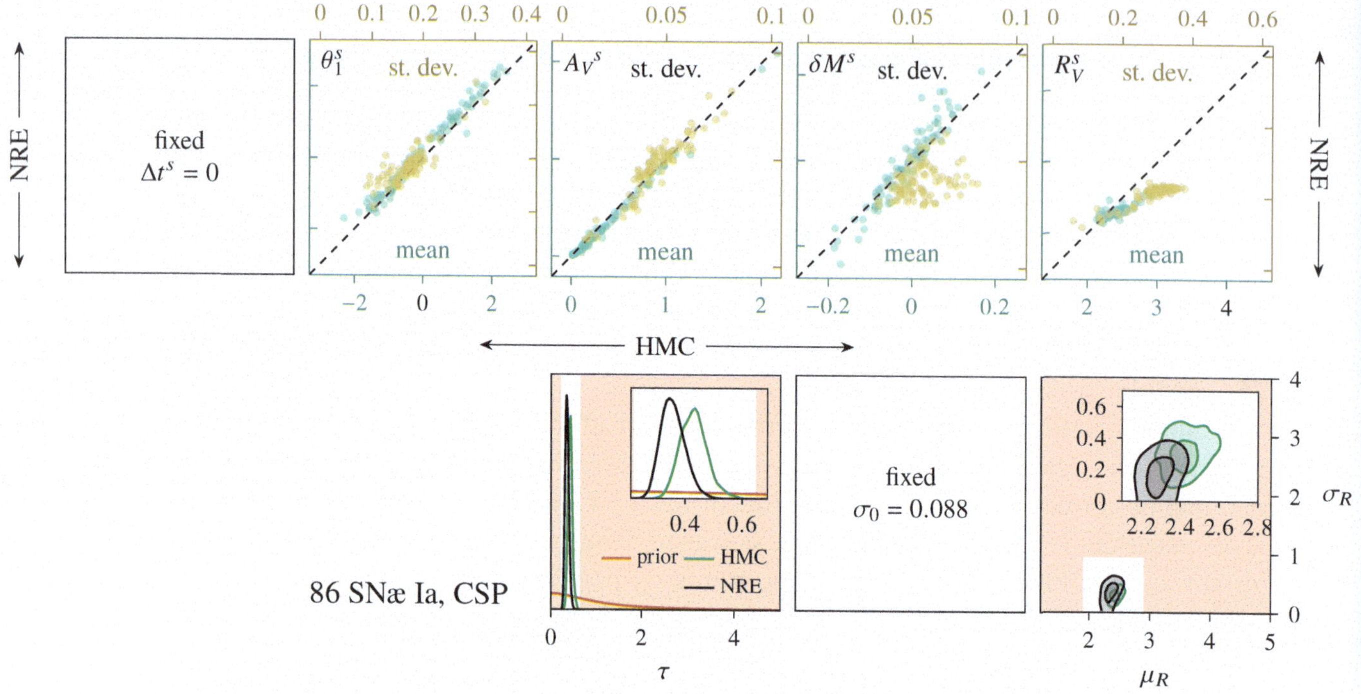

Figure 12.5: Inference results from the real data set. *Top*: comparison of marginal local-parameter posterior moments derived with NRE and HMC. *Bottom*: posteriors for the global parameters. See fig. 12.4 for more details.

Summary and outlook

We have presented detailed marginal neural simulation-based inference in the context of a hierarchical model of SN Ia light curves that incorporates realistic intrinsic (to each SN) and extrinsic (due to dust properties of the host galaxy) variability. By training a neural network to approximate the likelihood-to-evidence ratio with a training set of simulated light curves based on the Carnegie Supernova Project (CSP), we have derived marginal posteriors for the parameters of the populations of SNæ Ia and their hosts: the mean and standard deviation of dust-law parameters R_V^s, the average optical depth τ, and the residual scatter of SN Ia absolute magnitudes σ_0, and simultaneously inferred marginally the parameters of all $\approx$100 SNæ Ia. After validating the approach on simulated data, we have analysed the light curves of 86 real SNæ Ia observed by the CSP [298] and selected by Thorp & Mandel [510]. In both cases, we observe excellent agreement between our SBI results and

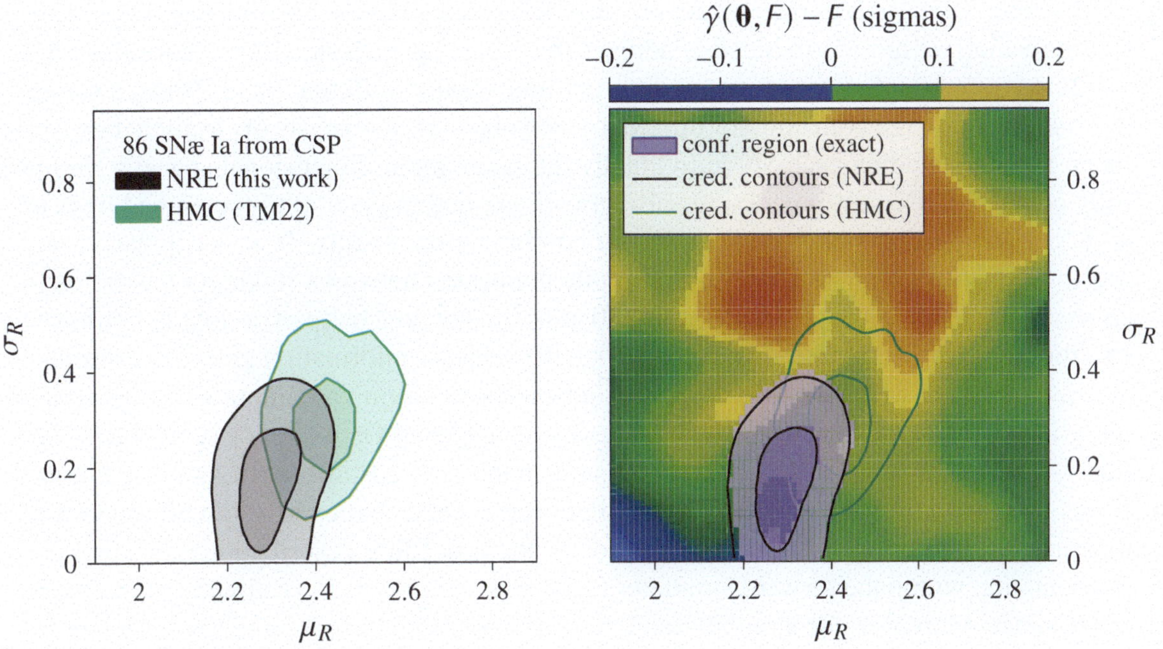

Figure 12.6: Population-level dust inference from TM22's subset of 86 CSP SNæ Ia. *Left*: a zoom-in of the respective panel in fig. 12.5, showing our NRE marginal posterior and that of TM22. *Right*: confidence regions (purple shade) with exact calibrated one- and two-sigma confidence (≈39 and ≈86 % in two dimensions). The coloured background depicts the "threshold" credibility offset by the prescribed (two-sigma) confidence (in terms of standard-Gaussian "sigmas"). Consult subsection 2.3.2 and SIDE-real's appendix A for the full details.

a baseline likelihood-based analysis as in Thorp & Mandel [510], Grayling et al. [185]. Concretely, posteriors for τ and σ_0 are in perfect agreement from the two methods, as well as the marginals for most local parameters (Δt^s, θ_1^s, A_V^s, δM^s). For the latter, SBI exhibits a slightly bigger uncertainty (by ≈10 %). Results for the dust-law parameters R_V^s and their population are also in good agreement, with only a small offset of about 1σ between NRE and HMC observed. As we illustrated in figs. 8.3 and 12.2, R_V^s have a minuscule impact on the data in comparison with the remaining variability, which makes them the hardest to infer and leads to hierarchy-dominated results: i.e. inference of one R_V^s depends on observations of all SNæ, regardless of the analysis methodology (likelihood- or simulation-based). In light of this extremely challenging learning task, the neural network exhibits excellent performance, having learned to extract and route the relevant information without access to the full high-dimensional likelihood but solely from training examples.

The precision and accuracy we achieve are largely due to the Super Tuple network architecture, which we have introduced to address the issue that each supernova in a survey has a different number of observations at different times and in different bands (thus, a survey is a *tuple*: an ordered collection of different-sized objects). It consists of: a single bespoke linear layer embedding each SN individually into a common-dimensional space; a *shared* fully connected SN post-processing sub-network applied in parallel to the embeddings of all SNæ; and a fully connected summariser combining the results. It is as expressive and fast to evaluate and train as conventional fully connected networks (taking a few hours to converge with training data generated in ~30 mins, in the same ballpark as highly optimised likelihood codes) but manages to fully extract the relevant information before overfitting.

In the present work, we have made a number of simplifying assumptions—e.g. fixing the Milky-Way extinction parameters and using a simplified Gaussian instrument description—that do not affect significantly the results in the low-redshift, fairly small-size, high-signal-to-noise case we consider. These do not represent a challenge to the SBI inference pipeline, i.e. do not require any modification to the procedure described here beyond implementation within the forward simulator (a framework for which we already presented in (SLiCsim)). Moreover, the simulator and network employed here are already sufficient for applications beyond hierarchical inference—namely, ground-breaking principled Bayesian model comparison in the presence of formidably numerous nuisance parameters as an answer to pressing questions regarding the distributions of SN Ia magnitudes and theirs hosts' dust properties, as we describe in the following chapter.

Lastly, in order to perform cosmological inference from near-future photometric-only data sets, we will need to account for two crucial probabilistic effects: redshift uncertainty and selection effects (including contamination). While the former can be addressed within the current framework—provided a suitable model for the process of measuring redshifts, as we will demonstrate in chapter 14—, the latter will require us to retire (improve upon) the Super Tuple architecture so as to be able to simulate and learn from examples with stochastic (*a priori* unknown) size: see subsection 4.2.1. We present our alternative solution—the conditioned deep set neural network—in chapter 15 with a simple model based on summary parameters but still envision the spirit of the Super Tuple to return when confronting stochastic collections of light curves, which will again need to be embedded into a common-dimensional space before being input into the deep set. In that setting, the fixed-input-size first layer of the Super Tuple will need to be replaced either with a more flexible linear embedding—i.e. a Gaussian process regression—or with a more sophisticated recurrent or attention-based neural network. These tools have already been applied to real data of *individual* SNæ (as we already presented) and are just waiting to be applied for hierarchical—and cosmological—inference.

Chapter 13

Simulation-based SN Ia Model selection

SimSIMS

SimSIMS is a first, brief,[270] and impactful application of the neural classification-based Bayesian model selection technique — which we presented in section 3.2 — to pressing questions regarding host-dependent SN Ia standardisation and dust extinction (subsections 8.3.3 and 8.3.4). Concretely, we address the *interplay* between

- the possibility of an offset between the intrinsic brightnesses of SNæ Ia hosted by low- and high-(stellar-)mass galaxies: a mass/magnitude step,[209] and
- different population models for the host dust, which may be influenced by other galaxy properties and give rise to an *apparent* (empirical) correlation with stellar mass.

We base this work on the hierarchical models of Thorp & Mandel [510, hereafter TM22] and the data they analyse: the same optical and NIR light curves of 86 non-peculiar SNæ Ia from the CSP [298] for which we performed parameter inference within *one* model[271] in SIDE-real (subsection 12.3.2). Here, we use essentially the same simulator — with the modifications presented below, which implement the different probabilistic hierarchies — to perform principled model comparison by deriving explicit posterior model probabilities (and hence, Bayes factors) after implicitly marginalising over 4000 nuisance parameters. The amortized nature of our technique allows us to validate it on simulated data and explore the dependence of its results on underlying parameter values, thus visualising and quantifying Occam's razor. When applied to the real CSP light curves, our method prefers a model with a *single dust law* and *no magnitude step*, disfavouring (based on SN Ia data alone) different dust laws for low- and high-mass hosts with odds in excess of 100:1.

---

[270] SimSIMS was presented at the NeurIPS's ML & the physical sciences workshop, which imposes a 4-page (pre-review) submission limit. Incidentally, the paper was also materialised in roughly 4 days, including setting up the model-selection pipeline in Clipppy and running it with the already-present simulator setup. *low-hanging fruit*

[271] In the nomenclature of this chapter, the model from SIDE-real is `M0, global`.

© The Author(s), under exclusive license to Springer Nature Switzerland AG 2026
K. Karchev, *Supernova Cosmology for the 21st Century*,
https://doi.org/10.1007/978-3-032-15072-1_13

13.1 The selection of models

We consider purely hierarchical modifications to the light-curve modelling framework we used in chapter 12, which is based on the BayeSN probabilistic SN Ia spectral timeseries parametrisation, to with dust extinction from the host and Milky Way, redshift, distance, and instrumental effects subsequently applied. We preserve the vast majority of the simulator settings from our previous parameter-inference application: i.e., we keep fixed the pre-trained M20 template, the MW dust law and optical depth, and distances derived from a fiducial flat ΛCDM cosmological model ($\Omega_{\mathrm{m}0} = 0.28$ and $\mathrm{H}_0 = 73.24\,\mathrm{km/s/Mpc}$ for $M_0 = -19.5$) and cosmological redshifts corrected for peculiar velocities from [89]. This still leaves 47 parameters *per SN* (42 of them describing the residual correlated light curve variability) for a total of more than 4000 for the analysed data set with 86 SNæ Ia.

In SimSIMS, we consider a further standardising covariate / "SN "-specific parameter: the host stellar mass, treating it simply as a fixed metadatum, adopting the values released with `SNANA`,[272] and ignoring their uncertainties when splitting into sub-populations across a *fixed*[273] threshold of $\log_{10}(M_*^{\mathrm{split}}/\mathrm{M}_\odot) = 10.5$, resulting in a 49/37 split.

Finally, we identify six distinct models, formed by assumptions for:

- the pre/absence of a magnitude step, ΔM, between[274] low- and high-mass hosts:

  `M0` : $$\delta M^s \sim \mathcal{N}\Big(0, \sigma_0^2\Big); \tag{13.1}$$

  `dM` : $$\delta M^s \sim \begin{cases} \mathcal{N}(0, \sigma_0^2) & \text{if } M_*^s \leq M_*^{\mathrm{split}}, \\ \mathcal{N}(\Delta M, \sigma_0^2) & \text{if } M_*^s > M_*^{\mathrm{split}}. \end{cases} \tag{13.2}$$

  On the additional free parameter we place a uniform prior

  $$\Delta M \sim \mathcal{U}(-0.2, 0.2), \tag{13.3}$$

  while keeping the original broad prior on σ_0 from table 12.1. This makes `M0` nested within `dM`, so the extents of these priors will directly influence the final Bayes factors, as we illustrated in fig. 3.2 and will explore in fig. 13.2.

---

data mining

[272] Their source in `SNANA` is listed once as "`??`" and then as "`XXX`" 💁‍♂️ … TM22 refer to Uddin et al. [517], whose estimates, as determined post-factum,[270] do not match the `SNANA` values… In fact, Uddin et al. [517, fig. 4], in turn, expose differences of up to 0.3 dex with previous analyses [380, 75]—none of which seem to be the `SNANA` source, and so the mystery continues… *Real data*

likelihood profiling

[273] As we remarked, M_*^{split} is usually *optimised* by looking for the threshold that results in greatest separation between the magnitudes of the two sub-samples: this corresponds to the *frequentist profile likelihood* approach, which ~~has no place~~ cannot easily be incorporated in a Bayesian model-selection framework, which relies on *marginal* likelihoods—impact appraisal pending.

[274] Our ΔM has opposite sign to TM22: here, $\Delta M < 0$ corresponds to brighter SNæ Ia in more massive hosts.

- the population of host dust-law parameters R_V^s (in all cases restricted to the range [0.5; 6] as in TM22):

$$\texttt{global}: \qquad R_V^s = \mu_R; \tag{13.4}$$

$$\texttt{local}: \qquad R_V^s \sim \mathcal{N}\left(\mu_R, \sigma_R^2\right); \tag{13.5}$$

$$\texttt{split}: \qquad R_V^s \sim \begin{cases} \mathcal{N}\left(\mu_R^{\text{lo}}, (\sigma_R^{\text{lo}})^2\right) & \text{if } M_*^s \leq M_*^{\text{split}}, \\ \mathcal{N}\left(\mu_R^{\text{hi}}, (\mu_R^{\text{hi}})^2\right) & \text{if } M_*^s > M_*^{\text{split}}. \end{cases} \tag{13.6}$$

The three dust models are thus also (*recursively*) nested. For the population parameters (in all of them) we use the same priors as in TM22 (essentially the same as in table 12.1).

моделёшка™ *(modelyoshka)*

13.2 Validation and exploration

We compose a neural-network classifier / model-posterior estimator from the data summariser of the Super Tuple in section 12.2 (cf. eqs. (12.3) to (12.5)) and a final linear transformation giving the unnormalised model probabilities (cf. eq. (3.4)):

$$\hat{\mathrm{r}}_m([\mathbf{d}^s]) \equiv \texttt{Linear}(\texttt{Summariser}([\texttt{SNHead}(\texttt{SNEmbed}_s(\mathbf{d}^s))]) \to N_{\text{mod}}). \tag{13.7}$$

We train it to optimise eq. (3.3), approximated via 96 000[275] examples from *each* of the 6 models (i.e., adopting a uniform model prior), using Adam [289] and a `OneCycle` learning-rate schedule [485]. Simulating the 576 000 mock survey realisation and training until convergence (for about 100 000 steps) on a single NVIDIA A-100 took about 1 h each.

After the upfront cost of training, neural simulation-based model comparison is amortised, and this opens up possibilities — all but impossible with traditional techniques — for verification and exploration/interpretation of its results, which we briefly explore now.

We plot in fig. 13.1 the refinedness and reliability of the trained classifier (see subsection 3.2.1 for the details), evaluated on held-out simulations. The prominent diagonal in fig. 13.1a indicates that *on average* the network assigns high probabilities to the true model from which data is generated. A faint "resonance" can be seen, suggesting "confusion" of `M0` and `dM`, but we remind the reader that, in general, the refinedness is a function both of the classifier's performance and of the intrinsic power of the data to distinguish between the models (in turn related to the broadness of the priors used within them). On the other hand, diagonal reliabilities (fig. 13.1b) are a more direct indicator of the classifier's good overall calibration, meaning that its results can indeed be interpreted as frequencies/probabilities.

[275] chosen so as to fit the full training set in GPU memory at once

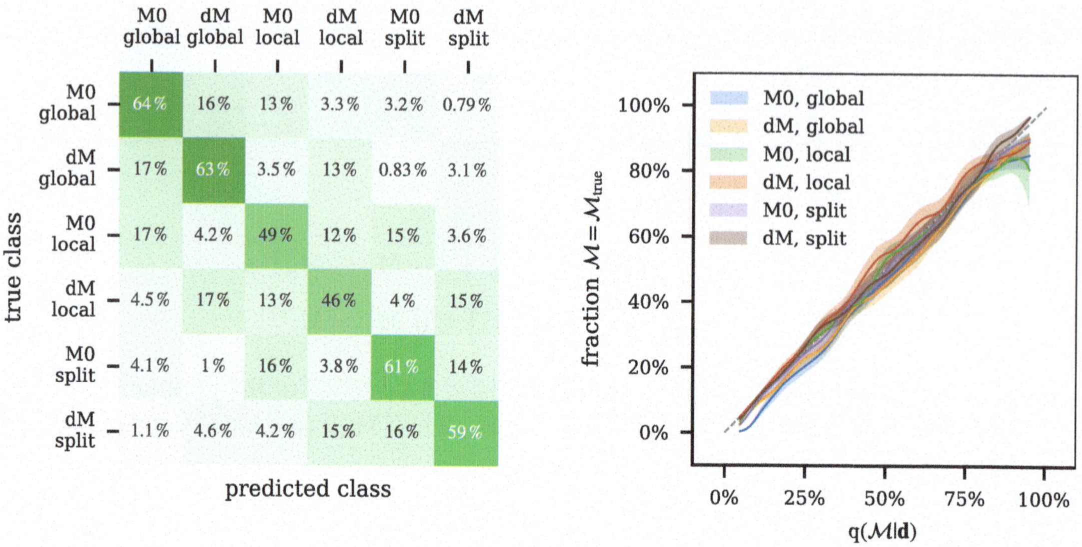

(a) Refinedness: each row shows the posterior over models (as labelled above), averaged over a collection of data simulated with the model indicated on the left. A prominent diagonal indicates *on average* confident and correct classification.

(b) Reliability: the fraction of the examples to which the NN assigned a given probability (abscissa) to a given model (colour) that indeed were generated with that model. Adherence to the diagonal signifies well calibrated probabilities.

Figure 13.1: Verification of amortised classifier-based model selection.

Moreover, owing to amortisation, we are able to explore Bayes factors (ratios of evidences, or equivalently, posterior odds for a uniform model prior) across a range of ground-truth parameters of simulated data. Figure 13.2, which compares nested models (`local` $\rightarrow$ `global` in μ_R–σ_R space and `dM` $\rightarrow$ `M0` in ΔM–σ_0 space), clearly demonstrates Occam's razor: data resulting from parameters sufficiently close to the location of the nested model ($\sigma_R = 0$ or $\Delta M = 0$) predominantly favour the simpler model (yellow/red regions). We also observe that, naturally, a step in magnitudes is harder to detect when their scatter (σ_0) is larger. A scatter in R_V^s (i.e. $\sigma_R > 0$) is also harder to detect when μ_R is large because, in that region, the effect on data is smaller due to the non-linear nature of the dust law.

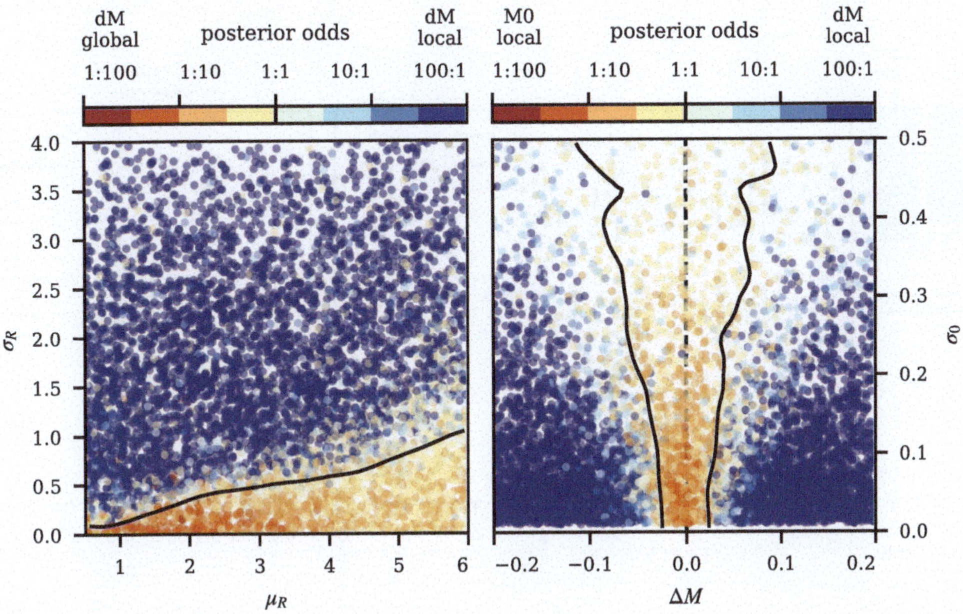

Figure 13.2: Occam's razor: Bayes factors (equivalent to posterior odds with a uniform model prior) for different simulated datasets as a function of the input parameters. In the left panel, we compare the hypothesis of a diversity in the host dust laws (`local`), which reduces to a single `global` dust law when $\sigma_R \to 0$, while marginalising over the unknown magnitude step and intrinsic sctter. On the right, we examine the preference for a non-zero magnitude step (`dM` *vs.* `M0`) as a function of its size and the intrinsic scatter, assuming a single but unknown (marginalised) R_V^s distribution. The solid black lines indicate parameters leading to equal posterior odds *on average over the chosen priors*: see fig. 3.2.

13.3 Results and controversy

Finally, we apply the trained classifier to the real CSP data and present the results (normalised model probabilities and $\log_{10}$ Bayes factors) for all 6 models in fig. 13.3. Highest probability is assigned to the simplest model, which supposes no split according to mass and a single dust *law* (characterised by a universal μ_R). No preference for a mass step is given, regardless of the dust model, while in general, a nonzero spread in R_V^s is mildly disfavoured (by a factor $\approx$2). A split in the dust-law *distribution* between low- and high-mass hosts is clearly disfavoured, regardless of the magnitude step, with a Bayes factor of $\approx$100, contrary to the conclusions of both Thorp & Mandel [510] and Brout & Scolnic [71].

In fig. 13.3, we also present posteriors (derived via NRE trained as in chapter 12 on *the same* simulations used for the model-comparison network), which support the conclusions stated above. In agreement with TM22, we find a magnitude step of $\Delta M \approx -0.05$, and

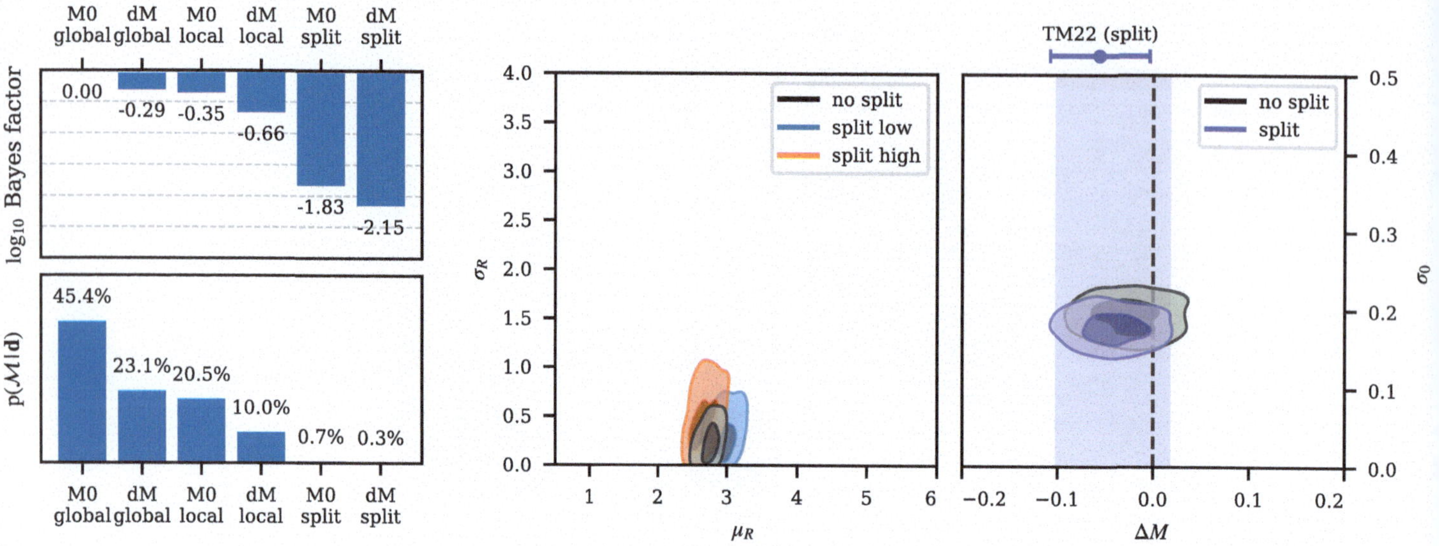

Figure 13.3: Results from the CSP data. *Left:* the posterior over models (bottom) and (top) $\log_{10}$ Bayes factor with respect to the highest-ranked model (no magnitude step, global dust law). *Right:* approximate marginal posteriors (1σ and 2σ HPD credible regions) from NRE, trained as in chapter 12 and reusing simulations from the model-selection run. The μ_R–σ_R plot compares the posteriors for $\left[\cdot^{\mathrm{low}}, \cdot^{\mathrm{high}}\right]$ from the `split` model with the result for a single dust-law distribution (`local`/"no split"). The ΔM–σ_0 plot compares posteriors from the same two models. The shaded strip there denotes the 95 % (2σ in one dimension) HPD region from the `split` model and is in excellent agreement with TM22 (2σ error bar above).

approximately 2σ away from 0, with the results only mildly affected by the dust model. We find a larger value $\sigma_0 \approx 0.2$ (cf. $\approx$0.1 in TM22) since this quantity in our analysis absorbs all residual variability present in the data, including peculiar velocity *uncertainties*, which we do not model explicitly. All of the global dust-parameter posteriors are in good agreement with TM22's fig. 8, and we obtain similar posteriors when treating low- and high-mass hosts separately as when we assume a single dust distribution (after marginalising over ΔM in all cases). This justifies the `split` dust model being strongly disfavoured, due to its considerably larger prior volume due to the two additional parameters.

Conclusion

Enabled by neural SBI, we have performed Bayesian model comparison on an unsolved problem in cosmology that requires realistic modelling of SN Ia light curves and marginalising over thousands of latent variables. A demonstration of Occam's razor, our results from low-redshift SN Ia data favour a global dust law and no magnitude step (with 45 % posterior probability up from 16.7 % *a priori*). The existence of a magnitude step or a distribution of R_V^s remain plausible (with posterior odds of approximately 1:2), while a split in global dust populations across $\log_{10}(M_*^{\text{split}}/\mathrm{M}_\odot) = 10.5$ is disfavoured with odds in excess of 100:1. These results are in contradiction with previous analyses even though we obtain similar "intermediate" results: this highlights the importance of performing end-to-end principled and rigorous analyses, instead of resorting to *ad hoc* procedures in order to handle the computational intractability of this high-dimensional and hierarchical model-comparison problem. We emphasise, however, that Bayesian model comparison is always dependent on the prior volumes considered, which includes the *fixing* (delta distribution) of components like cosmology, redshift and mass uncertainties, and peculiar velocities.

The scalability of our approach allows it to be applied to much larger data sets than demonstrated here, both present and future, with even more sophisticated Bayesian models (e.g. marginalising out the location of the mass split, or extending the mass standardisation beyond a binary split or with other host-related covariates), and more realistic simulators (self-consistently estimating redshifts and peculiar velocities, including selection effects and non-Ia contamination), ushering in[276] the era of principled simulation-based *pomp* fully Bayesian SN Ia cosmology.

[276] And indeed, soon after SimSIMS—unrelatedly or not—Ocampo et al. [390, 391], Goh et al. [177] published a number of cosmological model-comparison works employing NN classifiers (and containing no mention of SimSIMS, which was a first-of-its-kind study in the field 🥹). We, unperturbed, have instead moved on to usher in principled eras in other fields like exoplanetary atmosphere retrievals [331].

Chapter 14

SN Ia cosmology with TMNRE (scaling to 100 000)

SICRET

In SICRET, we take a step back from raw light curves, astrophysical modelling, and real data and look to the future: the aim of this chapter is to prove that a truncated marginal neural ratio estimation (TMNRE) analysis can deliver accurate and precise posteriors for cosmological parameters from future-sized data sets, which are expected to contain on the order of 10^5 SNæ Ia and thus fall beyond the computational capabilities of likelihood-based methods for all but the simplest models (cf. subsection 8.3.1). To this end, we generate mock data of different sizes (kept fixed during NRE training) from a deceptively simple Bayesian hierarchical model (BHM) inspired by BAHAMAS (itself based on the SALT light-curve parameters x_1, c). In this idealised case, for which traditional inference is valid, we use it to verify the precision of our SBI-derived constraints, which exhibit optimal scaling up to 100 000 SNæ Ia. After one minor — but critical — addition: photometric redshift uncertainty, we demonstrate that a model simplified so as to be solvable with sampling methods quickly goes catastrophically astray when applied to a large photometric survey while our SBI pipeline delivers consistently unbiased results.

From a technical side, in this study we use iterative truncation to systematically restrict the prior ranges of the 11 global parameters of the model (among them cosmology) and deliver maximally precise and still accurate results with a simple fully connected MLP network. Following global inference, we employ the methodology described in section 4.1 to marginally, but simultaneously, infer the 100 000 SN-specific standardised absolute brightnesses (which are the target of conventional fitting/standardisation analyses) and demonstrate that they too have the expected accuracy and precision. Finally, once again we validate and calibrate our final results — for the parameters of ΛCDM cosmology — as an added measure of their robustness.

© The Author(s), under exclusive license to Springer Nature Switzerland AG 2026
K. Karchev, *Supernova Cosmology for the 21st Century*,
https://doi.org/10.1007/978-3-032-15072-1_14

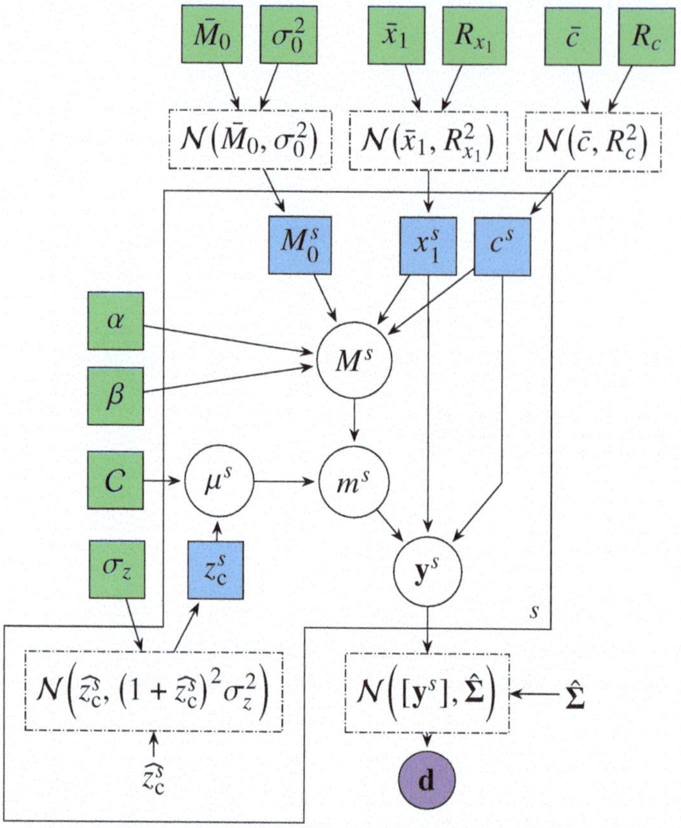

Figure 14.1: A simple BHM for SN Ia cosmology. Once again, blue / green squares are SN-specific / global parameters, and the purple circle is the data vector. In SICRET, the fixed model inputs are $\left[\widehat{z_c^s}\right]$ and $\hat{\boldsymbol{\Sigma}}$. Notice that the redshift-related "sub-graph" is inverted with respect to the usual "forward" flow, corresponding to using *external* redshift constraints.

14.1 Bayesian SN Ia cosmology with summaries

We begin by describing the simulator used for the present analysis (for generating both mock observations and training data): not a particularly daunting task since in this chapter, we model SNæ Ia only through summary statistics with analytic distributions, as per the tradition of Bayesian hierarchical SN Ia cosmology. Still, probabilistically, the model is similar to that presented before and consists of: a layer of parametrised population distributions, $\mathcal{O}(N_{\mathrm{SN}})$ SN-local parameters (redshifts, brightnesses, and light-curve parameters) exhibiting intrinsic scatter, and a simplified observational model accounting for "measurement" noise. It is depicted and detailed in its entirety in fig. 14.1 and table 14.1.

Table 14.1: SN Ia parameters, (hierarchical) priors and values used to generate mock data in SICRET. SN-specific parameters are sampled randomly, while the "observed" redshifts $\hat{z}^s_c$ and covariance matrix $\hat{\boldsymbol{\Sigma}}$ are taken from Pantheon (with replacement: see subsection 14.1.3). See also fig. 14.1 for a (directed) graphical representation of the model.

parameter		prior	mock value
latent redshift	z^s_c	$\mathcal{N}\left(\hat{z}^s_c, (1+\hat{z}^s_c)^2\sigma_z^2\right)$	~
measured redshift	$\hat{z}^s_c$	fixed	Pantheon
redshift uncertainty	σ_z^2	$\gamma^{-1}(0.0003, 0.0003)$	0.04^2
correlation coefficients	α	$\mathcal{U}(0,1)$	0.14
	β	$\mathcal{U}(0,4)$	3.1
abs. magnitude	M^s_0	$\mathcal{N}(\bar{M}_0, \sigma_0^2)$	~
mean abs. mag.	$\bar{M}_0$	$\mathcal{N}(-19.3, 2^2)$	−19.5
residual mag. scatter	σ_0^2	$\gamma^{-1}(0.003, 0.003)$	0.1^2
SALT "stretch"	x^s_1	$\mathcal{N}(\bar{x}_1, R^2_{x_1})$	~
x_1 prior mean	$\bar{x}_1$	$\mathcal{N}(0, 10^2)$	0
x_1 prior st. dev.	R_{x_1}	$\log\mathcal{U}(10^{-5}, 10^2)$	1
SALT "colour"	c^s	$\mathcal{N}(\bar{c}, R^2_c)$	~
c prior mean	$\bar{c}$	$\mathcal{N}(0, 1^2)$	0
c prior st. dev.	R_c	$\log\mathcal{U}(10^{-5}, 10^2)$	0.1
data covariance	$\hat{\boldsymbol{\Sigma}}$	fixed	Pantheon
DM density	Ω_{m0}	$\mathcal{U}(0,2)$	0.3
DE density	$\Omega_{\Lambda 0}$	$\mathcal{U}(0,2)$	0.7

14.1.1 SN Ia model: BAHAMAS

We represent the "physical" aspect of SNæ Ia through the BAHAMAS [347, 479] analytical hierarchical model, which represents a straightforward Bayesian extension to the orthodox (Phillips/Tripp/SALT) standardisation procedure. In it, the standardising covariates are elevated to SN-specific (often called "latent") random variables and assigned hierarchical prior distributions. Concretely, we will assume that each of a fixed number N_{SN} observed SNæ Ia has absolute magnitude[147] deterministically given by[277]

$$M^s = M_0^s - \alpha x_1^s + \beta c^s, \quad \text{with stochasticity transferred[278] to} \quad M_0^s \sim \mathcal{N}\left(\bar{M}_0, \sigma_0^2\right), \tag{14.1}$$

where the usual residual scatter is now understood as the standard deviation of the population of M_0^s around a mean/standard SN Ia absolute magnitude $\bar{M}_0$. The standardising covariates are assumed to follow similar normal population distributions

$$x_1^s \sim \mathcal{N}\left(\bar{x}_1, R_{x_1}^2\right) \quad \text{and} \quad c^s \sim \mathcal{N}\left(\bar{c}, R_c^2\right), \tag{14.2}$$

and their parameters are assigned uninformative priors: normal for the means, an inverse-gamma distribution for the residual scatter, and log-normal for the stretch and colour variances, as listed in table 14.1. Lastly, for the effect of cosmology (we use ΛCDM), we consider the trivial distance modulus (eq. (5.2)) calculated with the luminosity distance, assuming K-corrections have been applied (eq. (5.15)) to produce directly the "apparent brightness" m^s

$$m^s = M^s + \mu\left(z_c^s, C\right). \tag{14.3}$$

14.1.2 Observables

Up to this point, the present model is a mild extension of the toy we used in the previous application in chapter 10, with the addition of two extra linear local parameters. Where it differs significantly is in the model for observables. On the side of the redshifts, instead of explicitly modelling the measurements (after assuming a rate-based prior as per eq. (8.14), or an eq. (10.14)-like approximation thereof), we rely on an *external* analysis of independent auxiliary data (e.g. galaxy photometry) to have derived a Gaussian[279] *posterior*:

$$z^s \sim \mathcal{N}\left(\hat{z}_c^s, \left(1 + \hat{z}_c^s\right)^2 \sigma_z^2\right), \tag{14.4}$$

[277] the minus in front of α is conventional so that a positive correlation implies broader–brighter.

[278] In comparison with the previous $\mathcal{N}(\bar{M}_0 - \alpha x_1^s + \beta c^s, \sigma_0^2)$, this re-formulation brings the SALT statistical description closer to that of BayeSN ($M_0^s = \bar{M}_0 + \delta M^s$).

[279] We remind the reader that this is not a fully appropriate description of photometric redshift estimates, regardless of the size of the assumed uncertainty: see the discussion in subsection 8.3.2.

where $\widehat{z_c^s}$ is the estimated[280] cosmological redshift, and σ_z^2 is its variance scale. In the context of a subsequent cosmological analysis, this is interpreted as a *prior* for the cosmological redshift of each SN Ia: compare fig. 14.1 with figs. 10.1b and 15.2 and note the subtle differences between eq. (14.4) and eq. (10.15); accordingly, σ_z^2 is treated as a free global parameter (for which we assume an inverse-gamma prior).

For the SN Ia *observables* (summary parameters) *observables*

$$\mathbf{y}^s \equiv \left[m^s, x_1^s, c^s\right] \quad \rightarrow \quad \mathbf{d}^s = \left[\widehat{m}^s, \hat{x}_1^s, \hat{c}^s\right], \tag{14.5}$$

BAHAMAS considers a *non-factorised* sampling distribution[178]; i.e., instead of a collection of independent observations $\{\mathbf{d}^s\}$, the data is a vector $\mathbf{d} \equiv [\mathbf{d}^s]$ of length $3N_{\mathrm{SN}}$ formed by concatenating measurements of the three observables of all SNæ. Following the standard formalism, a joint multivariate normal sampling distribution is assumed:

$$\mathbf{d} \sim \mathcal{N}\left([\mathbf{y}^s], \hat{\boldsymbol{\Sigma}}\right) \tag{14.6}$$

where $\hat{\boldsymbol{\Sigma}}$ is a possibly dense $3N_{\mathrm{SN}}{\times}3N_{\mathrm{SN}}$ covariance matrix that—more or less faithfully—encodes the statistical and systematic uncertainties related to extracting estimates of $\mathbf{y}^s$ (hopefully, unbiased: whence the use of the concatenated parameters as the distribution's mean) from the underlying raw data. It, like the $\left[\widehat{z_c^s}\right]$, is part of the present BHM's (fixed) settings but is in reality *derived from data*: e.g. the statistical contribution to $\hat{\boldsymbol{\Sigma}}$ is formed by the uncertainties in fits to the individual light curves, which extract estimators $\widehat{x_0^s}$, $\widehat{x_1^s}$, $\widehat{c^s}$. In a holistic SBI pipeline, therefore, even $\hat{\boldsymbol{\Sigma}}$ would need to be stochastically simulated, which presents an enormous bottleneck due to the necessity to *perform* the fits for every simulated data set. As we discuss in SICRET—and demonstrated in chapter 12—, the appropriate path forward is to bypass the fitting stage with an end-to-end inference network trained on comprehensive simulations.

14.1.3 Survey specification: mimicking Pantheon

The model/simulator requires two inputs: the vector of redshift estimates, $\left[\widehat{z_c^s}\right]$, and the observational covariance, $\hat{\boldsymbol{\Sigma}}$. When simulating mock, we built these based on the then state-of-the-art Pantheon compilation [474], so that the mock SNæ Ia has a realistic distribution of redshifts[281] and reasonably sized observational uncertainties, appropriate for each SN's

280 Since this is not a random variable in our model—indeed, it is a fixed input to the simulator (fig. 14.1)—, we adopt the previous estimator-like notation: an un-leap.

281 Since future large surveys will observe SNæ with a different distribution of z_c from Pantheon, owing to their different selection probabilities, and since the statistical power of a SN Ia sample for cosmological parameter inference depends strongly on this distribution, our posteriors from 10^5 SNæ Ia should be taken as an indication rather than prediction of possible future constraints.

redshift. We select N_{SN} supernovæ from Pantheon (with replacement when N_{SN} is larger than the compilation size[282]) and concatenate their reported redshifts into $[\widehat{z}^s_{\mathrm{c}}]$ and stack diagonally their *individual* parameter covariance matrices $[\hat{\boldsymbol{\Sigma}}^s]$ (each of size 3×3) to form a *block-diagonal* $\hat{\boldsymbol{\Sigma}}$ for this proof of concept. Thus, we can define SN-specific metadata

$$\mathbf{a}^s_{\mathrm{SN}} = \left[\hat{\boldsymbol{\Sigma}}^s, \widehat{z}^s_{\mathrm{c}}\right] \tag{14.7}$$

with which to identify the [283] the individual SNæ.

14.2 Super Massive hierarchical TMNRE

The goal of SICRET is to demonstrate and validate the scalability of our combined global and local inference procedure (presented in section 4.1). In spirit, this is a similar task to the one we performed in SIDE-real (chapter 12), with the simplifying difference that now the input data **d** consists of summary parameters and can be decomposed into a collection of same-sized elements: $\mathbf{d} \to [\mathbf{d}^s]$, unlike the tuple of light curves we considered before.[284] Correspondingly, the inference network we use—depicted and detailed in fig. 14.2 and table 14.2—is even simpler than the Super Tuple: its defining characteristic now irrelevant, the bespoke embedding layer `SNEmbed`$_s$ is simply removed, and the three "measurements" of each supernova directly serve as the "embeddings":

$$\mathbf{d}^s \equiv \left[\widehat{m}^s, \hat{x}^s_1, \hat{c}^s\right] \to \mathbf{f}^s, \tag{14.8}$$

which are featurised into $\mathbf{d}^s$ = `SNHead`($\mathbf{d}^s$) as in eq. (12.4). The feature array is then summarised using the same fully-connected $\mathbf{S}$ = `Summariser`($[\mathbf{d}^s]$) component. Ineffi-
~~elegance~~ cient and inelegant as this solution might admittedly be, it is *simple* and delivers excellent
simplicity results, as we will show shortly. And even with 100 000 SNæ/ feature vectors of dimension 32 and summary size of 256 neurons, i.e. approximately a billion parameters, it is far from the sizes of modern-day AI systems. Still, in the next chapter, we will finally depart from the fully-connected paradigm and, with the conditioned deep set, be able to handle varying-sized input with a parametrisation independent of the survey size.

---

[282] Nominally, the Pantheon data release contains 1048 SNæ Ia, but for two of them ("`16232`" and "`PTF10bjs`") the reported parameter covariances are not positive definite 👤… so we only use 1046 💁‍♂️

[283] Note that this is not possible with a dense $\hat{\boldsymbol{\Sigma}}$ since in that case it is in fact *impossible* to separate the objects, e.g. due to the possibility of re-parameterisation / linear transformation of **d**, which scrambles the information of any set of delineated "objects".

array [284] The term for an ordered collection of equal-sized objects is *array*, or "масив" (massive) in Bulgarian, so I call the network used here Super Massive—which it is indeed due to its fully connected `Summariser` layer.

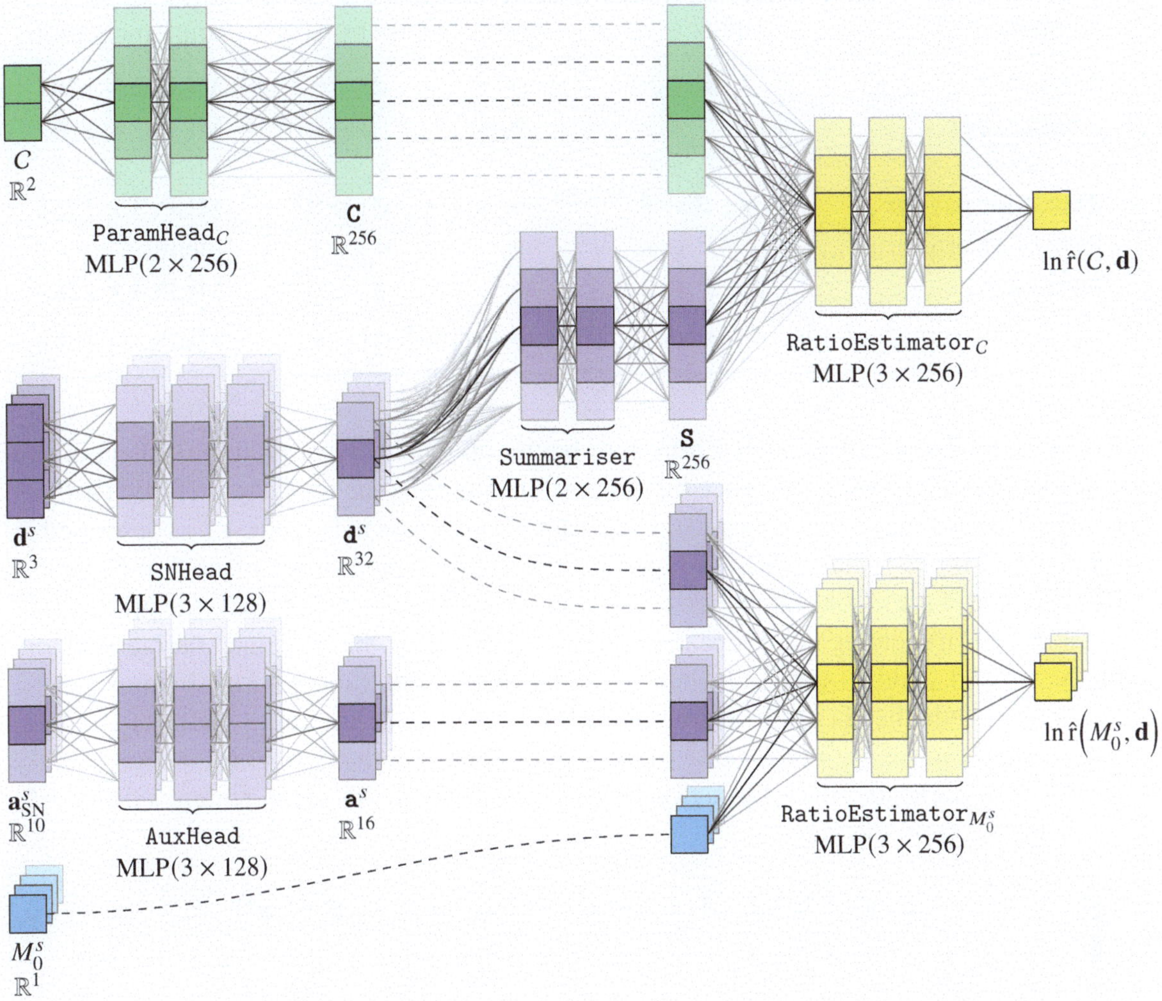

Figure 14.2: The Super Massive inference network — named after the dense / fully connected layer $[\mathbf{d}^s]$ → Summariser — as used in the last truncation stage, in which the (global) cosmological parameters C are inferred simultaneously with the 100 000 standardised absolute brightnesses $\left[M_0^s\right]$. Refer to fig. 12.3 for details on the depiction and note the omission (here) of the light-curve embedder SNEmbed and the connection between **S** and the local-parameter ratio estimator.

Table 14.2: Details about the components of the Super Massive NN. See the caption of table 12.2 for details about the notation and fig. 14.2 for a graphical depiction.

component	inputs ∈ space	→	output ∈ space	implementation
$\mathtt{SNHead}$	$\mathbf{d}^s \in \mathbb{R}^3$	$\rightarrow$	$\mathbf{d}^s \in \mathbb{R}^{32}$	MLP(3 × 128)
$\mathtt{Summariser}$	$[\mathbf{d}^s]_{s=1}^{N_{\mathrm{SN}}} \in \mathbb{R}^{32 \times N_{\mathrm{SN}}}$	$\rightarrow$	$\mathbf{S} \in \mathbb{R}^{256}$	MLP(2 × 256)
$\mathtt{ParamHead}_{\gamma_g}$	$\boldsymbol{\theta}_g \in \mathbb{R}^m$	$\rightarrow$	$\boldsymbol{\gamma}_g \in \mathbb{R}^{256}$	MLP(2 × 256)
$\mathtt{RatioEstimator}_{\gamma_g}$	$\boldsymbol{\gamma}_g, \mathbf{S} \in \mathbb{R}^{256+256}$	$\rightarrow$	$\ln \hat{r}(\boldsymbol{\gamma}_g, \mathbf{d}) \in \mathbb{R}^1$	MLP(3 × 256)
$\mathtt{RatioEstimator}_{M_0}$	$M_0^s \in \mathbb{R}^1$	$\rightarrow$	$M_0^s \in \mathbb{R}^1$	Identity
$\mathtt{AuxHead}$	$\hat{\boldsymbol{\Sigma}}^s, \widehat{z_{\mathrm{c}}^s} \in \mathbb{R}^{9+1}$	$\rightarrow$	$\mathbf{a}^s \in \mathbb{R}^{16}$	MLP(3 × 128)
$\mathtt{RatioEstimator}_{M_0}$	$M_0^s, \mathbf{d}^s, \mathbf{a}^s \in \mathbb{R}^{1+32+16}$	$\rightarrow$	$\ln \hat{r}\left(M_0^s, \mathbf{d}\right) \in \mathbb{R}^1$	MLP(3 × 256)

Since the (mock) data analysed in this chapter are much larger than the collection we considered in previously, the procedure for performing inference is more involved (still, its core philosophy remains unchanged). Concretely, it proceeds in two phases, as described in section 4.1. In the initial phase, we employ iterative truncation (implemented in CLIPPPY) for inference of the global parameters. This proceeds in stages, starting with the priors listed in table 14.1, truncating them repeatedly[285] after evaluating the re-trained ratio estimators on the target $\mathbf{d}_{\mathrm{o}}$, and terminating when none of the parameter ranges shrink by more than a factor of 2. Only then is a ratio estimator for local parameters $\mathtt{RatioEstimator}_{\lambda_g}$ trained; moreover, we make use of the observation from section 4.1 that global truncation already approximately encodes the information in $\mathbf{S}$ that is relevant in this second inference phase;[286] so we omit it from the inputs:

$$\text{local:} \quad \ln \hat{r}(\boldsymbol{\lambda}_g^s, \mathbf{d}) = \mathtt{RatioEstimator}_{\lambda_g}(\boldsymbol{\lambda}_g^s, \not{\mathbf{S}}, \mathbf{d}^s, \mathbf{a}^s). \tag{14.9}$$

The last architectural modifications from the network used in chapter 12 — besides the component/layer sizes — are purely cosmetic (and may have thus remained unnoticed in the re-production of eq. (12.7) into eq. (14.9)): we featurise the SN-specific metadata $\mathbf{a}_{\mathrm{SN}}^s \rightarrow \mathbf{a}^s \equiv \mathtt{AuxHead}(\mathbf{a}_{\mathrm{SN}}^s)$ in parallel with the $\mathtt{SNHead}$ from eq. (12.4) and do not featurise the latent parameters ($\mathtt{ParamHead} \rightarrow$ Identity) this time around.[287]

[285] We use simple "rectangular" truncation preserving 99.99 % of the approximate posterior mass (eq. (2.26)).

[286] This assumes that inference of a given group of local parameters of interest $\boldsymbol{\lambda}_g$ is not hierarchy-dominated, as was the case of R_V^s in chapter 12.

black magic

[287] These intuitive choices are motivated by the author's experimentation with *black magic*, i.e. NN design.

14.3 Experiments and results

In this section, we demonstrate and validate our inference procedure with mock data. We first elaborate on the application of TMNRE to infer global model parameters and produce calibrated frequentist confidence regions in the space of cosmological parameters by analysing a data set of 100 000 mock SNæ Ia generated with the full forward model described above. We also demonstrate ultra high-dimensional marginal NRE on the latent parameters: specifically, on the standardised absolute magnitudes M_0^s. Then, we investigate the scaling properties of our method by comparing its results to those of MCMC sampling for mock data generated by a *simplified tractable* model, which, however, is a poor description of *real* data (i.e. of data generated by the full nonlinear model).

14.3.1 Simultaneous inference from 100 000 SNæ Ia

We apply TMNRE to mock data sets containing between 10^3 and 10^5 SNæ Ia, all with the same global parameter values listed in table 14.1 but with different latent realisations. For each sample size, we generate a distinct survey specification: a collection of observed redshifts, $\hat{\mathbf{z}}$, an observational covariance, $\hat{\mathbf{\Sigma}}$, and a target data $\mathbf{d}_\mathrm{o}$. We then train a Super Massive inference network for groups representing each individual global parameter in fig. 14.1, except that we infer the $C \rightarrow C_{\Lambda\mathrm{CDM}} \equiv [\Omega_{\mathrm{m}0}, \Omega_{\Lambda 0}]$ jointly. We train with 640 000 simulated examples in each truncation stage, randomly generated *on demand* so as to avoid overfitting,[288] and re-initialise the network for each stage. The full procedure for 10^5 SNæ Ia took around 12 h (≈2 h per stage), with the runtime dominated by the *simulator* and a memory footprint largely due to the network's dense summarising layer. *online training*

Results from the trained networks in each stage ($\hat{\mathrm{r}}(\boldsymbol{\gamma}_g, \mathbf{d}_\mathrm{o}) \times \mathrm{p}(\boldsymbol{\gamma}_g)$ *numerically evaluated on grids*) are illustrated in figs. 14.3 and 14.4 for the data set with 10^5 SNæ Ia. At first, not all parameters are constrained: e.g. α and β are learnt appreciably only after two truncations, when the training data variance is reduced enough by constraining the rest of the parameters. On the other hand, the obvious approximate degeneracy in the cosmological parameters ($q_0 = \Omega_{\mathrm{m}0}/2 - \Omega_{\mathrm{de}0}$) is immediately picked up by the network, as indicated by the narrow strip in the first panel of fig. 14.4. Note that the posteriors in the initial stages are much larger than later ones even though training has largely converged (as evidenced by the loss stabilising): this is due to the limited network capacity and demonstrates the need and utility of truncation. *NRE posterior on a grid*

[288] Hence, we do not include dropout as in the summariser of the Super Tuple. Unfortunately, this strategy is only applicable to very fast analytic probabilistic programs and much less so to more sophisticated "physical" simulators of e.g. light curves.

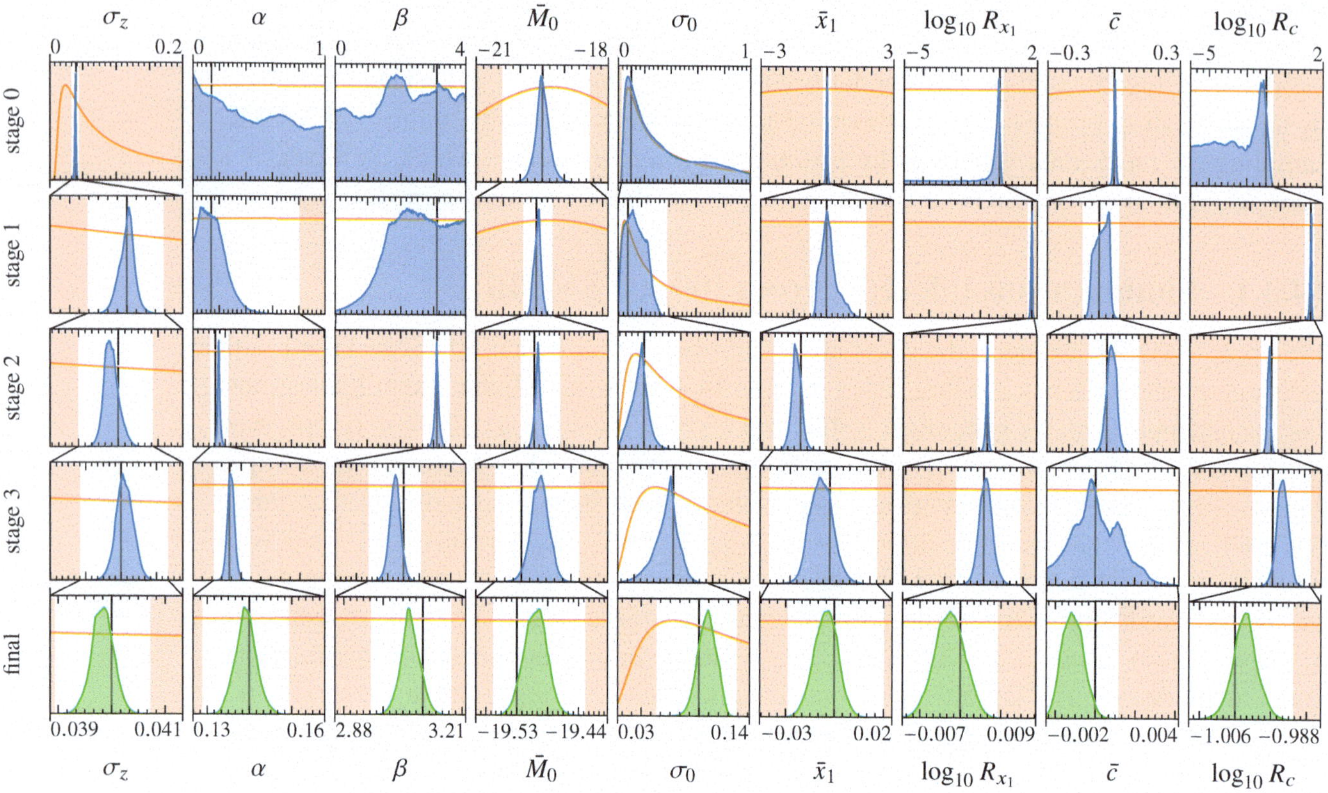

Figure 14.3: Approximate posteriors (blue and finally green filled curves) for the non-cosmological global parameters at sequential stages of truncation (each row is a stage) in the analysis of 10^5 SNæ Ia. The orange line shown the prior density (the same across stages), which gets truncated to the unshaded region (containing 99.99 % of the estimated probability mass) for the following stage. In all plots, the vertical line denotes the value used to produce the target mock observations (from table 14.1), and the approximate posteriors and priors are independently arbitrarily normalised to aid presentation.

The final cosmological posterior we obtain by training with the final truncated priors for all global parameters but only inferring the cosmological parameters in order to utilise the full flexibility of the data pre-processor network. After performing this procedure with all-sized target data, we derive the TMNRE results presented in fig. 14.5, where we also compare with constraints derived by a traditional likelihood-based analysis using a *simplified version* of the model intended to make it tractable for MCMC, relying on linear propagation of redshift uncertainties (eq. (8.13)): see `mphotoz` in subsection 14.3.2. This approximation significantly biases the cosmological results (as we anticipated in the discussion of Eddington bias), with the effect's severity increasing with the sample size. Furthermore, propagating uncertainties only linearly artificially enlarges the expected scatter, which leads to overconfident results (evidenced by the tightly constraining, yet biased, MCMC posteriors in fig. 14.5) when these variations are not present in the data. In contrast, marginal NRE places no restrictions on the model and produces unbiased posteriors with correct uncertainties[289] even from 100 000 SNæ Ia.

Calibrated confidence regions for the cosmological parameters (as described and already demonstrated for dust-population inference in fig. 12.6) are shown in fig. 14.6. We note that, even though the approximation is in some regions conservative and in others under-covering by a few tens per cent (at 68 % nominal credibility), this corresponds to only small inaccuracies in the size of the inferred posterior, which would not affect scientific conclusions and can in any case be calibrated away.

Marginal inference of 100 000 SN-specific parameters is performed after global truncation is finalised, simultaneously with the retraining of the final cosmological posterior. After training the local-parameter $\texttt{RatioEstimator}_{\lambda_g}$, we use simulator samples (the ones used in training) to represent the prior and re-weight them using the neural ratio estimate: see the note in section 4.1. We can then calculate moments or build histograms to represent the marginal NRE posteriors.

We present marginal results for the standardised absolute magnitudes, $\gamma_g \to M_0$, of $N_{\mathrm{SN}} = 10^5$ SN Ia simulated with spectroscopic redshifts (i.e. setting $\sigma_z = 0$) since data with photometric redshift uncertainty is too weakly constraining in the parameters of an individual SN Ia: the marginal posteriors nearly coincide with the constrained effective prior. In this case, as describe below, we can also perform the global inference with MCMC and then use the samples to obtain latent-variable posteriors, which we consider as ground truth. In fig. 14.7, we compare their moments to moments of the marginal NRE posteriors.

[289] We verify the correctness of NRE results with analyses of mock data generated *from* the linearised BAHAMAS (`mphotoz`), for which MCMC produces appropriate (ground-truth) posteriors. In fig. 9 of SICRET, we observe a similar level of similarity between the two methods as we did in fig. 12.4, even from 100 000 SNæ Ia.

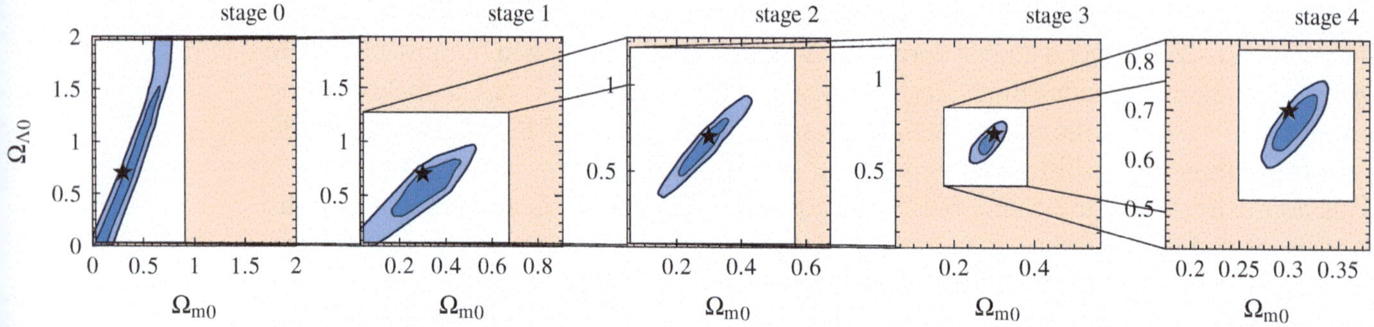

Figure 14.4: Approximate posterior density and 68 % and 95 % HPD contours for the cosmological parameters at sequential stages of truncation in the analysis of 10^5 SNæ Ia. The rectangular boxes delineate the selected range for the following stage, which bounds the irregular HPD region with 99.99 % estimated credibility. The prior used is always flat across the depicted range, and green lines denote the values used to produce the mock data.

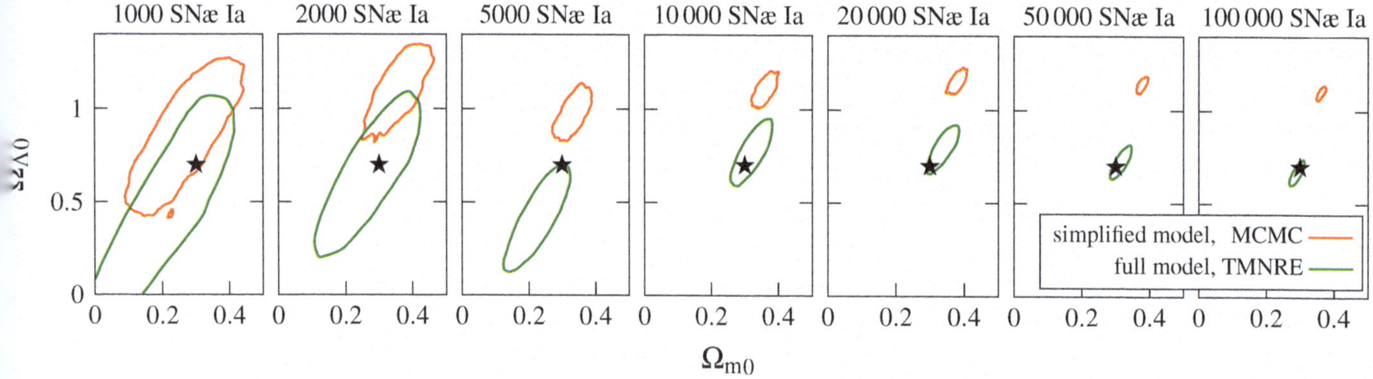

Figure 14.5: Marginal posteriors for cosmological parameters (90 % HPD credible regions) for increasing SN Ia sample size: from TMNRE (with the full generative model) in green and from MCMC (with a linearised model, required to make the problem tractable for sampling methods) in red. A star marks the values used to simulate mock data.

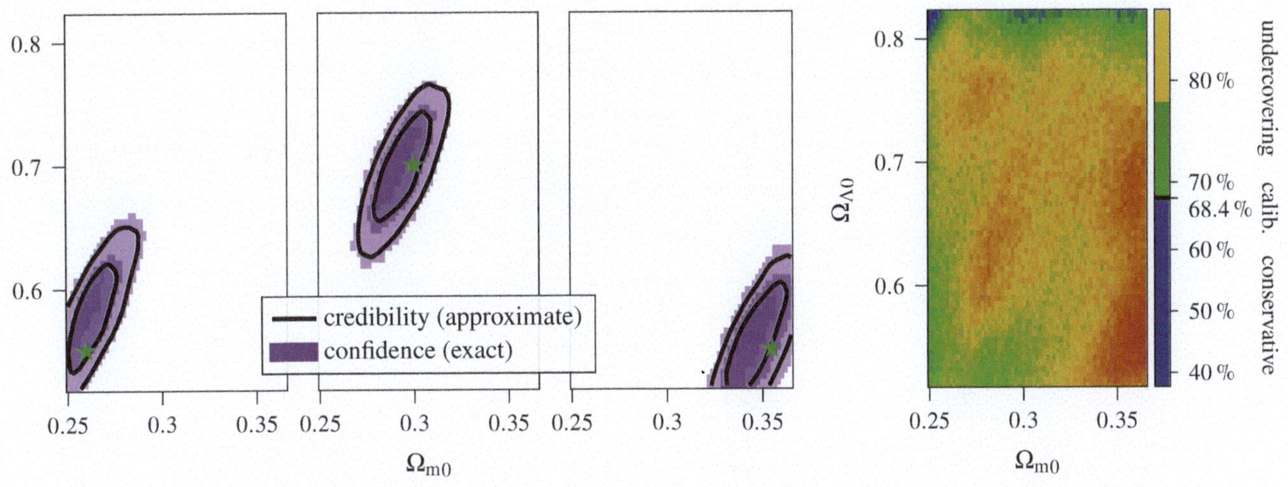

(a) (b)

Figure 14.6: (a) Calibrated constraints from 10^5 mock SNæ Ia with different input cosmological parameters (indicated by a star). Black contours delineate the 68 % and 96 % and 95 per cent (approximate) credible regions from the last stage of TMNRE (whose constrained prior range is the extent of the plots), while the respective calibrated (exact) confidence regions are shown in purple. (b) Threshold nominal credibility* required for 68.4 % confidence: cf. fig. 12.6, where the same quantity is expressed in relative sigmas.

*See SICRET's fig. A1 for its derivation from the calibration plots of *individual pixels*.

NRE learns to correctly identify the location and size of the latent-variable posteriors. There are small deviations from the ground truth for extreme values: where the posterior (and thence, the data) falls in the tail of the prior, and so the network has seen few similar examples during training, and when the posterior is tightly constraining (i.e. has a small standard deviation). In the latter case, the NRE estimate is again conservative.

14.3.2 Validation of scaling properties

In this subsection, we investigate quantitatively the scaling (with N_{SN}) of cosmological constraints derived as above, verifying that it follows the behaviour of traditional analyses. In order to not bias the comparison (i.e. so that we can consider the MCMC posteriors correct unlike in fig. 14.5), we generate mock data from a simplified/linearised model (described fully in SICRET, appendix B [following 347, 479]) that we then use for MCMC inference and training NREs. To this end, we consider two settings:

- `specz`: we imitate spectroscopic redshift estimation by setting $\sigma_z \to 0$; in this case, the layer of SN-specific parameters is analytically exactly marginalisable [347, 479];

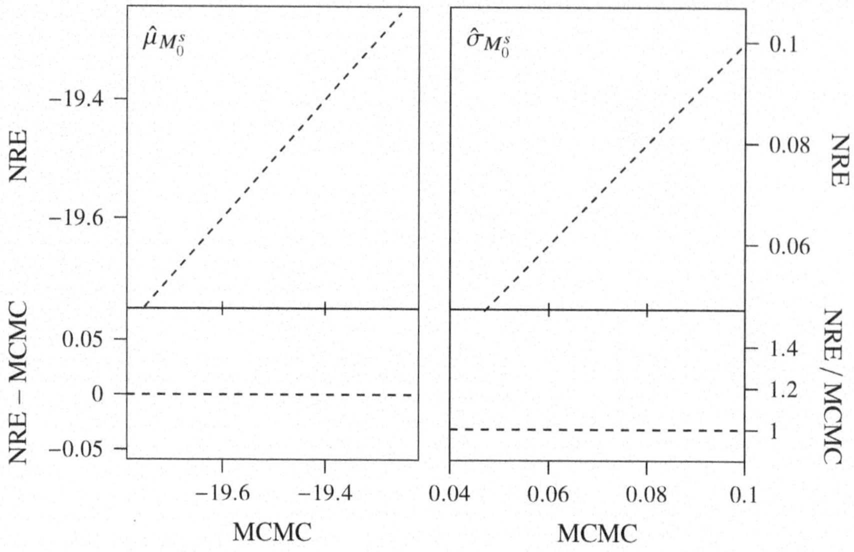

Figure 14.7: Comparison of marginal posterior moments (left: mean, and right: standard deviation) for the standardised absolute magnitude, $\{M_0^s\}$,* derived with NRE and MCMC. The values themselves are plotted in the top row, while the bottom panels show the difference of means and ratio of standard deviations. The analysed mock data contains 10^5 SNæ Ia with spectroscopic redshifts as described in the text. All 10^5 NRE posteriors were produced by one single inference network, whereas the MCMC results required an initial global sampling step and lengthy post-processing (see SICRET, appendix B1). The clustering of posterior variances at specific values is due to the sampling with replacement of the Pantheon observational covariances when generating mock data.

*Since we omit the connection between the global summary $\mathbf{S}$ and the local-parameter ratio estimator, inference of the SN-specific properties is fully permutation-invariant (provided the relevant metadata $\mathbf{a}_{\mathrm{SN}}^s$), so we write them as a {set} rather than an [array].

- `mphotoz`: “photometric-like” redshift uncertainties, modelled *implicitly* as an increase of the distance modulus uncertainty / residual scatter as in eq. (8.13); assuming this, the redshifts are *not introduced* in the first place, and the “remaining” SN-specific parameters can once again be analytically marginalised (see SICRET, appendix B), while σ_z is still an unknown parameter to be inferred / marginalised.

We consider, in turn, increasing sample sizes between $N_{\mathrm{SN}} = 10^3$ and 10^5 SNæ Ia, spanning the range from current to near-future surveys. In each case, we generate 10 data realisations with the same global parameters (from table 14.1) but with random SN-specific properties. We perform the MCMC analysis using the affine-invariant sampler implemented in `emcee` (and interfaced through CLIPPPY and `phytorch`'s auto-differentiation for uncertainty propagation to *the original* probabilistically programmed forward model), sampling all 11 global parameters (10 for `specz`) for 1000 steps (discarding the first 200 as burn-in) with 50 chains. For each sample size N_{SN} and data type (`specz`or `mphotoz`) we then train a neural ratio estimator for the cosmological parameters only. We imitate the last stage of iterative truncation by constraining the prior of each global parameter so that it contains (to within 5σ) the MCMC posteriors from all 10 analysed data sets. Once the inference network is trained, the 10 NRE posteriors can easily be evaluated as above.

The behaviour of NRE's precision and accuracy[290] across the range of SN Ia sample sizes is presented in fig. 14.8, alongside "ground-truth" inference with MCMC We confirm that NRE posteriors are consistently larger than their MCMC counterparts by up to a few tens per cent per parameter. Still, both MCMC and NRE posterior sizes clearly scale as $1/\sqrt{N_{\mathrm{SN}}}$ per parameter, while the offset of the mean is proportional to the standard deviation for both MCMC and NRE analyses, as expected in a purely Gaussian model. This means that the inference network succeeds in combining the information from the large number of observed objects. Overall, NRE achieves comparable accuracy and precision to traditional inference across the sample sizes considered, and we do not observe signs of bias.

Summary and conclusion

We have presented a proof of concept for scalable and model-agnostic Supernova Ia cosmology using TMNRE (SICRET). After demonstrating the systematic bias that can be introduced by oversimplifying aspects of the data-generation process, e.g. uncertainties in measured redshift, we have presented SBI posteriors for cosmological parameters from 10^5 mock SNæ Ia simulated from the BAHAMAS model. Exploiting the local amortisation of TMNRE inference, we have derived regions with exact frequentist confidence, which — in combination with validation of its Bayesian coverage properties and detailed comparison[289] with ground-truth likelihood-based analyses on mock data from a simplified tractable model — strengthen our trust in the inference procedure. Moreover, we have presented a method to simultaneously infer the latent parameters of all 10^5 SNæ Ia with a single small neural network that takes advantage of the truncation of global-parameter prior ranges. We have verified the marginal posteriors we derive for standardised absolute magnitudes against ground-truth MCMC (in the case of spectroscopic redshifts). Finally, we have shown that the inference network is able to extract the relevant information even from a large numbers of SNæ Ia realistically expected of future observational campaigns.

[290] We measure precision through the determinant of the approximate-posterior covariance: size $\equiv \sqrt{|\langle CC^\top \rangle|}$, and accuracy with the offset of the mean from the ground-truth: bias$^2 \equiv |\langle C \rangle - C_{\mathrm{true}}|^2$, where averages are over $q(C \mid \mathbf{d})$. Even though these two quantities have the same dimensions for our 2D cosmological posteriors, we caution against comparing them directly since the former represents a volume in parameter space, while the latter is the square of a length.

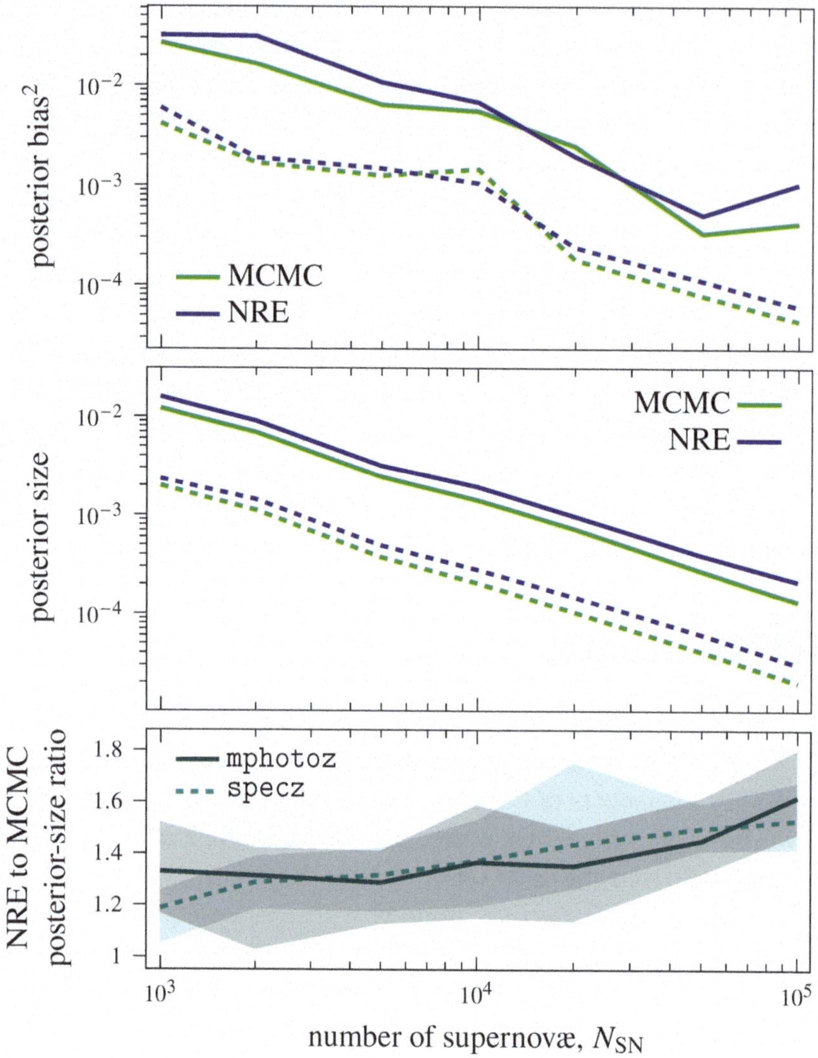

Figure 14.8: Comparison of the accuracy (top) and precision[290] of NRE (blue) and MCMC (green) posteriors for data of different sizes generated and analysed with the linearised marginal BAHAMAS. The units in the top two panels are $[C]^2$, and the general trend is $\propto 1/N_{SN}$ (i.e. $1/\sqrt{N_{SN}}$ *per dimension* in the 2D space of cosmological parameters), as expected from Gaussian theory. The bottom panel shows the ratio of posterior sizes from the two methods (i.e. of the respective lines in the middle panel). Everywhere solid lines pertain to `mphotoz`data/analyses and dashed lines to `specz`. All plots have been averaged over 10 data realisations, and in the bottom panel, the shaded areas additionally show the range of values across the different mock data sets.

Chapter 15

Ratio Estimation for SN selection Effects (now you detect it, now you don't)

RESSET

RESSET[291] presents the last substantial methodological advance required for the application of SBI to *realistic* cosmological SN Ia samples: the ability to account for sample selection (and contamination, although we do not explicitly demonstrate this here). In it, we address, on one hand, the arbitrariness of the "state-of-the-art" bias-correction procedure, which can lead—in the best case—to deceptive confidence and otherwise to wildly biased results; on the other hand, we overcome the issues that plague *any* hierarchical likelihood-based inference strategy: namely, the requirement for its probabilistic exlicitation and for joint sampling of an exponentially growing parameter space—even if one is ultimately interested in $\mathcal{O}(1)$ cosmological parameters.

STAR NRE

To this end, we introduce *set-based truncated autoregressive neural ratio estimation (STAR NRE)*, a simulation-based approach that makes use of a conditioned deep set NN and combines efficient high-dimensional global inference with sub-sampling-based truncation in order to scale to very large survey sizes while training on sets with stochastic cardinality. Applying it to a simplified SN Ia model that consists of standardised brightnesses and redshifts with Gaussian uncertainties and a selection procedure based on the expected LSST sensitivity, we demonstrate precise and unbiased inference of cosmological parameters and the redshift evolution of the volumetric SN Ia rate from ≈100 000 mock SNæ Ia. Our method bypasses the latent layer and delivers marginal results, imposing no restrictions on the simulator's output size (in fact, it naturally extracts useful information from it) and the nature of individual objects. It can thus handle an arbitrarily complicated selection and classification procedure and be applied to complex data like light curves in the future.

[291] which stands, alternatively, for "retribution for the set!" after the initial ~~failed~~ sub-optimal[268] application of the *un*conditioned deep set architecture to the fixed-structure nested-"sets" data analysed in SIDE-real

© The Author(s), under exclusive license to Springer Nature Switzerland AG 2026
K. Karchev, *Supernova Cosmology for the 21st Century*,
https://doi.org/10.1007/978-3-032-15072-1_15

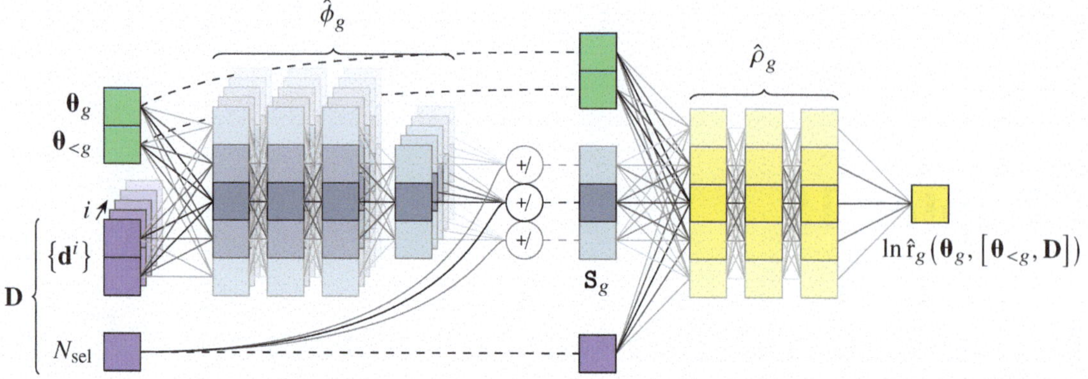

Figure 15.1: Conditioned deep set-based ARNRE implementing eq. (15.1) with MLPs.

15.1 Conditioned deep set: the little NN that could

As we already established, selection effects are most straight*forward*ly accounted for by a simulator that produces faithful mock catalogues of *selected* objects. This requires, on one hand, treating their count as explicitly informative and, on the other, being able to learn from simulated collections of different sizes, as we depict in fig. 4.3. Our solution for both, inspired[119] by the conditional structure of the forward model itself, is the conditioned deep set: a minimal NN architecture that can be implemented with standard fixed-input–output-size components as we described schematically in eq. (4.13).

In this section, we describe the particular implementation that we use in RESSET, which is a slight extension to eq. (4.13) for use in the autoregressive framework we describe shortly. Each autoregressive global-parameter[292] neural ratio estimator (cf. eq. (2.17)) is implemented as a *separate* conditioned deep set: since the featuriser takes the parameters as input, we cannot reuse the summary across groups g as before. Moreover, in the interest of numerical performance, instead of summing, we average the featurised set elements and append the cardinality to the output.[293] The final NN structure, depicted in fig. 15.1, is thus

$$\ln \hat{r}_g\big(\boldsymbol{\theta}_g, [\boldsymbol{\theta}_{<g}, \mathbf{D}]\big) = \hat{\rho}_g\big(\boldsymbol{\theta}_g, \boldsymbol{\theta}_{<g}, \mathbf{S}_g, N_{\rm sel}\big) \quad \text{with} \quad \mathbf{S}_g \equiv \frac{1}{N_{\rm sel}} \sum_{\mathbf{d}\in\mathbf{D}} \hat{\phi}_g\big(\boldsymbol{\theta}_g, \mathbf{d}\big). \tag{15.1}$$

As per tradition, all featurisers, $\hat{\phi}_g$, and post-processors, $\hat{\rho}_g$, are implemented as MLPs with three layers of 128 or 256 neurons, respectively, and ReLU non-linearities preceded

[292] We do not consider SN-specific parameter inference… yet, so we will use $\boldsymbol{\theta}_g$ to mean $\boldsymbol{\gamma}_g$.

[293] It is good practice to keep inputs to fully connected NN layers close to zero with order-unity scatter, which is the typical range of nonlinearities. However, the output of the summation operator trivially scales with the cardinality of the input set, so we artificially extract this source of variations.

by layer normalisation[294] [28]. For added expressivity, parameters passed to $\hat{\rho}_g$ are first embedded in 32 dimensions by a small MLP with two hidden layers of 64 neurons.

15.2 STAR NRE and fuzzy business

Finally, we present a strategy that combines the scalability and flexibility of conditioned deep set NNs with the ARNRE formulation,[295] which allows high-dimensional sequential truncation and thus alleviates the burden of simulation-based inference with "big data" both in terms of training time and memory required for storing and processing simulations. The scheme relies on the intuition that inference from one small subset of the data is, in general, less constraining than a full analysis, yet unbiased if the sub-sampling is random—much like intermediate results during truncation. Therefore, we introduce the similar in spirit "sub-sampling stages", in each of which we train with increasingly bigger sets before reaching the size of the target data.

Strategy 4 (STAR NRE)**.** When analysing a given data set of N_{sel} objects with a simulator tuned to generate total populations of size $N_{\text{tot}} \sim \text{p}(N_{\text{tot}} \mid \boldsymbol{\gamma})$ (resulting in "selected" samples with $N_{\text{sel}} \sim \text{p}(N_{\text{sel}} \mid \boldsymbol{\gamma})$ from eq. (4.6)), we initially simulate much smaller example/mock sets $\mathfrak{D}$ of *nominal* size $\approx N_{\text{sel}}/k$, either by randomly sub-sampling[296] $\mathbf{D}$ or tuning the simulator so as to generate linearly smaller populations, which can often be achieved through a setting[297] that modifies *random subset*

$$\langle N_{\text{tot}} \rangle(\boldsymbol{\gamma}) \to \langle N_{\text{tot}} \rangle(\boldsymbol{\gamma})/k \tag{15.2}$$

(cf. eq. (4.7)); in this case, it is important that the setting controls only the total number of objects and not (the distribution of) their properties.

[294] The use of layer- instead of batch normalisation is beneficial for two reasons: first, it is slightly faster since it does not need to track running statistics; and second, because it does not depend on the input data, we can reuse network components across training tasks, i.e. truncation and cardinality stages. Still, we perform one-off data-set "whitening" by shifting and re-scaling all NN inputs (parameters and data) by their respective means and standard deviations from the training set.

[295] The autoregressive nature of the ratio estimator is not important for the present discussion: it is implemented so as to enable precise *joint* constraints for the global parameters of interest in the interest of truncating *nuisance* parameters, which was also the original motivation for its introduction by Anau Montel et al. [14]. Further discussion on joint *vs.* marginal inference from (sub)sets (which refers to the benefits of ensuring conditional independence) can be found in RESSET, appendix B.1.

[296] in fact; partitioning, by assigning each element of $\mathbf{D}$ to a random subset $\mathfrak{D}$ with equal probability, which effectively transforms a small number of large simulations into a larger number of smaller examples

[297] For astronomical transient surveys, good candidates are the sky area covered and the survey duration, which we introduce in eq. (15.7); in fact, their product forms the proportionality constant in eq. (8.14), which is the actual handle we use.

Once we have generated a… set of examples $\{\mathfrak{D}_i\}_{i=1}^{N_{\text{train}}}$, we proceed to train on them a conditioned deep set-based ARNRE as usual and evaluate it,[298] independently, on k subsets forming a partition of the original data:

$$\mathbf{D}_\mathrm{o} \to \left\{\mathfrak{D}_{\mathrm{o},j}\right\}_{j=1}^{k} \quad \text{with} \quad \cup_j \mathfrak{D}_{\mathrm{o},j} = \mathbf{D}_\mathrm{o} \quad \text{and} \quad \mathfrak{D}_{\mathrm{o},j} \cap \mathfrak{D}_{\mathrm{o},i\neq j} = \emptyset. \tag{15.3}$$

fuzzy business

This results in k "sub-posteriors" (posterior-to-marginal-prior ratio estimates multiplied by $\prod_g \mathrm{p}(\boldsymbol{\theta}_g)$ as in eq. (2.17)), which we *average* into a *"fuzzy"* joint posterior estimate

$$\sum_{j=1}^{k} \frac{\mathrm{p}(\boldsymbol{\theta} \,|\, \mathfrak{D}_{\mathrm{o},j})}{k} \approx \frac{1}{k} \sum_{j=1}^{k} \prod_g \hat{\mathrm{r}}_g\left(\boldsymbol{\theta}_g, \left[\boldsymbol{\theta}_{<g}, \mathfrak{D}_{\mathrm{o},j}\right]\right) \mathrm{p}(\boldsymbol{\theta}_g), \tag{15.4}$$

which is naturally wider than $\mathrm{p}(\boldsymbol{\theta} \,|\, \mathbf{D}_\mathrm{o}) \overset{?[299]}{=} \prod_j \mathrm{p}(\boldsymbol{\theta} \,|\, \mathfrak{D}_{\mathrm{o},j})$ and rough (due to sub-sampling variance), hence "fuzzy". We use its contours only to define a truncation region (see eqs. (2.25) and (2.26))[300] and then re-train the ratio estimator on constrained samples, initialising it with the trained network parameters from the previous stage. If the constraints using the current k do not improve significantly, we switch to a *smaller* k, i.e. to generating bigger training sets. In this case, again, we re-use the previous NN parameters. As a result of this iteration between restricting the parameter space from which examples are simulated (i.e. the usual truncation scheme) and increasing the size of example data sets used for training, the two components of the deep set network—the featuriser and post-processor—are quickly and inexpensively pre-trained on small simulations and only fine-tuned on bigger more computationally and memory-intensive ones. We note that the final results are always derived from a network that is trained and evaluated on full-size (not sub-sampled) data.

15.3 Proper modelling of SNæ Ia for cosmology

This section presents a simplified model for cosmological inference from SNæ Ia in the presence of selection effects. For a change,[301] we consider the possibility of evolving dark

[298] Deriving joint posteriors from an ARNRE in practice requires MCMC since re-weighting samples or evaluating the ratio estimator on a grid—the strategies we relied on in chapters 12 and 14—are not appropriate for the high-dimensional setting that ARNRE usually targets. Instead, we resort to CLIPPPY's `emcee` integration.

[299] This equality is only valid if the individual objects—and hence the subsets $\{\mathfrak{D}_j\}$ are independent, conditionally on $\boldsymbol{\theta}$ cf. footnote 295.

[300] The procedure of "fuzzy" truncation is fully elaborated in RESSET, appendix B: see especially fig. 9 and note the further complication related to sampling from the truncated *prior*, for which we once again resort to `emcee`, with a crude bound estimate implemented through `dynesty`.

[301] It does away with the ~~first~~second-order degeneracy seen in fig. 14.4 and makes the biases introduced by unprincipled selection-effects modelling clearer to illustrate; besides, it represents a scientifically more interesting question that goes beyond simply retrieving precise values within a given parametrisation of nature.

Table 15.1: SN Ia parameters, (hierarchical) priors and values used to generate mock data in SIDE-real. Notice that unlike tables 12.1 and 14.1, there are no redshift-related *inputs* (metadata). See also fig. 15.2 for a (directed) graphical representation of the model.

parameter		(hyper) prior	mock value / range	
DM density ($z_c = 0$)	Ω_{m0}	$\mathcal{U}(0, 1)$	0.3	
DE density ($z_c = 0$)	Ω_{de0}	$1 - \Omega_{m0}$	0.7	
DE EOS	w_{de}	$\mathcal{U}(-2, -0.5)$	−1	
SN Ia rate at $z_c = 0$	R_0	$\mathcal{N}\left(\begin{bmatrix}2.5\\1.5\end{bmatrix}, \begin{bmatrix}0.5^2 & -0.24\\-0.24 & 0.6^2\end{bmatrix}\right)$	2.5	$\times 10^{-5}\,\mathrm{h}_{70}^3/\mathrm{Mpc}^3/\mathrm{yr}$
low-z rate exponent	β		1.5	
high-z rate exponent	γ	fixed	−0.5	
rate "break"	z_{break}	fixed	1	
true redshift	z_c^s	$\mathrm{Pois}(\mathrm{d}\langle N_{tot}\rangle/\mathrm{d}z_c)$	$\in [0; \infty)^{\otimes N_{tot}}$	
redshift estimate	$\widehat{z_c}^s$	$\mathcal{N}\left(z_c^s, (1 + z_c^s)^2 \sigma_z^2\right)$	$\in [0; \infty)^{\otimes N_{tot}}$	
redshift uncertainty	σ_z	$\mathcal{U}(0, 0.06)$	0.04	
observed magnitude	$\widehat{m}^s$	$\mathcal{N}(M_0 + \mu^s, \sigma_0^2)$	$\in (-\infty; \infty)^{\otimes N_{tot}}$	
mean abs. mag.	M_0	$\mathcal{U}(-20, -19)$	−19.5	
mag. scatter / noise	σ_0	$\mathcal{U}(0, 0.2)$	0.1	
detection indicator	S^s	$\mathrm{p}(S^s \mid z_c^s, \widehat{m}^s)$	$\in \{\mathfrak{m}, \mathfrak{s}\}^{\otimes N_{tot}}$	

energy: a cosmic fluid with constant equation of state (EOS) $w_0 \neq 1$ (in general), but constrain the spatial curvature to $\Omega_{k0} = 0 \implies \Omega_{de0} = 1 - \Omega_{m0}$. Beyond that, the forward model we use here (depicted graphically in fig. 15.2 and detailed in table 15.1) is an extension of our first demonstration from chapter 10—we adopt the same data-sampling distribution for (photometric-like) uncertain *cosmological* redshift estimates and standardised brightnesses / distance moduli[302] from eqs. (10.15) and (10.16)—in two respects: we present a more involved description of the SN Ia rate—which gives rise to the redshift distribution—and, titularly, a nontrivial selection procedure.

[302] After SICRET,[278] we return to the original formulation from eq. (8.9), in which σ_0 is realised as observational noise on $M_0 + \mu^s$ (which RESSET writes as $\bar{M} + \mu^s$) rather than a population scatter of M_0^s (eq. (14.1)).

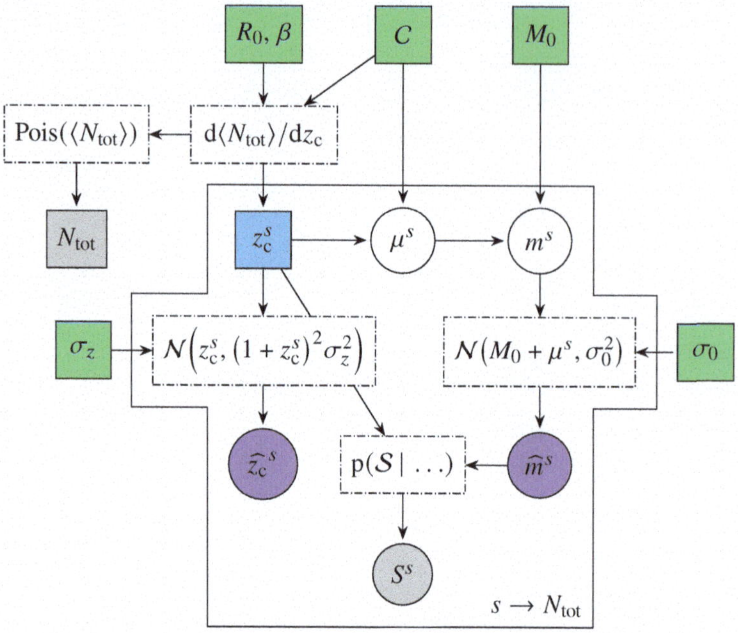

Figure 15.2: A simple BHM for SN Ia cosmology with sample selection. This is the forward formulation, which represents the total population and is easy to implement as a simulator (cf. fig. 4.3). Considering only "selected" SNæ Ia ($S^s \to \mathfrak{s}$) re-configures it so that selection effects are implicitly accounted for.

15.3.1 Volumetric SN Ia rate

survey volume = sky area × duration

Following from eq. (8.14), the expected redshift distribution[303] of *all* (including undetected) transients of a given type within the *surveyed volume*,[297] ΩT, where Ω is the survey *sky area* (in steradian or deg^2), and T is the survey *duration* (in Earth years), is:

$$\frac{\mathrm{d}\langle N_{\mathrm{tot}}\rangle}{\mathrm{d}z_c} = \Omega T \times \frac{R(z_c)}{1+z_c}\frac{\partial V_c(z_c, C)}{\partial z_c} \tag{15.5}$$

and depends on the cosmological model. For SNæ Ia, we use a comoving volumetric rate as in PLASTICC (table 2):

$$R(z_c) = R_0 \times \begin{cases} (1+z_c)^{\beta}, & z_c \leq z_{\mathrm{break}}; \\ (1+z_c)^{\gamma}(1+z_{\mathrm{break}})^{\beta-\gamma}, & z_c > z_{\mathrm{break}}; \end{cases} \tag{15.6}$$

with slope $\gamma = -0.5$ above $z_{\mathrm{break}} = 1$ [231, after 454, 182]. In practice, the high-redshift rate will not influence our analysis since LSST is not expected to detect SNæ Ia above z_{break}.

[303] technically, the rate function of an inhomogeneous Poisson process

At low redshift, we use the estimates of Dilday et al. [128][304] from SDSS:

$$\begin{aligned} R_0 &= (2.5 \pm 0.5) \times 10^{-5}\, \mathrm{h}_{70}^3/\mathrm{Mpc}^3/\mathrm{yr},\ ^{305} \\ \beta &= 1.5 \pm 0.6, \\ \text{with } \operatorname{corr}(R_0, \beta) &= -0.8 \end{aligned} \tag{15.7}$$

as a 2-dimensional correlated Gaussian prior. The rate parameters will be much better constrained with future large samples, as we show in section 15.4—assuming the general functional form of eq. (15.7) is correct. In chapter 16, we will finally transcend this arbitrariness and tie the rate of SN Ia occurrence with the evolution of galaxies and stellar populations.

Given particular values for R_0, β, and cosmology, one can integrate[306] eq. (15.7) to derive the expected size of the "total"[306] population, obtain a realisation $N_{\mathrm{tot}} \sim \operatorname{Pois}(\langle N_{\mathrm{tot}} \rangle)$ (in accordance with eq. (4.7)) and then sample that many redshifts in proportion to eq. (8.14). After simulating redshift and magnitude measurements according to the sampling distributions in table 15.1/fig. 15.2, the total population is fully fleshed out, and it only remains to determine the collection of *selected* objects, as we describe next.

15.3.2 LSST-like selection probability

In practice, a SN Ia is detected using difference imaging and selected for inclusion in cosmological analyses based on the quality of its light curve (number of observations in different bands pre/post peak and quality of a fit with e.g. SALT). Since we are building a toy model that only includes the SN's redshift and apparent brightness at peak, our detection criterion will be correspondingly simplified. For the probability of detection & selection (which we call *selection* for short), we adopt a simple criterion based on the expected LSST `fiveSigmaDepth` for the band in which the SN peaks based on its redshift: at $z_c = 0$, the peak of SNæ Ia is generally in the blue part of the spectrum; we assume at 4385 Å: the effective wavelength of the B band. The `fiveSigmaDepth` depends on a variety of factors [124, 552] and in simulations shows significant variations; therefore, for each supernova, we compare $\widehat{m}^s$ to a random simulated `fiveSigmaDepth`;[307] and if $\widehat{m}^s$ is lower (brighter),

[304] Dilday et al. [128, subsection 6.4.1] give $2.6^{+0.6}_{-0.5}$, but we use 2.5 as in PLASTICC and a symmetric uncertainty for simplicity.

[305] Notice the scaling by $\mathrm{h}_{70} \equiv \mathrm{H}_0/70\,\mathrm{km/s/Mpc}$, which makes predictions (inference) about total number counts, which depend on H_0, independent of the particular value used in simulations (analyses).

[306] up to a sufficiently high upper bound that under no *a priori* allowed γ, a SN can be conceivably detected with a higher redshift; we integrate up to $z_c = 2$

[307] We use the `baseline_v2` run of the LSST's `rubin_sim` [553], but the survey details do not affect this simplified model.

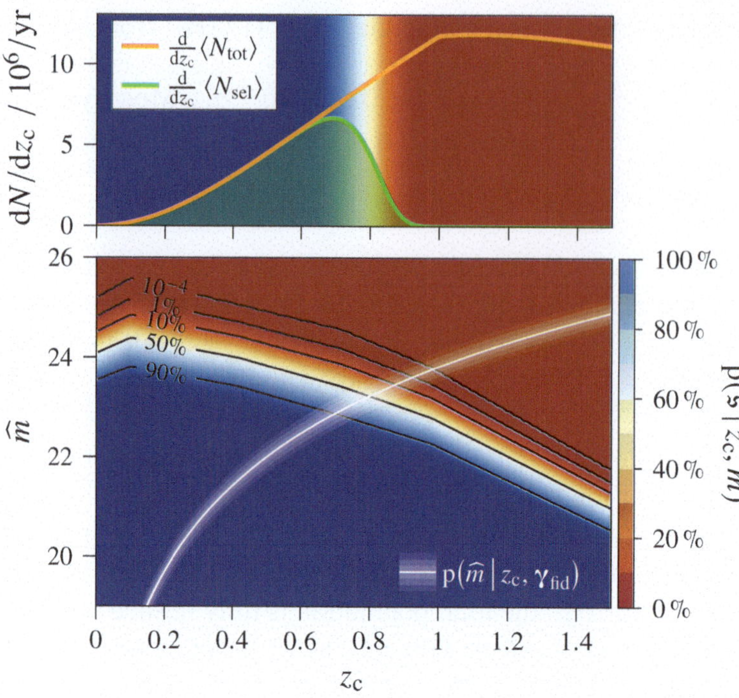

Figure 15.3: *Bottom:* The selection efficiency adopted in this work (colour axis) as a function of the SN's *true* redshift and measured magnitude with some threshold contours as labelled. A white line indicates the Hubble diagram under the fiducial cosmological model, with up to $3\sigma_0$ uncertainty/scatter around it. *Top:* Number (density per unit redshift) of SNæ Ia under the fiducial cosmological and rate models as a function of their *true* redshift: expected total number in orange and expected selected count in green. The backdrop depicts their ratio $\mathrm{p}(\mathfrak{s} \,|\, z_\mathrm{c}, \boldsymbol{\gamma}_\mathrm{fid}) \equiv \int \mathrm{p}(\mathfrak{s} \,|\, z_\mathrm{c}, \widehat{m})\, \mathrm{p}(\widehat{m} \,|\, z_\mathrm{c}, \boldsymbol{\gamma}_\mathrm{fid})\, \mathrm{d}\widehat{m}$, i.e. the average along all $\widehat{m}$ of the bottom plot weighted by the white shaded region.

the SN is selected. We thus assume, for the purposes of this toy model, that while the wavelength/band of the peak changes, its magnitude does not; i.e. we ignore the necessity for K-corrections.

Due to the stochasticity of the depth simulator (which reflects a variety of observational conditions), it defines a complicated selection probability, $\mathrm{p}(\mathfrak{s}^i \,|\, z_\mathrm{c}, \widehat{m})$ that depends on $\widehat{m}$ and z_c—notice, the *true* redshift rather than the noisy estimate $\widehat{z_\mathrm{c}}$. We illustrate it, in comparison with the fiducial population of SNæ Ia, in fig. 15.3. Thus, in general, $\mathrm{p}(\mathfrak{s}^s \,|\, \widehat{z_\mathrm{c}}^s, \widehat{m}^s) \neq 1$, even for data on objects that ended up being selected; instead, calculating it requires averaging $\mathrm{p}(\mathfrak{s}^s \,|\, z_\mathrm{c}^s, \widehat{m}^s)$ over the *posterior* $\mathrm{p}(z_\mathrm{c} \,|\, \{\widehat{z_\mathrm{c}}^s, \widehat{m}^s\})$, which makes it practically intractable even in this extremely simplified scenario because of the significant redshift uncertainty assumed. However, the selection procedure can be easily realised in a forward simulator: given a collection of (true) redshifts and magnitudes generated as described above, each object is stochastically detected/selected with probability $\mathrm{p}(\mathfrak{s}^s \,|\, z_\mathrm{c}^s, \widehat{m}^s)$; after all, the simulator always has access to the necessary latent variables.

For real observations, the selection procedure will be far more complex [see e.g. 303, subsection 11.2.1], based on criteria like number of detections in a variety of bands, efficiency of obtaining reliable redshift estimates, correctly classifying transients using machine-learning models, etc., and will be only representable through simulations: the essential impediment to likelihood-based techniques, which our simplified model already exhibits and which we naturally overcome through stochastic-cardinality SBI.

15.3.3 Mock data

To demonstrate our inference procedure, we generate a mock data set $\mathbf{D}_\mathrm{o}$ from the model described above with parameter values as listed in table 15.1. We adjust[297] the survey volume[308] so that 100 000 SNæ Ia are *expected* to be detected under fiducial values; for our particular stochastic realisation, the precise number is $N_\mathrm{sel} = 105\,287$.

This mock data set was depicted in the illustration of Malmquist and Eddington biases in fig. 8.4. We remind the reader that due to the latter—which arises from the combination of a varying rate (non-constant redshift distribution) and significant redshift-estimation uncertainties—not even the complete (total) population follows the cosmological model (i.e. the white line in fig. 15.3) when binned according to "observed"[309] redshifts $\hat{z}^s$, leading to a significant bias in *naïve*—or *naïvely* bias corrected—fits when the data is constraining enough, as we demonstrated in fig. 8.6. *naïveté*

15.4 Unbiased results

Following STAR NRE,

1.1. we start training with examples from 1/50 of the total survey size (i.e. we adjust ΩT so that $\langle N_\mathrm{obs} \,|\, \boldsymbol{\Upsilon}_\mathrm{fid} \rangle \approx 2000$),
1.2. truncate the parameter space, preserving only values consistent with a "fuzzy" posterior (eq. (15.4)) formed from analysing 50 disjoint subsets (eq. (15.3)) of the mock data $\mathbf{D}_\mathrm{o}$,
1.3. and fine-tune the network on new, targeted examples.

---

[308] $\Omega T \approx 1600\,\mathrm{deg}^2 \cdot \mathrm{yr}$, although the correspondence $\Omega T \leftrightarrow N_\mathrm{sel}$ from this toy example will not hold for more complicated selection procedures; for comparison, the LSST will cover roughly 20 000 deg^2, a smaller fraction of the SN population will pass the selection criteria employed in practice.

[309] In fact, the concept itself of an "observed redshift" is dependent on the assumption of Gaussianity, whose inapplicability we already brought up (repeatedly).

2. Then, we increase the survey size tenfold ($\langle N_{\text{obs}} \mid \gamma_{\text{fid}} \rangle \approx 20\,000$) and repeat, *starting with the previously trained network.*
3. Finally, we repeat using the full survey size, this time doing one extra step of truncation to ensure properly converged results.

For each of these (7) stages, we generate 64 000 training examples (with 6400 more used for validation), which takes <30 min, and train on 4 (8 for the last stages) NVIDIA A100 GPUs using the Adam optimiser [289] until convergence, usually achieved withing 3 h (per stage).

The results from the final stage for each survey size (cardinality stage) are shown in fig. 15.4. As expected, they are progressively more concentrated around the parameters from which the mock data were generated. To appraise the amount of information extracted by our inference procedure (i.e. the strength of the constraints it produces), in the inset of fig. 15.4, we compare the marginal posterior for the cosmological parameters from the final stage of STAR NRE to alternative approaches:

- On one hand, we perform fully hierarchical Bayesian analysis (with latent redshifts marginalised numerically via integration on a grid) of *complete* data, i.e. that has no magnitude-based selection and all SNæ Ia up to a given cutoff *true* redshift are included.[310] While complete data is naturally more constraining due to the additional presence of high-redshift objects and correspondingly higher statistics, such an analysis would in practice be limited to a low redshift of about $z_{\text{c}} \approx 0.6$ (in our setup: see fig. 15.3) with the infeasible requirement of selecting the objects based on *true* redshift (so that the analysis is computationally tractable). STAR NRE manages to extract information beyond the completeness limit, effectively correcting the selection bias and delivering constraints comparable to those from the similarly sized complete data (up to $z_{\text{c}} \approx 0.8$, where the selection probability drops to 50 %).
- We also compare our result with a traditional χ^2 fit to "de-biased" selected data—importantly, with the fiducial model used for calculating bias corrections matching the one from which the mock data were simulated—and analysed with redshift uncertainties propagated linearly to magnitudes, again assuming the fiducial model and *disregarding* the intrinsic SN Ia rate. Even if the resulting constraints appear more stringent that ours, this is largely due to the optimal choice (made before analysing the data) of a fiducial cosmology close to the true model, which, as we demonstrated in fig. 8.6, is indispensable for obtaining systematically unbiased posteriors even for the simplest SN model such as ours. In reality, the selection procedures and bias corrections are far more complicated, making bias-corrected fits extremely prone to systematic bias.

[310] This true-redshift cutoff is still a kind of selection criterion, but—as opposed to a cutoff in *measured* redshift—the correction associated with it is a simple re-normalisation of the redshift prior, which we account for by calculating the size of the population below the cutoff and including it in the hierarchical likelihood.

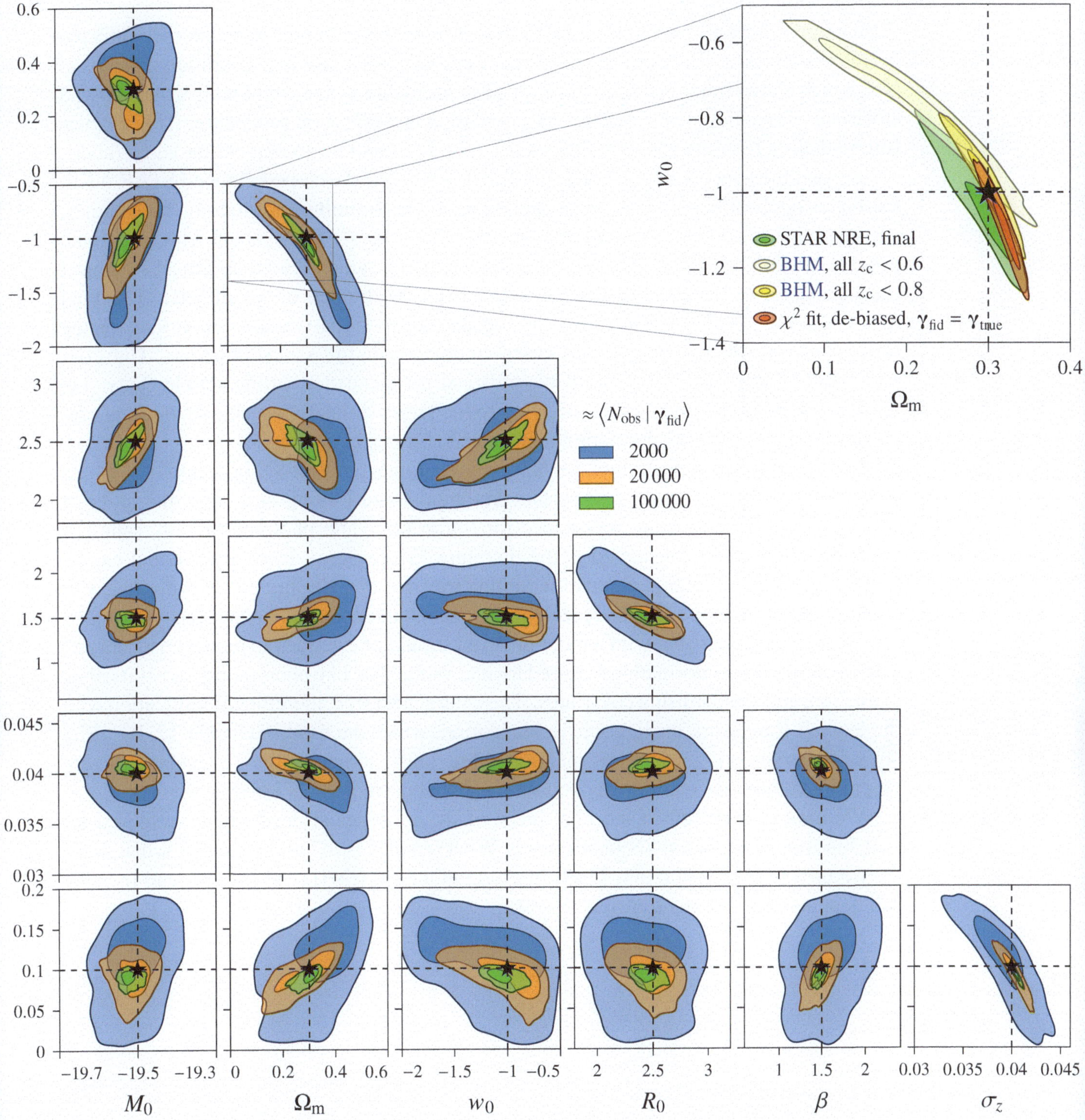

Figure 15.4: Two-dimensional projections of the joint 7-dimensional "fuzzy" posteriors (39 and 86 % HPD credibility) in successive cardinality stages (in different colours as indicated in the legend). *Inset:* comparison of our final result with alternative likelihood-based analyses of complete sub-samples as well as the full selected mock data de-biased with fiducial parameters set to their true values.

Conclusion

We have presented a simulation-based analysis framework to handle large data sets affected by selection biases. The cornerstone of our approach is the use of a conditioned deep set NN, which allows training data to have varying cardinality as a result of realistic end-to-end stochastic simulation of the underlying population and selection procedure. This enables full modelling freedom and straightforward inference contrary to non-set-based methods.

Beyond the network architecture, the strategy we have developed, dubbed STAR NRE, facilitates applications to very large catalogues by pre-training on random sub-samples, whose size is gradually increased in "sub-sampling stages", interspersed with traditional prior truncation. This benefits the deep-set featuriser, which learns an object-by-object transformation of the individual data and can thus be trained on small sets. The final result, however, is derived after fine-tuning the network (particularly important for the post-processor, which aggregates *all* observed objects) on full-size simulations. Thus, our methodology imposes no condition of statistical independence on the set elements: i.e. allows even marginal inference, and owing to the target simulations achieves optimal precision all while implicitly correcting any conceivable selection biases.

For our current demonstration, we have used a simple model of SN Ia cosmology, which nevertheless, still captures the important Malmquist and Eddington biases, the latter of which will be of increasing importance for future large and photometric-only samples, yet is currently rarely considered in the literature. Moreover, it incorporates (and can be used to extract information about) the volumetric rate of SNæ Ia and its evolution with redshift, which will be the centrepiece of the next chapter, where the analysis will follow the exact same strategy but with a more involved simulator extended to model host data and extract all auxiliary information required for analysing SNæ like photometric redshifts, dust extinction, and standardising covariates. The natural next step, therefore, is unification with the explicit framework for light-curve simulation, which we used in chapter 12, by using any of the existing LC featurisers mentioned previously to handle their irregular structure before input into the conditioned deep set for global cosmological inference.

We have thus presented the final piece of methodological development required for scalable simulation-based SN Ia inference. What remains is to build a simulator that faithfully represents our vision of a The future: grand unified SN (Ia) cosmology and reap the full benefits of upcoming transient surveys.

Chapter 16

One with the environment: combined inference of galaxy redshift and SNæ

CIGaRS [271, published post-submission] is a proof of the concept of unified inference of SN Ia and host properties in an interconnected Bayesian hierarchical model as in fig. 8.7. This chapter presents its initial implementation, which combines an extension of our SICRET description of SALT-like SN Ia standardisation with the `Prospector` [313, 536] full spectral model of galaxies via two "interaction" channels: self-consistent dust modelling via Simple-BayeSN to transfer the physically motivated galaxy extinction onto the SN Ia colour; and a rate of SN Ia occurrence tied to the host's stellar population via a phenomenological (parametrised) delay-time distribution (DTD) indicative of the progenitor scenario(s).

The observables we consider are meant to represent data from a photometric-only survey like the LSST: on one side, we will assume noisy estimates of the SN Ia[311] light-curve parameters, as if derived from independent fits (or—better—by a bespoke LC-summariser NN or estimated *marginally* as in chapters 12 and 14); and, on the side of the[312] hosts, only broadband photometry in the LSST *ugrizy* filters. Notice: *no explicit/external redshift estimates*; instead, we directly model the data from which they are typically[313] extracted.

SN Ia cosmology without explicit redshifts

Our goal in this setting will be twofold: inferring the cosmological parameters and constraining the DTD, whereby we target jointly the constituents and evolution of our Universe and the astrophysics of stellar populations. To this end, we will once again employ S(T)AR NRE, exactly as described in chapter 15.

---

[311] Again, non-Ia contamination can easily be included, provided an explicit model for its observables.

[312] It will be interesting to consider also host (mis-)association, which will be easy to realise in the forward model by simply associating the probable host(s)'s photometry with the particular SN's LC/summaries.

[313] In the present implementation, this is only the host observables, which we (and the community in general) have determined provide superior photo-z constraints—when available; but a fuller analysis will be able to extract redshift information also from the SNæ's light curves, e.g. with Photo-zSNthesis or traditional fitters.

© The Author(s), under exclusive license to Springer Nature Switzerland AG 2026
K. Karchev, *Supernova Cosmology for the 21st Century*,
https://doi.org/10.1007/978-3-032-15072-1_16

16.1 Properer modelling of SNæ Ia and their hosts for cosmology

causality

coincidence

A grand unified description of SN Ia cosmology (e.g. fig. 8.7) is a diptych formed by the largely independent representations of host and transient data, joined by a *causal connection* relating the occurrence and intrinsic properties of SNæ to the stellar populations from which they originate, as well as by the *coincidence* of extrinsic effects to their light: most prominently, dust extinction. Below, we present a toy model that incorporates schematically these connections and follows them through to (mock) observables (save for the dependence of the SN population properties on the host, which would include e.g. evolution and local-environment modelling and inter-host localisation).

16.1.1 Galaxies and their photometry

utility

Comprehen~~ding~~sive galaxy modelling [366] is far beyond the scope of this thesis; instead, we will adopt the utilitarian approach and resort to an *off-the-shelf* forward model for galactic spectral flux: `Speculator-α` [11], which satisfies two important criteria:

- it is fast: `Speculator` is a NN emulation framework (which resorts to 50-component PCA for initial compression / final decompression, i.e. it resembles a SN Ia template); we have transcribed in `PyTorch` and incorporated as a `ConstantSource` in SLiCsim its version trained on `Prospector-α` [313] simulations of stellar population synthesis (SPS); this allows quickly generating mock galaxy photometry with minimal additional effort;

stellar population

- it has a convenient Bayesian parametrisation [536, `Prospector-β`], which includes the "hooks" to which to attach the DTD (an age-binned *distribution of stellar ages* **SFH**) and an extinction law which to affect SNæ hosted by the galaxy: `Prospector` uses the Kriek & Conroy [297] dust model;[314] controlled by a parameter δ, which we convert into an R_V via the latter's definition in eq. (8.8). We also identify galaxy-modellers' τ_2 as A_V.

We have implemented `Prospector-β` as a probabilistic program in `Pyro`, and while the details are not crucial, it is important to note that it is *not* hierarchical, i.e. does not include global/population parameters, and the distribution of the various properties (importantly to us, masses and dust) are *a priori independent*. To "fix" this and implement the expected/observed causal connections, we include the empirical relations of Alsing et al.

[314] Which is different from F99 — otherwise it wouldn't have a distinct name... —, and a LC-based SN+host model must consider this in order to be self-consistent; moreover, it must consider that the average dust which affects the integrated host photometry might not be representative of that at the location of the supernova, which must be taken into account as we discussed in relation to the local SN environment.

[12, eqs. (14) and (15)]:

$$A_V^h \sim \mathcal{N}\left(0.2 + 0.5\,\mathrm{ReLU}\left(\lg \mathrm{SFR}^h\right), 0.2^2\right), \tag{16.1}$$

$$\delta^h \sim \mathcal{N}\left(-0.095 + 0.111 A_V^h - 0.0066 (A_V^h)^2, 0.4^2\right), \tag{16.2}$$

where the star-formation rate (SFR) is the most-recent (current) component of $\mathbf{SFH}^h$. Moreover, the marginal distribution $\mathrm{p}(z_\mathrm{c}, M_*)$ (which is the first step of `Prospector-β` forward sampling, similarly to the SN Ia rate) is *independent of cosmology*[315] (cf. the dashed line in fig. 8.7). This can be relaxed by modelling large-scale structure formation from the first Principles of physical cosmology.

For this demonstration, we sample a bank of 1 000 000 i.i.d. galaxy parameter vectors

$$\mathbf{g}^h \equiv \left[z_\mathrm{c}^h, \mathbf{SFH}^h, R_V^h, A_V^h, \ldots\right] \tag{16.3}$$

from `Prospector-β` and generate the respective *absolute ugrizy* photometry (`Redshifted` but not yet affected by `CosmologicalDistance`) with `Speculator-α`. When compiling a final mock survey, we choose a random C and calculate apparent brightnesses $\mathbf{m}(C)$ (using the transverse comoving distance). Then, we add noise (optimistically, 0.01 mag) and evaluate detectability of each host—for the particular realisation of cosmology and noise—by comparing with the "5σ co-add depth" from Bianco et al. [45]. This represents a sharp (deterministic) selection criterion $S^h = S\left(\widehat{\mathbf{m}}^h\right)$ that determines the sample of *selected* galaxies in each simulator run, within which[316] we proceed to generate SNæ Ia.

16.1.2 SNæ Ia in hosts

Each galaxy's $\mathbf{SFH}^h = \left[\mathrm{SFH}_i^h\right]_{i=1}^7$ is an array of masses, each associated with a given age t_i (with respect to the moment the galaxy is observed), that represent the total amount of stars formed at different times. The SN Ia DTD expresses what fraction of them explode as SNæ Ia per unit time interval; that is, the rate (in the observer's frame) of SN Ia occurrence *from a specific galaxy's specifically old* population is

$$R_i^h = T \times \frac{\mathrm{SFH}_i^h \times \mathrm{DTD}(t_i)}{1+z}. \tag{16.4}$$

[315] `Prospector-β` does rely on an assumed WMAP [219] $T_\mathrm{age}(z_\mathrm{c})$, however, for building the history of star formation, but we do not expect this to have an effect on the simulations relevant to SNæ Ia or to our inference.

[316] This is an implicit selection criterion *on the SNæ*: we only consider, in other words, SNæ for which we can observe and uniquely identify the host (with certainty), which is usually not a stronger requirement than the cuts typically applied to cosmological SN Ia samples.

Following Heringer et al. [211, 212], we will assume a power law (set to zero before 100 Myr to allow for the creation of a white dwarf):

$$\mathrm{DTD}(t) = A \times (t/\mathrm{Gyr})^s \mathrm{M}_\odot^{-1}\,\mathrm{yr}^{-1}; \tag{16.5}$$

$$\log_{10} A \sim \mathcal{N}\left(-12.15, 0.1^2\right), \tag{16.6}$$

$$s \sim \mathcal{N}\left(-1.34, 0.2^2\right). \tag{16.7}$$

The double-degenerate formation scenario, in which the two WDs lose energy through gravitational radiation corresponds to an exponent $s = -1$, is in slight tension with current results. Thus, a precise determination of s—or, better, of a more flexible functional form (e.g. containing a weighted combination of multiple components for the different possible formation channels)—offers a unique vantage point onto SNæ Ia as physical phenomena rather than mere tools for distance measurement.

In our forward model, we evaluate eq. (16.4) for *all* stellar populations of *all* galaxies that were observed,[316] and, given the survey duration,[297] obtain Poisson realisations for the number of SNæ Ia that explode *in a specific galaxy's specifically old* population: summing those within and across the galaxies gives, respectively, N^h_{SN} (from fig. 8.7) and N_{tot} (the size of the total population). This procedure allows us to unique identify the "local environment" from which each simulated SN originated and to, possibly, assign different properties based on that information, although we do not do this at present.

Instead, for each of the N_{tot} SNæ, we generate observables according to Simple-BayeSN. This is very similar to BAHAMAS and the model we used in chapter 14 but contains three modifications concerning colour. The first is that, as in any BayeSN, extinction due to dust—which results in *external* reddening—is explicitly accounted for by splitting the

intrinsic colour

apparent colour into an *intrinsic* c_{int} and the host-dust colour excess [345, eq. (5)]:

$$c^s = c^s_{\mathrm{int}} + A_V^{h(s)} \Big/ R_V^{h(s)}, \tag{16.8}$$

where the dust parameters used for each SN refer to those used to simulate its host's light, achieving self-consistent modelling. Correspondingly, the (Phillips/Tripp) colour–magnitude standardisation coefficient β refers only to c_{int}, while the dimming effect of extinction is again explicitly implemented (assuming magnitudes are in the B band) [345, eq. (4)]:

$$M^s = M^s_0 - \alpha x_1 + \beta c_{\mathrm{int}} + \left(R_V^{h(s)} + 1\right) A_V^{h(s)} \Big/ R_V^{h(s)}. \tag{16.9}$$

And lastly, Simple-BayeSN allows for a correlation between x_1 and c_{int} via a new "colour-standardisation" coefficient α_c (empirically found to be close to zero) [345, eq. (20)]:

$$c^s_{\mathrm{int}} \sim \mathcal{N}\left(\bar{c} + \alpha_c x^s_1, \sigma_c^2\right). \tag{16.10}$$

After applying observational noise (here we use the simple $\hat{\boldsymbol{\Sigma}}^s = 0.01^2\mathbb{1}$, but a starting point towards sophistication of this modelling aspect are the *priors* on observational (co)-variances listed in Boyd et al. [62]) and a detection/selection procedure (the LSST-inspired one from subsection 15.3.2), we have finished the SN Ia side of the forward model. Finally, we build a mock survey as a set whose elements combine each SN Ia's observables (and, when it is not trivial/fixed, the metadata $\hat{\boldsymbol{\Sigma}}^s$)[317] and the photometry of their host:

$$\mathbf{D} = \left\{ \left(\widehat{m}^s, \hat{x}_1^s, \hat{c}^s, \cancel{\hat{\boldsymbol{\Sigma}}^s}, \widehat{\mathbf{m}}^{h(s)} \right) \right\}_{s=1}^{N_{\text{sel}}}. \tag{16.11}$$

16.2 Preliminary results and outlook

As a preliminary proof of concept (and so that the analysis can more or less be completed on time for inclusion in this thesis), we generate a "small" sample of, on expectation, 1000 (in "reality", 1080) SNæ Ia. To analyse them, we train ARNRE to infer $\boldsymbol{\theta}_g \in [M_0, [A, s], \alpha, \beta, \alpha_c, [\Omega_{m0}, \Omega_{\Lambda 0}], \sigma_0]$ from 64 000 simulated surveys (all at the same T setting as the mock and with the x_1 and c population parameters, $\bar{x}_1$, σ_{x_1}, $\bar{c}$, σ_c fixed to the values in table 14.1 for simplicity) with the same conditioned deep set as in section 15.1. We show results from the first stage (no truncation yet...) in fig. 16.1 and note that they can definitely benefit from an extended and more careful training, which we expect to significantly improve the least constrained (currently) parameters: C and σ_0.

Overall, we receive fig. 16.1 with *optimism* as it shows that the network is able to learn—at least to a certain extent—all parameters; most significantly *optimism*

- A and s, which describe the DTD and explicitly connect the two panes of our diptych;
- M_0 and β, whose estimation requires taking into account both the SN Ia luminosities and colours *and* inferring the $R_V^{h(s)}$, $A_V^{h(s)}$ from the host photometry;
- cosmology, for which the results resemble the initial stages in fig. 14.4, so we conjecture that the NN has indeed learnt to extract redshifts from $\mathbf{m}^{h(s)}$—lest it has uncovered a method for cosmological inference without redshifts.

In the same framework, we can easily integrate non-Ia contaminants with *phisically* grounded rates and correlations with host properties, as well as any parametrised model for population evolution and/or local-environment dependence and extrinsic effects like peculiar velocities and weak lensing due to large-scale structure. Incorporating, finally, an explicit forward model for the light curves of all transients we are considering and for their summarisation using classifying/summarising NNs will allow the ultimate principled and verifiable all-in-one analysis of LSST's hard earned massive data.

[317] Recall that in end-to-end SBI, systematics can always be included in the forward model instead of a dense/non-trivial covariance since the SNæ are exchangeable: this is, after all, why we represent data as a set.

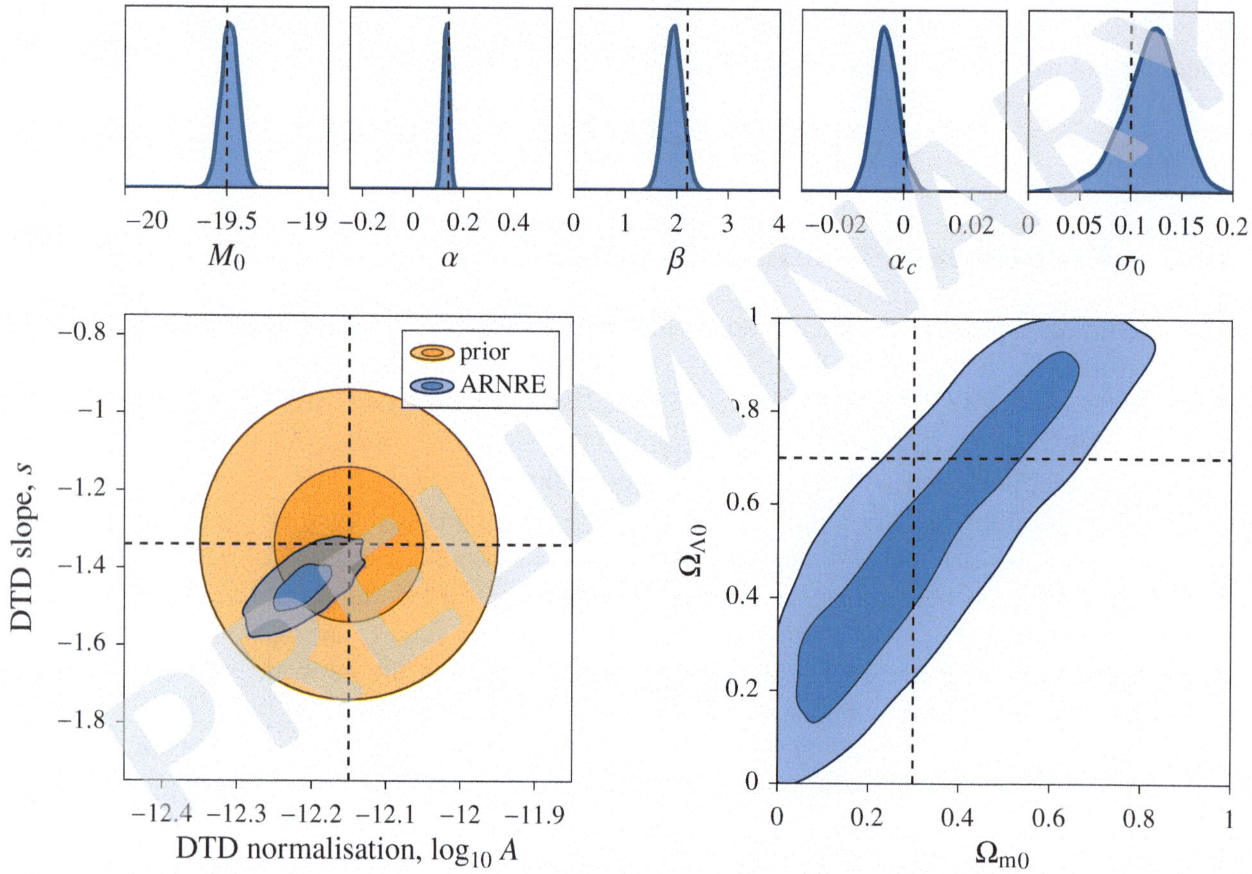

PRELIMINARY

Figure 16.1: First application of SBI to a (lesser) unified SNæ-Ia-and-hosts model (mock LC-summary data of 1080 SNæ Ia with identified and detected hosts (*ugrizy* broadband photometry) analysed with ARNRE and no truncation… yet; the priors are uniform across the extents of each plot, except for the DTD parameters; our results are *joint* over these 9 global variables (with the rest kept fixed), but we show only these projections as representing scientific interest). After a simple extension of the simulator and re-using our existing inference framework, we are able to constrain all parameters (although further training is advised) in a single independent analysis that does not rely on explicit redshift measurements, exposing a promising avenue for the future of SN Ia cosmology.

Epilogue

As a large language model, I can certainly summarise the thesis in the style of the author:

In this thesis, we have applied and extended the framework of simulation-based inference (SBI) in the field of supernova cosmology, with a focus on principled and scalable Bayesian analysis that can evade the pitfalls of current and future data and deliver no-compromise results for cosmological parameters. After surveying the different flavours of neural SBI, we have described several extensions to the method of truncated marginal neural ratio estimation in the setting of a large hierarchical model, as well as a method for calibrating its output and deriving constraints with a guaranteed confidence level.

We have then summarised the historical and cosmological background of SN Ia research and demonstrated their exponential expansion. In anticipation of future instruments, expected to discover and obtain data for orders of magnitude more transients than currently available, we have outlined a number of methodological challenges and highlighted how failing to take certain statistical and modelling effects properly into consideration—e.g. host dust extinction, all sources of redshift and their uncertainties, and sample selection—can lead to significantly biased results. We have argued that SBI is *the* methodology of choice for addressing them and elucidated out vision of a grand unified analysis of SN Ia and galaxy data that incorporates all conceivable physical details.

To this end, we have developed modern high-performance codes for physics and probabilistic modelling in general and SN Ia (and adjacent-data) simulation in particular and applied them to a series of analyses of real and mock data with increasing sophistication and scale. We have targeted the light curves of $\approx$100 low-redshift SNæ Ia and inferred simultaneously *all* their object-specific parameters *and* their populations. We have then performed the first principled Bayesian model comparison in the field, addressing the interdependence of magnitude standardisation and dust extinction and deriving results contrary to previous findings. Moreover, we have demonstrated the scalability of our framework to $\approx$100 000 SNæ Ia using a novel set-based truncation strategy making use of a minimal but powerful conditioned deep set neural network. Finally, we have shown a glimpse of our unified inference setup, which we envision to be the baseline for SN Ia cosmology in the future.

© The Editor(s) (if applicable) and The Author(s), under exclusive license to Springer Nature Switzerland AG 2026

K. Karchev, *Supernova Cosmology for the 21st Century*,
https://doi.org/10.1007/978-3-032-15072-1

See ya later, simulator!

A supernova explosion
over the Miramare castle;
painting in the style of Van Gogh
– DALL·E

Appendices

Chapter 17

Simulation-based HIerarchical Truncated inference

Truncation was an essential part of the analyses we have presented in part IV as it allowed us to obtain sufficiently precise results with finite network and training data. We even described strategies 2 to 4, which relate truncation to various hierarchical settings, concretely:

- strategy 2: truncating "nuisance" parameters in the Bayesian supermodel (fig. 3.1), where of interest is the high-level indicator variable $\mathcal{M}$;
- strategy 3: truncating global parameters to facilitate object-specific inference;
- strategy 4: truncating global parameters while increasing the constraining power of the data (its size).

In them, we never considered (out loud) truncating the local parameters of the i.i.d. (exchangeable) objects we analysed, even though we did place constraints on them in figs. 12.5 and 14.7. In fact, within the (stochastic-cardinality) simulators we use for catalogue-based inference (chapters 15 and 16), we *cannot* truncate local parameters since these are *not defined a priori*, i.e. before conditioning on the observed data (particularly its size) and placing labels (e.g. the metadata we used in SICRET, notably absent in RESSET) on the objects.

In this appendix, we demonstrate the danger that truncation of object-specific parameters (whenever well defined, i.e. in fixed-N_{obj} models) poses to population inference in high-dimensional settings (i.e. in the presence of many i.i.d. observations and hence local parameters). Assume a hierarchical model that factorises as

$$\mathrm{p}(\boldsymbol{\gamma}, \{\boldsymbol{\lambda}^i\}, \{\mathbf{d}^i\}) = \left[\prod_{i=1}^{N_{\text{obj}}} \mathrm{p}(\mathbf{d}^i \mid \boldsymbol{\lambda}^i, \boldsymbol{\chi})\, \mathrm{p}(\boldsymbol{\lambda}^i \mid \boldsymbol{\gamma})\right] \mathrm{p}(\boldsymbol{\gamma}), \tag{17.1}$$

i.e. where the global parameters that we are interested in do not influence the final sampling

© The Editor(s) (if applicable) and The Author(s), under exclusive license to Springer Nature Switzerland AG 2026
K. Karchev, *Supernova Cosmology for the 21st Century*,
https://doi.org/10.1007/978-3-032-15072-1

distribution but only the population of $\boldsymbol{\lambda}^i$. Now assume that we have constrained $\boldsymbol{\lambda}^i$ to some truncation regions $T_{\boldsymbol{\lambda}^i}$ (not necessarily the same) such that $\mathrm{p}(\mathbf{d}^i \,|\, \boldsymbol{\lambda}^i) \approx 0$ outside them for each i. By defining truncated priors[318]

$$\tilde{\mathrm{p}}(\boldsymbol{\lambda}^i \,|\, \boldsymbol{\gamma}) \equiv \frac{\mathrm{p}(\boldsymbol{\lambda}^i \,|\, \boldsymbol{\gamma}) \times \mathbb{1}(\boldsymbol{\lambda}^i \in T_{\boldsymbol{\lambda}^i})}{c^i(\boldsymbol{\gamma})} \quad \text{with} \quad c^i(\boldsymbol{\gamma}) \equiv \int_{T_{\boldsymbol{\lambda}^i}} \mathrm{p}(\boldsymbol{\lambda}^i \,|\, \boldsymbol{\gamma}) \, \mathrm{d}\boldsymbol{\lambda}^i, \tag{17.2}$$

we can write the *original* model as

$$\mathrm{p}(\boldsymbol{\gamma}, \{\boldsymbol{\lambda}^i\}, \{\mathbf{d}^i\}) = \underbrace{\frac{\prod_{i=1}^{N_\text{obj}} c^i(\boldsymbol{\gamma})}{\mathbb{1}(\{\boldsymbol{\lambda}^i \in T_{\boldsymbol{\lambda}^i}\})}}_{\text{correction}} \times \underbrace{\left[\prod_{i=1}^{N_\text{obj}} \mathrm{p}(\mathbf{d}^i \,|\, \boldsymbol{\lambda}^i) \frac{\mathrm{p}(\boldsymbol{\lambda}^i \,|\, \boldsymbol{\gamma}) \times \mathbb{1}(\boldsymbol{\lambda}^i \in T_{\boldsymbol{\lambda}^i})}{c^i(\boldsymbol{\gamma})}\right] \mathrm{p}(\boldsymbol{\gamma})}_{\tilde{\mathrm{p}}(\boldsymbol{\gamma}, \{\boldsymbol{\lambda}^i\}, \{\mathbf{d}^i\})}, \tag{17.3}$$

where $\tilde{\mathrm{p}}(\boldsymbol{\gamma}, \{\boldsymbol{\lambda}^i\}, \{\mathbf{d}^i\})$ is our truncated simulator, and $c(\boldsymbol{\gamma}) \equiv \prod_{i=1}^{N_\text{obj}} c^i(\boldsymbol{\gamma})$ is a correction factor[319] that accounts for the re-normalisation of the probabilities in the middle hierarchical layer and *must* be taken into account for global inference, *e.g.* by using it in the NRE optimisation objective while training.

re-weighted NRE training (hanc marginis exiguitas non caperet)

The preceding statement and eq. (17.3) must ring a bell: it is the same, in spirit, as eq. (4.10). Realising that selection is truncation, Anau Montel & Weniger [13] used the truncation formalism to represent selection *in their simulator*, estimating all the necessary corrections. However, we were less enthusiastic about this and demonstrated in fig. 4.4 how the infer-as-usual–correct strategy can be detrimental to the SBI approach due to large cancellations that require excessively (i.e. unrealistically) precise estimates. Conversely, in fig. 17.1, we illustrate the similar effect in the arguably more interpretable setting of high-dimensional object-specific truncation.

Using the usual Gaussian toy model $\lambda^i \sim \mathcal{N}(0, \sigma)$ we generate N_obj draws, truncate in a region ${\lambda^i}_\text{o} \pm 3\varepsilon$ around the realised values, and plot the resulting "correction" $c(\sigma)$. Both the "unconstrained" and "well-constrained" cases result in trivial modifications:

$$\varepsilon \to \infty \implies c(\sigma) \to \text{const}, \tag{17.4}$$

$$\varepsilon \to 0 \implies c(\sigma) \to \mathrm{p}(\{{\lambda^i}_\text{o}\} \,|\, \sigma). \tag{17.5}$$

In the "interesting" case of similar noise and scatter and for large samples, truncation introduces a significant shift between the two terms in eq. (17.3) just as in fig. 4.4: we quantify

[318] Clipppy's ConUnDis can do this in parallel over all i.i.d. λ^i, provided they are one-dimensional.

[319] We assumed above that the problematic denominator $\mathbb{1}(\{\boldsymbol{\lambda}^i \in T_{\boldsymbol{\lambda}^i}\})$ cancels with a similarly vanishing $\mathrm{p}(\mathbf{d}^i \,|\, \boldsymbol{\lambda}^i)$ to prevent divergences.

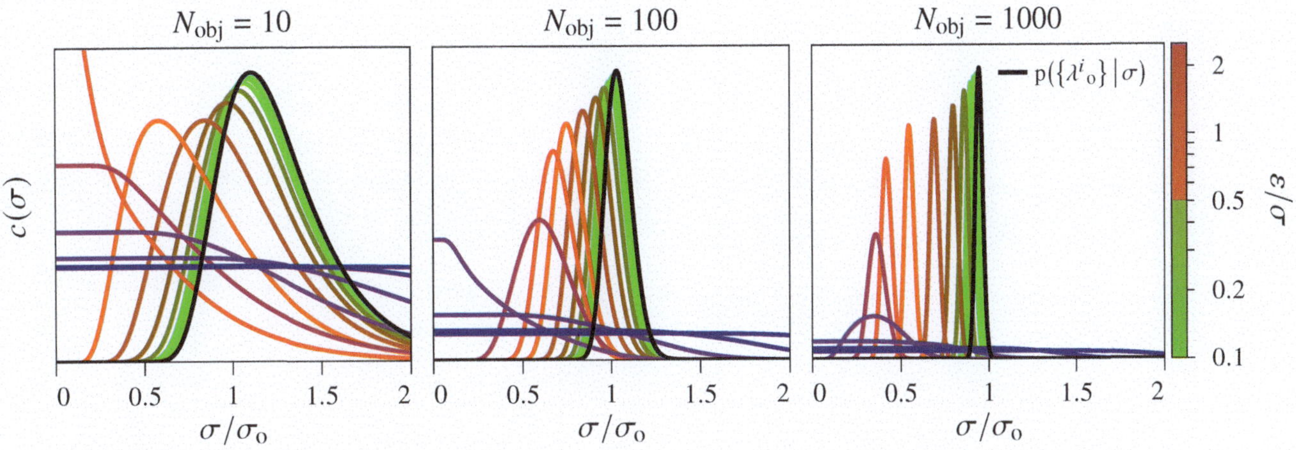

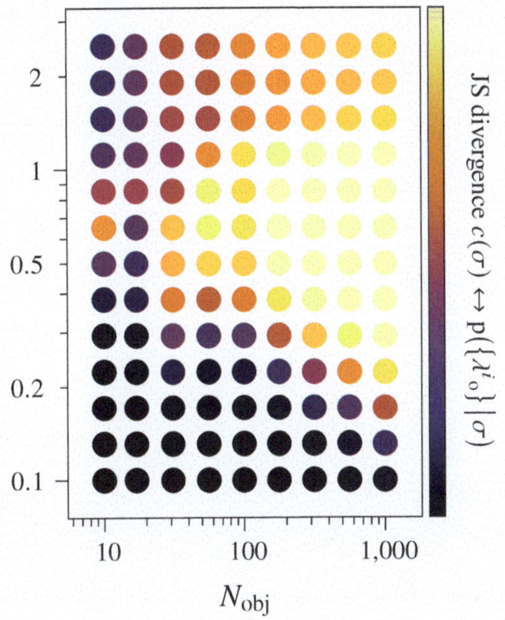

Figure 17.1: The danger of hierarchical truncation illustrated with a Gaussian toy model. *Top:* the correction factor as a function of the global parameter (the population standard deviation) compared to the likelihood of the sampled values, which is the target of SBI. For intermediate noise values, the two are significantly different, which means that results with truncated simulations would need to be severely—and precisely—corrected. *Left:* The JS divergence between the two (normalised across $\sigma \in [0; 2]$) for different noise/scatter and sample size; brighter is higher (more different), and the "noise" comes from sampling variance.

it via the JS divergence for different N_{obj} and ε/σ. The bottom line is that, exactly as we discussed in relation to handling selection, the necessity of precise estimation over many orders of magnitudes all but precludes successful SBI with large numbers of *independently* truncated local parameters.

Abbreviations

ΛCM least common multiple
ΛCDM Λ–cold dark matter
SN supernova
SN Ia type Ia supernova

ABC approximate Bayesian computation
AD automatic differentiation
ADU analog-to-digital unit
ARNRE autoregressive neural ratio estimation

BAO baryon acoustic oscillations
BCE binary cross-entropy
BHM Bayesian hierarchical model
BIC Bayesian information criterion
BNN Bayesian neural network

CC SN core-collapse supernova
CCA canonical correlation analysis
CCD charge-coupled device
CDF cumulative distribution function
CDM cold dark matter
CfA Harvard-Smithsonian Center for Astrophysics
CMB cosmic microwave background
CNN convolutional neural network
CPU central processing unit
CSP Carnegie Supernova Project

DAG directed acyclic graph
DD double-degenerate
DDPM de-noising diffusion probabilistic model
DE dark energy
DEH$_0$VILS Dark Energy, H_0, and peculiar velocities using Infrared Light from supernovæ
DES Dark Energy Survey
DESI Dark Energy Spectroscopic Instrument
DLR directional light radius
DM dark matter
DRP distance to random point
DTD delay-time distribution

ELASTICC Extended LSST Astronomical Time-series Classification Challenge
ELBO evidence lower bound
ELT Extremely Large Telescope
EOS equation of state
ESSENCE Equation of state: SupErNovæ trace Cosmic Expansion

FLRW Friedman(n)–Lemaître–Robertson–Walker

GAN generative adversarial network

© The Editor(s) (if applicable) and The Author(s), under exclusive license to Springer Nature Switzerland AG 2026
K. Karchev, *Supernova Cosmology for the 21st Century*,
https://doi.org/10.1007/978-3-032-15072-1

GOTO Gravitational-wave Optical Transient Observer
GP Gaussian process
GPU graphics processing unit
GR general theory of relativity
GW gravitational wave

HL highest-likelihood
HMC Hamiltonian Monte Carlo
HPD highest posterior density

i.i.d. independent and identically distributed
IMNN information-maximising neural network
IR infrared

JLA joint light-curve analysis
JS divergence Jensen–Shannon divergence

KDE kernel density estimation
KL divergence Kullback–Leibler divergence

LC light curve
LLM large language model
LMC Langevin Monte Carlo
LOSS Lick Observatory supernova search
LSS large-scale structure
LSST Legacy Survey of Space and Time
LTB Lemaître–Tolman–Bondi

MALA Metropolis-adjusted Langevin algorithm
MAP maximum *a posteriori*
MCMC Markov chain Monte Carlo
MDN mixture density network
MH Metropolis–Hastings
MJD modified Julian day
ML machine learning
MLE maximum likelihood estimator
MLP multi-layer perceptron
MOPED Massively Optimised Parameter Estimation and Data compression
MSE mean squared error
MVN multivariate normal
MW Milky Way

NDE neural density estimation
NF normalising flow
NIR near infrared
NJE neural joint estimation
NLE neural likelihood estimation
NLL negative log-likelihood
NLP natural-language processing
NN neural network
NPE neural posterior estimation
NPSE neural posterior score estimation
NRE neural ratio estimation
NS nested sampling
NUTS no-u-turn sampler

P–P plot probability–probability plot
Pan-STARRS Panoramic Survey Telescope and Rapid Response System
PCA principal component analysis
PDF probability density function
PLASTICC Photometric LSST Astronomical Time-series Classification Challenge
PMVN partial multivariate normal
POP parity-odd power
PP probabilistic programming

RAISIN SN Ia in the IR
ReLU rectified linear unit
RNN recurrent neural network
RST Rubin Space Telescope

SBI simulation-based inference
SD single-degenerate

SDSS Sloan Digital Sky Survey
SED spectral energy distribution
SFR star-formation rate
SLSN superluminous supernova
SNLS SuperNova Legacy Survey
SNR signal-to-noise ratio
SPCC supernova photometric classification challenge
SPS stellar population synthesis
STAR NRE set-based truncated autoregressive neural ratio estimation

TMNRE truncated marginal neural ratio estimation

VI variational inference
VMIM variational mutual-information maximisation

WD white dwarf
WFIRS Wide-Field Infrared Survey
WMAP Wilkinson Microwave Anisotropy Probe

ZTF Zwicky Transient Facility

Symbols

Statistics and inference

$\mathcal{U}(a, b)$ uniform distribution on $[a; b]$
$\mathcal{N}(\mu, \sigma^2)$ normal distribution with mean μ and variance σ^2
$\mathrm{Pois}(x)$ Poisson distribution with rate parameter x

$\mathbf{d}$ data array
$\mathbf{D}$ data set (unordered, includes information in its cardinality)
$\mathfrak{D}$ a random subset of $\mathbf{D}$
$\mathbf{s}(\mathbf{d})$ summary statistics

$\boldsymbol{\psi}$ all parameters of a Bayesian hierarchical model
$\boldsymbol{\theta}$ parameters inferred in a given study ("of interest")
$\boldsymbol{\nu}$ nuisance parameters (not of interest, including global and object-specific)
$\boldsymbol{\gamma}$ global parameters in a hierarchical model
$\boldsymbol{\lambda}^i$ local (specific to object i) parameters in a hierarchical model
$\mathbf{a}$ metadata (object/pointing-specific settings)
$\boldsymbol{\theta}_{\mathrm{fid}}$ fiducial parameter values

$\boldsymbol{\theta}_g$ one of a number of groups of parameters of interest
$\mathrm{q}(\boldsymbol{\theta} \mid \mathbf{d})$ (possibly neural) posterior estimator
$\mathrm{r}(\boldsymbol{\theta}, \mathbf{d})$ joint/marginal = likelihood/evidence = posterior/prior ratio
$\hat{\mathrm{r}}(\boldsymbol{\theta}, \mathbf{d})$ (possibly neural) r estimator
$T(\mathbf{d}_{\mathrm{o}})$ truncation region
$\tilde{\mathrm{p}}_{T(\mathbf{d}_{\mathrm{o}})}(\boldsymbol{\theta})$ truncated prior

$\mathbf{F}$ Fisher matrix/information
$\mathcal{I}$ mutual information
$\mathcal{H}$ entropy
γ (nominal) Bayesian credibility

© The Editor(s) (if applicable) and The Author(s), under exclusive license to Springer Nature Switzerland AG 2026

K. Karchev, *Supernova Cosmology for the 21st Century*,
https://doi.org/10.1007/978-3-032-15072-1

F frequentist confidence

Cosmology

C the cosmological parameters
$H_0 \equiv H(t=0)$ Hubble parameter and its present value (the Hubble "constant")
Ω_{r0} relative present-day radiation density
Ω_{m0} relative present-day total matter density
Ω_{de0} relative present-day dark energy density
w_0 equation of state of dark energy
$q_0,\ j_0$ present-day deceleration and jerk parameters

z total redshift (including cosmological and peculiar velocity)
z_c cosmological redshift
$D_c(z_c, C)$ radial comoving distance
$D_M \equiv D_c \operatorname{sinc}\left(\sqrt{k D_c^2}\right)$ transverse comoving distance
$D_L(z_c, C) \equiv D_M/(1+z_c)$ luminosity distance
$\mu(z_c, C) \equiv -2.5 \log_{10}\left[(D_L/10\,\mathrm{pc})^2\right]$ distance modulus

SNæ

M_0 standard absolute magnitude of SNæ Ia
σ_0 residual magnitude scatter after standardisation
x_1, c SALT "stretch" and "colour" parameters
$\alpha,\ \beta$ stretch- and colour-related standardisation coefficients
A_V extinction in the V band
R_V total-to-selective extinction ratio: $A_V/(A_B - A_V)$

`FLUXCAL` & `FLUXCALERR` calibrated "flux" (signal) and "noise"

R (comoving) volumetric rate (in the rest frame) of SN Ia occurrence

$\mathbf{g}$ parameters of the host galaxy
M_* stellar mass of the host galaxy
ΔM magnitude step across the split in host stellar mass
$M_\odot$ solar mass: 2×10^{30} kg

Bibliography

[1] Abril-Pla O., et al., 2023, PeerJ Computer Science, 9, e1516

[2] Adam A., Perreault-Levasseur L., Hezaveh Y., 2022a, Pixelated Reconstruction of Gravitational Lenses Using Recurrent Inference Machines (arXiv:2207.01073), doi:10.48550/arXiv.2207.01073

[3] Adam A., Coogan A., Malkin N., Legin R., Perreault-Levasseur L., Hezaveh Y., Bengio Y., 2022b, Posterior Samples of Source Galaxies in Strong Gravitational Lenses with Score-Based Priors (arXiv:2211.03812), doi:10.48550/arXiv.2211.03812

[4] Agol E., Hernandez D. M., Langford Z., 2021, Monthly Notices of the Royal Astronomical Society, 507, 1582

[5] Aitken A. C., Silverstone H., 1942, Proceedings of the Royal Society of Edinburgh Section A: Mathematics, 61, 186

[6] Allam T., McEwen J. D., 2024, RAS Techniques and Instruments, 3, 209

[7] Alsing J., Wandelt B., 2018, Monthly Notices of the Royal Astronomical Society, 476, L60

[8] Alsing J., Wandelt B., 2019, Monthly Notices of the Royal Astronomical Society, 488, 5093

[9] Alsing J., Wandelt B. D., Feeney S. M., 2018, arXiv:1808.06040 [math, stat]

[10] Alsing J., Charnock T., Feeney S., Wandelt B., 2019, Monthly Notices of the Royal Astronomical Society, 488, 4440

[11] Alsing J., et al., 2020, The Astrophysical Journal Supplement Series, 249, 5

[12] Alsing J., Peiris H., Mortlock D., Leja J., Leistedt B., 2023, The Astrophysical Journal Supplement Series, 264, 29

[13] Anau Montel N., Weniger C., 2022, Detection Is Truncation: Studying Source Populations with Truncated Marginal Neural Ratio Estimation (arXiv:2211.04291), doi:10.48550/arXiv.2211.04291

[14] Anau Montel N., Alvey J., Weniger C., 2024, Monthly Notices of the Royal Astronomical Society, 530, 4107

[15] Anderson J. R., Peterson C., 1987, Complex Systems, 1, 995

[16] Andrieu C., Djurić P. M., Doucet A., 2001, Signal Processing, 81, 19

[17] Aristotle 1901, Posterior Analytics. Blackwell, Oxford

[18] Arnett W. D., 1969, The Astrophysical Journal, 157, 1369

[19] Arnett W. D., 1982, The Astrophysical Journal, 253, 785

© The Editor(s) (if applicable) and The Author(s), under exclusive license to Springer Nature Switzerland AG 2026

K. Karchev, *Supernova Cosmology for the 21st Century*,
https://doi.org/10.1007/978-3-032-15072-1

[20] Ashton G., et al., 2022, Nature Reviews Methods Primers, 2, 39

[21] Astier P., et al., 2006, Astronomy and Astrophysics, 447, 31

[22] Astropy Collaboration 2013, Astronomy and Astrophysics, 558, A33

[23] Astropy Collaboration 2018, The Astronomical Journal, 156, 123

[24] Augustine M. T., 2024, A Survey on Universal Approximation Theorems (arXiv:2407.12895), doi:10.48550/arXiv.2407.12895

[25] Autenrieth M., Wright A. H., Trotta R., van Dyk D. A., Stenning D. C., Joachimi B., 2024a, Improved Weak Lensing Photometric Redshift Calibration via StratLearn and Hierarchical Modeling (arXiv:2401.04687), http://arxiv.org/abs/2401.04687

[26] Autenrieth M., van Dyk D. A., Trotta R., Stenning D. C., 2024b, Statistical Analysis and Data Mining: The ASA Data Science Journal, 17, e11643

[27] Avelino A., Friedman A. S., Mandel K. S., Jones D. O., Challis P. J., Kirshner R. P., 2019, The Astrophysical Journal, 887, 106

[28] Ba J. L., Kiros J. R., Hinton G. E., 2016, Layer Normalization (arXiv:1607.06450), doi:10.48550/arXiv.1607.06450

[29] Baade W., 1926, Astronomische Nachrichten, 228, 359

[30] Baade W., 1938, The Astrophysical Journal, 88, 285

[31] Baes M., Camps P., Van De Putte D., 2017, Monthly Notices of the Royal Astronomical Society, 468, 927

[32] Barbary K., 2016, Extinction, Zenodo, doi:10.5281/zenodo.804967

[33] Barbary K., et al., 2016, SNCosmo, Zenodo, doi:10.5281/zenodo.592747

[34] Barber D., Agakov F., 2003, in Advances in Neural Information Processing Systems. MIT Press, https://papers.nips.cc/paper_files/paper/2003/hash/a6ea8471c120fe8cc35a2954c9b9c595-Abstract.html

[35] Barker G. F., 1887, Proceedings of the American Philosophical Society, 24, 166

[36] Bayes Mr., 1763, Philosophical Transactions of the Royal Society of London Series I, 53, 370

[37] Beaumont M. A., 2019, Annual Review of Statistics and Its Application, 6, 379

[38] Benito M., Karchev K., Leane R. K., Põder S., Smirnov J., Trotta R., 2024, Journal of Cosmology and Astroparticle Physics, 2024, 038

[39] Bernardo R. C., Grandón D., Levi Said J., Cárdenas V. H., 2023, Physics of the Dark Universe, 40, 101213

[40] Beskos A., Roberts G., Stuart A., 2009, The Annals of Applied Probability, 19, 863

[41] Beskos A., Pillai N., Roberts G., Sanz-Serna J.-M., Stuart A., 2013, Bernoulli, 19, 1501

[42] Betancourt M., 2016, Identifying the Optimal Integration Time in Hamiltonian Monte Carlo (arXiv:1601.00225), doi:10.48550/arXiv.1601.00225

[43] Betancourt M. J., Girolami M., 2013, Hamiltonian Monte Carlo for Hierarchical Models (arXiv:1312.0906), doi:10.48550/arXiv.1312.0906

[44] Betoule M., et al., 2014, Astronomy and Astrophysics, 568, A22

[45] Bianco F. B., Jones L., Ivezić Ž., Ritz S., 2022, Technical Report PSTN-054, Updated Estimates of the Rubin System Throughput and Expected LSST Image Depth. Vera C. Rubin Observatory

[46] Bingham E., et al., 2019, The Journal of Machine Learning Research, 20, 973

[47] Binney J., Tremaine S., 1987, Galactic Dynamics. Princeton University Press

[48] Bishop C. M., 1994, Monograph, Mixture Density Networks, `https://publications.aston.ac.uk/id/eprint/373/`. Aston University, Birmingham, `https://publications.aston.ac.uk/id/eprint/373/`

[49] Bishop C. M., 2006, Pattern Recognition and Machine Learning. Springer Science and Business Media LLC

[50] Blanco-Mallo E., Morán-Fernández L., Remeseiro B., Bolón-Canedo V., 2023, Pattern Recognition, 141, 109646

[51] Blei D. M., Kucukelbir A., McAuliffe J. D., 2017, Journal of the American Statistical Association, 112, 859

[52] Bloom J. S., et al., 2012, The Astrophysical Journal, 744, L17

[53] Bohlin R. C., Gilliland R. L., 2004, The Astronomical Journal, 127, 3508

[54] Böhm V., Feng Y., Lee M. E., Dai B., 2021, Astronomy and Computing, 36, 100490

[55] Bond-Taylor S., Leach A., Long Y., Willcocks C. G., 2022, IEEE Transactions on Pattern Analysis and Machine Intelligence, 44, 7327

[56] Bondi H., 1947, Monthly Notices of the Royal Astronomical Society, 107, 410

[57] Boone K., 2019, The Astronomical Journal, 158, 257

[58] Boruah S. S., Hudson M. J., Lavaux G., 2020, Monthly Notices of the Royal Astronomical Society, 498, 2703

[59] Boruah S. S., Hudson M. J., Lavaux G., 2021, Monthly Notices of the Royal Astronomical Society, 507, 2697

[60] Bounissou S., Thatte N., Zieleniewski S., Houghton R. C. W., Tecza M., Hook I., Neichel B., Fusco T., 2018, Monthly Notices of the Royal Astronomical Society, 478, 3189

[61] Box G. E. P., 1979, in Launer R. L., Wilkinson G. N., eds, , Robustness in Statistics. Academic Press, pp 201–236, doi:10.1016/B978-0-12-438150-6.50018-2

[62] Boyd B. M., Grayling M., Thorp S., Mandel K. S., 2024, Accounting for Selection Effects in Supernova Cosmology with Simulation-Based Inference and Hierarchical Bayesian Modelling (arXiv:2407.15923), doi:10.48550/arXiv.2407.15923

[63] Bradbury J., et al., 2018, JAX: Composable Transformations of Python+NumPy Programs, `http://github.com/google/jax`

[64] Brahe T., 1573, Tychonis Brahe Dani De nova et nvllivs ævi memoria privs visa stella iam pridem anno à nato Christo 1572, mense Nouembrj primùm conspecta, contemplatio mathematica: cui præter exactam eclipsis lvnaris, huius, anni, pragmatian, et elegantem in Vraniam elegiam, epistola quo[que] dedicatoria accessit : in qua, noua & erudita conscribendi diaria metheorologica methodus, vtrius[que] astrologiæ studiosis. Impressit Lavrentivs Benedictj

[65] Branch D., Patchett B., 1973, Monthly Notices of the Royal Astronomical Society, 161, 71

[66] Branch D., Fisher A., Nugent P., 1993, The Astronomical Journal, 106, 2383

[67] Brehmer J., Cranmer K., Louppe G., Pavez J., 2018a, Physical Review D, 98, 052004

[68] Brehmer J., Cranmer K., Louppe G., Pavez J., 2018b, Physical Review Letters, 121, 111801

[69] Brehmer J., Louppe G., Pavez J., Kyle Cranmer 2020, Proceedings of the National Academy of Sciences, 117, 5242

[70] Brout D., Riess A., 2023, in , Hubble Constant Tension. arXiv (arXiv:2311.08253), `http://arxiv.org/abs/2311.08253`

[71] Brout D., Scolnic D., 2021, The Astrophysical Journal, 909, 26

[72] Brout D., et al., 2022a, The Astrophysical Journal, 938, 110

[73] Brout D., et al., 2022b, The Astrophysical Journal, 938, 111

[74] Burns C. R., et al., 2011, The Astronomical Journal, 141, 19

[75] Burns C. R., et al., 2018, The Astrophysical Journal, 869, 56

[76] Burrows A., Vartanyan D., 2021, Nature, 589, 29

[77] Cacciatori S. L., Gorini V., Re F., 2024, in Streit-Bianchi M., Gorini V., eds, , New Frontiers in Science in the Era of AI. Springer Nature Switzerland, Cham, pp 217–251, doi:10.1007/978-3-031-61187-2_13

[78] Cai X., McEwen J. D., Pereyra M., 2022, Proximal Nested Sampling for High-Dimensional Bayesian Model Selection (arXiv:2106.03646), doi:10.48550/arXiv.2106.03646

[79] Calcino J., Davis T., 2017, Journal of Cosmology and Astroparticle Physics, 2017, 038

[80] Calzetti D., Armus L., Bohlin R. C., Kinney A. L., Koornneef J., Storchi-Bergmann T., 2000, The Astrophysical Journal, 533, 682

[81] Campagne J.-E., et al., 2023, JAX-COSMO: An End-to-End Differentiable and GPU Accelerated Cosmology Library (arXiv:2302.05163), doi:10.48550/arXiv.2302.05163

[82] Campeau-Poirier È., Perreault-Levasseur L., Coogan A., Hezaveh Y., 2023, Time Delay Cosmography with a Neural Ratio Estimator (arXiv:2309.16063), doi:10.48550/arXiv.2309.16063

[83] Carlson B. C., 1977, Special Functions of Applied Mathematics. Academic Press, New York; San Francisco

[84] Carlson B. C., 1995, Numerical Algorithms, 10, 13

[85] Carlson B. C., 1999, Journal of Symbolic Computation, 28, 739

[86] Carlson B. C., 2022, NIST Digital Library of Mathematical Functions

[87] Carpenter B., Hoffman M. D., Brubaker M., Lee D., Li P., Betancourt M., 2015, The Stan Math Library: Reverse-Mode Automatic Differentiation in C++ (arXiv:1509.07164), doi:10.48550/arXiv.1509.07164

[88] Carpenter B., et al., 2017, Journal of Statistical Software, 76, 1

[89] Carrick J., Turnbull S. J., Lavaux G., Hudson M. J., 2015, Monthly Notices of the Royal Astronomical Society, 450, 317

[90] Chandrasekhar S., 1957, An Introduction to the Study of Stellar Structure. Courier Corporation

[91] Charnock T., Lavaux G., Wandelt B. D., 2018, Physical Review D, 97, 083004

[92] Chen S., Guo W., 2023, Mathematics, 11, 1777

[93] Chen R. T. Q., Rubanova Y., Bettencourt J., Duvenaud D., 2019, Neural Ordinary Differential Equations (arXiv:1806.07366), doi:10.48550/arXiv.1806.07366

[94] Chen J.-F., Wang Y.-C., Zhang T., Zhang T.-J., 2023, Physical Review D, 107, 063517

[95] Chen R., et al., 2024, Evaluating Cosmological Biases Using Photometric Redshifts for Type Ia Supernova Cosmology with the Dark Energy Survey Supernova Program (arXiv:2407.16744), `http://arxiv.org/abs/2407.16744`

[96] Chevallier M., Polarski D., 2001, International Journal of Modern Physics D, 10, 213

[97] Chianese M., Coogan A., Hofma P., Otten S., Weniger C., 2020, Monthly Notices of the Royal Astronomical Society, 496, 381

[98] Choromanski K. M., et al., 2020, in International Conference on Learning Representations. `https://openreview.net/forum?id=Ua6zuk0WRH&utm_campaign=NLP%20News&utm_medium=email&utm_source=Revue%20newsletter`

[99] Coelho R. C. V., Calvão M. O., Reis R. R. R., Siffert B. B., 2015, European Journal of Physics, 36, 015007

[100] Colgate S. A., 1979, The Astrophysical Journal, 232, 404

[101] Collaboration P., et al., 2020, Astronomy and Astrophysics, 641, A6

[102] Collaboration L. D. E. S., et al., 2022, DESC DC2 Data Release Note (arXiv:2101.04855), doi:10.48550/arXiv.2101.04855

[103] Conley A., et al., 2011, The Astrophysical Journal Supplement Series, 192, 1

[104] Coogan A., et al., 2022, Efficient Template Bank Generation with Differentiable Waveforms (arXiv:2202.09380), doi:10.48550/arXiv.2202.09380

[105] Cook S. R., Gelman A., Rubin D. B., 2006, Journal of Computational and Graphical Statistics, 15, 675

[106] Cramer H., 1922, Mathematical Methods Of Statistics. Princeton University Press, Princeton, `http://archive.org/details/in.ernet.dli.2015.149716`

[107] Cranmer K., Pavez J., Louppe G., 2016, arXiv:1506.02169 [physics, stat]

[108] Cranmer K., Brehmer J., Louppe G., 2020, Proceedings of the National Academy of Sciences, 117, 30055

[109] DES Collaboration et al., 2024, The Astrophysical Journal, 973, L14

[110] DESI Collaboration et al., 2024, DESI 2024 VI: Cosmological Constraints from the Measurements of Baryon Acoustic Oscillations, doi:10.48550/arXiv.2404.03002

[111] Dabrowski M., Stelmach J., 1986, The Astronomical Journal, 92, 1272

[112] Dai M., Kuhlmann S., Wang Y., Kovacs E., 2018, Monthly Notices of the Royal Astronomical Society, 477, 4142

[113] Dai M., Jones D. O., Kenworthy W. D., Kessler R., Pierel J. D. R., Foley R. J., Jha S. W., Scolnic D. M., 2023, The Astrophysical Journal Supplement Series, 267, 1

[114] Dalmasso N., Izbicki R., Lee A., 2020, in Proceedings of the 37th International Conference on Machine Learning. PMLR, pp 2323–2334, `https://proceedings.mlr.press/v119/dalmasso20a.html`

[115] Dalmasso N., Masserano L., Zhao D., Izbicki R., Lee A. B., 2022, Likelihood-Free Frequentist Inference: Confidence Sets with Correct Conditional Coverage (arXiv:2107.03920), doi:10.48550/arXiv.2107.03920

[116] Darmois G., 1945, Revue de l'Institut International de Statistique / Review of the International Statistical Institute, 13, 9

[117] Davis T. M., et al., 2011, The Astrophysical Journal, 741, 67

[118] Davis T. M., Hinton S. R., Howlett C., Calcino J., 2019, Monthly Notices of the Royal Astronomical Society, 490, 2948

[119] Dawid A. P., 1982, Journal of the American Statistical Association, 77, 605

[120] DeGroot M. H., Fienberg S. E., 1983, Journal of the Royal Statistical Society. Series D (The Statistician), 32, 12

[121] Deckers M., et al., 2024, ZTF SN Ia DR2: The Secondary Maximum in Type Ia Supernovae (arXiv:2406.19460), http://arxiv.org/abs/2406.19460

[122] Deistler M., Goncalves P. J., Macke J. H., 2022, in Advances in Neural Information Processing Systems. https://openreview.net/forum?id=QW98XBAqNRa

[123] Delaunoy A., Hermans J., Rozet F., Wehenkel A., Louppe G., 2022, arXiv:2208.13624

[124] Delgado F., Saha A., Chandrasekharan S., Cook K., Petry C., Ridgway S., 2014, in Modeling, Systems Engineering, and Project Management for Astronomy VI. SPIE, pp 422–445, doi:10.1117/12.2056898

[125] Devroye L., Györfi L., Lugosi G., 1996, A Probabilistic Theory of Pattern Recognition, corrected edition edn. Springer, New York

[126] Dhawan S., Thorp S., Mandel K. S., Ward S. M., Narayan G., Jha S. W., Chant T., 2023, Monthly Notices of the Royal Astronomical Society, 524, 235

[127] Di Valentino E., et al., 2021, Classical and Quantum Gravity, 38, 153001

[128] Dilday B., et al., 2008, The Astrophysical Journal, 682, 262

[129] Do A., et al., 2024, Hawai'i Supernova Flows: A Peculiar Velocity Survey Using Over a Thousand Supernovae in the Near-Infrared (arXiv:2403.05620), http://arxiv.org/abs/2403.05620

[130] Donoso-Oliva C., Becker I., Protopapas P., Cabrera-Vives G., Vishnu M., Vardhan H., 2023, Astronomy and Astrophysics, 670, A54

[131] Doppler C., 1842, Proceedings of the Royal Bohemian Society of Sciences, Prague (PartV), 465, 482

[132] Draine B. T., 2003, Annual Review of Astronomy and Astrophysics, 41, 241

[133] Duane S., Kennedy A. D., Pendleton B. J., Roweth D., 1987, Physics Letters B, 195, 216

[134] Durkan C., Murray I., Papamakarios G., 2020, in Proceedings of the 37th International Conference on Machine Learning. PMLR, pp 2771–2781, https://proceedings.mlr.press/v119/durkan20a.html

[135] Durmus A., Moulines É., Pereyra M., 2018, SIAM Journal on Imaging Sciences, 11, 473

[136] Dyson F. W., 1913, Monthly Notices of the Royal Astronomical Society, 73, 334

[137] Eddington A. S., 1913, Monthly Notices of the Royal Astronomical Society, 73, 359

[138] Edwards T. D. P., Wong K. W. K., Lam K. K. H., Coogan A., Foreman-Mackey D., Isi M., Zimmerman A., 2023, Ripple: Differentiable and Hardware-Accelerated Waveforms for Gravitational Wave Data Analysis (arXiv:2302.05329), doi:10.48550/arXiv.2302.05329

[139] Einstein A., 1905, Annalen der Physik, 322, 132

[140] Einstein A., 1916, Annalen der Physik, 354, 769

[141] Einstein A., 1917, Sitzungsberichte der Königlich Preussischen Akademie der Wissenschaften, pp 142–152

[142] Eisenstein D. J., 1997, An Analytic Expression for the Growth Function in a Flat Universe with a Cosmological Constant (arXiv:astro-ph/9709054), doi:10.48550/arXiv.astro-ph/9709054

[143] Elsemüller L., Schnuerch M., Bürkner P.-C., Radev S. T., 2023, A Deep Learning Method for Comparing Bayesian Hierarchical Models (arXiv:2301.11873), doi:10.48550/arXiv.2301.11873

[144] Fakhouri H. K., et al., 2015, The Astrophysical Journal, 815, 58

[145] Fechner G. T., 1860, Elemente der Psychophysik. Breitkopf u. Härtel

[146] Feige B., 1992, Astronomische Nachrichten, 313, 139

[147] Ferrand G., Warren D. C., Ono M., Nagataki S., Röpke F. K., Seitenzahl I. R., 2019, The Astrophysical Journal, 877, 136

[148] Ferrand G., et al., 2021, The Astrophysical Journal, 906, 93

[149] Fesen R. A., Kremer R., Patnaude D., Milisavljevic D., 2012, The Astronomical Journal, 143, 27

[150] Fink D., 1997, Technical report, A Compendium of Conjugate Priors, `https://www.semanticscholar.org/paper/A-Compendium-of-Conjugate-Priors-Fink/2a5d43d7f96312455c463116117911da2b7fadc9`. Montana State Univeristy, `https://www.semanticscholar.org/paper/A-Compendium-of-Conjugate-Priors-Fink/2a5d43d7f96312455c463116117911da2b7fadc9`

[151] Fisher R. A., 1922, Philosophical Transactions of the Royal Society of London. Series A, Containing Papers of a Mathematical or Physical Character, 222, 309

[152] Fisher C., et al., 2024, Monthly Notices of the Royal Astronomical Society, 535, 27

[153] Fitzpatrick E. L., 1999, Publications of the Astronomical Society of the Pacific, 111, 63

[154] Fitzpatrick E. L., Massa D., 2007, The Astrophysical Journal, 663, 320

[155] Foley R. J., 2012, The Astrophysical Journal, 748, 127

[156] Foley R. J., et al., 2013, The Astrophysical Journal, 767, 57

[157] Foley R. J., et al., 2018, Monthly Notices of the Royal Astronomical Society, 475, 193

[158] Foreman-Mackey D., Hogg D. W., Lang D., Goodman J., 2013, Publications of the Astronomical Society of the Pacific, 125, 306

[159] Fréchet M., 1943, Revue de l'Institut International de Statistique / Review of the International Statistical Institute, 11, 182

[160] Friedman A., 1922, Zeitschrift für Physik, 10, 377

[161] Friedman A., 1999, General Relativity and Gravitation, 31, 1991

[162] Gagliano A., Contardo G., Mackey D. F., Malz A. I., Aleo P. D., 2023, First Impressions: Early-Time Classification of Supernovae Using Host Galaxy Information and Shallow Learning (arXiv:2305.08894), doi:10.48550/arXiv.2305.08894

[163] Gagnon-Hartman S., Ruan J., Haggard D., 2023, Monthly Notices of the Royal Astronomical Society, p. stad069

[164] Galan A., Vernardos G., Peel A., Courbin F., Starck J.-L., 2022, Using Wavelets to Capture Deviations from Smoothness in Galaxy-Scale Strong Lenses (arXiv:2207.05763), doi:10.48550/arXiv.2207.05763

[165] Ganeshalingam M., et al., 2010, The Astrophysical Journal Supplement Series, 190, 418

[166] Garnett R., 2023, Bayesian Optimization, 1st edition edn. Cambridge University Press, Cambridge, United Kingdom ; New York, NY

[167] Ge H., Xu K., Ghahramani Z., 2018, in Proceedings of the Twenty-First International Conference on Artificial Intelligence and Statistics. PMLR, pp 1682–1690, https://proceedings.mlr.press/v84/ge18b.html

[168] Geffner T., Papamakarios G., Mnih A., 2023, Compositional Score Modeling for Simulation-based Inference (arXiv:2209.14249), doi:10.48550/arXiv.2209.14249

[169] Gelfand A. E., Smith A. F. M., 1990, Journal of the American Statistical Association, 85, 398

[170] Gelman A., Gilks W. R., Roberts G. O., 1997, The Annals of Applied Probability, 7, 110

[171] Geman S., Geman D., 1984, IEEE Transactions on Pattern Analysis and Machine Intelligence, PAMI-6, 721

[172] Gerardi F., Feeney S. M., Alsing J., 2021, Physical Review D, 104, 083531

[173] Gibbons J. D., Chakraborti S., 2010, Nonparametric Statistical Inference, 5th edition edn. Chapman and Hall/CRC, Boca Raton

[174] Ginolin M., et al., 2024a, ZTF SN Ia DR2: Environmental Dependencies of Stretch and Luminosity of a Volume Limited Sample of 1,000 Type Ia Supernovae (arXiv:2405.20965), http://arxiv.org/abs/2405.20965

[175] Ginolin M., et al., 2024b, ZTF SN Ia DR2: Colour Standardisation of Type Ia Supernovae and Its Dependence on Environment (arXiv:2406.02072), doi:10.48550/arXiv.2406.02072

[176] Girolami M., Calderhead B., 2011, Journal of the Royal Statistical Society Series B: Statistical Methodology, 73, 123

[177] Goh L. W. K., Ocampo I., Nesseris S., Pettorino V., 2024, Distinguishing Coupled Dark Energy Models with Neural Networks (arXiv:2411.04058), doi:10.48550/arXiv.2411.04058

[178] Goobar A., Johansson J., Carracedo A. S., 2024, Strongly Lensed Supernovae: Lessons Learned (arXiv:2406.13519), doi:10.48550/arXiv.2406.13519

[179] Goodfellow I., Pouget-Abadie J., Mirza M., Xu B., Warde-Farley D., Ozair S., Courville A., Bengio Y., 2014, in Advances in Neural Information Processing Systems. Curran Associates, Inc., https://proceedings.neurips.cc/paper_files/paper/2014/hash/5ca3e9b122f61f8f06494c97b1afccf3-Abstract.html

[180] Goodman J., Weare J., 2010, Communications in Applied Mathematics and Computational Science, 5, 65

[181] Gordon T. A., Agol E., 2022, The Astronomical Journal, 164, 111

[182] Graur O., et al., 2014, The Astrophysical Journal, 783, 28

[183] Gray N., 2002, Mathematics of Computation, 71, 311

[184] Grayling M., Popovic B., 2024, BayeSN and SALT: A Comparison of Dust Inference Across SN Ia Light-curve Models with DES5YR (arXiv:2410.13747), doi:10.48550/arXiv.2410.13747

[185] Grayling M., Thorp S., Mandel K. S., Dhawan S., Uzsoy A. S., Boyd B. M., Hayesn E. E., Ward S. M., 2024, Scalable Hierarchical BayeSN Inference: Investigating Dependence of SN Ia Host Galaxy Dust Properties on Stellar Mass and Redshift (arXiv:2401.08755), http://arxiv.org/abs/2401.08755

[186] Green P. J., 1995, Biometrika, 82, 711

[187] Green P. J., 2003, in Hjort N. L., Richardson S., eds, Oxford Statistical ScienceNo. 27, Highly Structured Stochastic Systems, 1st edn, Oxford University Press, pp 179–206

[188] Greene K. L., Cyr-Racine F.-Y., 2022, Journal of Cosmology and Astroparticle Physics, 2022, 002

[189] Gu A., et al., 2022, The Astrophysical Journal, 935, 49

[190] Gull S. F., 1988, in Erickson G. J., Smith C. R., eds, , Maximum-Entropy and Bayesian Methods in Science and Engineering: Foundations. Springer Netherlands, Dordrecht, pp 53–74, doi:10.1007/978-94-009-3049-0_4

[191] Gupta R. R., et al., 2016, The Astronomical Journal, 152, 154

[192] Guy J., Astier P., Nobili S., Regnault N., Pain R., 2005, Astronomy and Astrophysics, 443, 781

[193] Guy J., et al., 2007, Astronomy and Astrophysics, 466, 11

[194] Guy J., et al., 2010, Astronomy and Astrophysics, 523, A7

[195] Halmos P. R., Savage L. J., 1949, The Annals of Mathematical Statistics, 20, 225

[196] Hamuy M., et al., 1993, The Astronomical Journal, 106, 2392

[197] Hamuy M., Phillips M. M., Suntzeff N. B., Schommer R. A., Maza J., Aviles R., 1996, The Astronomical Journal, 112, 2391

[198] Hanbury Brown R., Hazard C., 1952, Nature, 170, 364

[199] Handley W. J., Hobson M. P., Lasenby A. N., 2015, Monthly Notices of the Royal Astronomical Society, 453, 4384

[200] Harris C. R., et al., 2020, Nature, 585, 357

[201] Harrison E., 1987, Darkness at Night: A Riddle of the Universe, first edition. annotated. edn. Harvard University Press, Cambridge, Mass.

[202] Hastings W. K., 1970, Biometrika, 57, 97

[203] Hayes D. S., Pasinetti L. E., Philip A. G. D., eds, 1985, Calibration of Fundamental Stellar Quantities: Proceedings of the 111th Symposium of the International Astronomical Union Held at Villa Olmo, Como, Italy, May 24–29, 1984. Springer Netherlands, Dordrecht, doi:10.1007/978-94-009-5456-4

[204] Hearin A. P., Chaves-Montero J., Becker M. R., Alarcon A., 2021, The Open Journal of Astrophysics, 4, 10.21105/astro.2105.05859

[205] Hearin A. P., Ramachandra N., Becker M. R., DeRose J., 2022, The Open Journal of Astrophysics, 5, 3

[206] Hearin A. P., Chaves-Montero J., Alarcon A., Becker M. R., Benson A., 2023, Monthly Notices of the Royal Astronomical Society, 521, 1741

[207] Heavens A. F., Jimenez R., Lahav O., 2000, Monthly Notices of the Royal Astronomical Society, 317, 965

[208] Heavens A., Panter B., Jimenez R., Dunlop J., 2004, Nature, 428, 625

[209] Heavens A., Fantaye Y., Mootoovaloo A., Eggers H., Hosenie Z., Kroon S., Sellentin E., 2017, Marginal Likelihoods from Monte Carlo Markov Chains (arXiv:1704.03472), doi:10.48550/arXiv.1704.03472

[210] Heinrich L., Mishra-Sharma S., Pollard C., Windischhofer P., 2024, Transactions on Machine Learning Research

[211] Heringer E., Pritchet C., Kezwer J., Graham M. L., Sand D., Bildfell C., 2017, The Astrophysical Journal, 834, 15

[212] Heringer E., Pritchet C., van Kerkwijk M. H., 2019, The Astrophysical Journal, 882, 52

[213] Hermans J., Begy V., Louppe G., 2020, in III H. D., Singh A., eds, Proceedings of Machine Learning Research Vol. 119, Proceedings of the 37th International Conference

on Machine Learning. PMLR, pp 4239–4248, https://proceedings.mlr.press/v119/hermans20a.html

[214] Hermans J., Delaunoy A., Rozet F., Wehenkel A., Begy V., Louppe G., 2022, Transactions on Machine Learning Research

[215] Hicken M., et al., 2009, The Astrophysical Journal, 700, 331

[216] Hicken M., et al., 2012, The Astrophysical Journal Supplement Series, 200, 12

[217] Hill R., et al., 2018, Monthly Notices of the Royal Astronomical Society, 481, 2766

[218] Hillebrandt W., Kromer M., Röpke F. K., Ruiter A. J., 2013, Frontiers of Physics, 8, 116

[219] Hinshaw G., et al., 2013, The Astrophysical Journal Supplement Series, 208, 19

[220] Hinton G. E., Srivastava N., Krizhevsky A., Sutskever I., Salakhutdinov R. R., 2012, Improving Neural Networks by Preventing Co-Adaptation of Feature Detectors (arXiv:1207.0580), doi:10.48550/arXiv.1207.0580

[221] Hinton S. R., et al., 2019, The Astrophysical Journal, 876, 15

[222] Hlozek R., et al., 2012, The Astrophysical Journal, 752, 79

[223] Hložek R., et al., 2020, Results of the Photometric LSST Astronomical Time-series Classification Challenge (PLAsTiCC) (arXiv:2012.12392), doi:10.48550/arXiv.2012.12392

[224] Ho J., Jain A., Abbeel P., 2020, in Advances in Neural Information Processing Systems. Curran Associates, Inc., pp 6840–6851, https://proceedings.neurips.cc/paper/2020/hash/4c5bcfec8584af0d967f1ab10179ca4b-Abstract.html

[225] Hoffman M. D., Gelman A., 2014, Journal of Machine Learning Research, 15, 1593

[226] Hogg D. W., 2000, arXiv e-prints, pp astro–ph/9905116

[227] Hogg D. W., Villar S., 2024, Is Machine Learning Good or Bad for the Natural Sciences? (arXiv:2405.18095), doi:10.48550/arXiv.2405.18095

[228] Holz D. E., Hughes S. A., Schutz B. F., 2018, Physics Today, 71, 34

[229] Hotelling H., 1936, Biometrika, 28, 321

[230] Hou Yip K., Changeat Q., Al-Refaie A., Waldmann I., 2022, To Sample or Not To Sample: Retrieving Exoplanetary Spectra with Variational Inference and Normalising Flows (arXiv:2205.07037), doi:10.48550/arXiv.2205.07037

[231] Hounsell R., et al., 2018, The Astrophysical Journal, 867, 23

[232] Hoyle F., Fowler W. A., 1960, The Astrophysical Journal, 132, 565

[233] Hsiao E. Y., Conley A., Howell D. A., Sullivan M., Pritchet C. J., Carlberg R. G., Nugent P. E., Phillips M. M., 2007, The Astrophysical Journal, 663, 1187

[234] Huang C.-W., Krueger D., Lacoste A., Courville A., 2018, in Proceedings of the 35th International Conference on Machine Learning. PMLR, pp 2078–2087, https://proceedings.mlr.press/v80/huang18d.html

[235] Hubble E., 1929, Proceedings of the National Academy of Science, 15, 168

[236] Hubble E., 1936, The Astrophysical Journal, 84, 517

[237] Huterer D., 2020, The Astrophysical Journal, 904, L28

[238] Inserra C., et al., 2018, Astronomy and Astrophysics, 609, A83

[239] Inserra C., et al., 2021, Monthly Notices of the Royal Astronomical Society, 504, 2535

[240] Ivezić Ž., et al., 2019, The Astrophysical Journal, 873, 111

[241] Izbicki R., Lee A., Schafer C., 2014, in Proceedings of the Seventeenth International Conference on Artificial Intelligence and Statistics. PMLR, pp 420–429, `https://proceedings.mlr.press/v33/izbicki14.html`

[242] Jamieson D., Li Y., de Oliveira R. A., Villaescusa-Navarro F., Ho S., Spergel D. N., 2022, Field Level Neural Network Emulator for Cosmological N-body Simulations (arXiv:2206.04594), doi:10.48550/arXiv.2206.04594

[243] Järvenpää M., Gutmann M. U., Pleska A., Vehtari A., Marttinen P., 2019, Bayesian Analysis, 14, 595

[244] Jaynes E. T., 1968, IEEE Transactions on Systems Science and Cybernetics, 4, 227

[245] Jaynes E. T., 2003, Probability Theory: The Logic of Science, annotated edition edn. Cambridge University Press, Cambridge, UK ; New York, NY

[246] Jeans J., 2009, The Growth of Physical Science. Cambridge Library Collection - Physical Sciences, Cambridge University Press, Cambridge, doi:10.1017/CBO9780511694387

[247] Jeffrey N., Wandelt B. D., 2020, Solving High-Dimensional Parameter Inference: Marginal Posterior Densities & Moment Networks (arXiv:2011.05991), doi:10.48550/arXiv.2011.05991

[248] Jeffrey N., Wandelt B. D., 2023, Evidence Networks: Simple Losses for Fast, Amortized, Neural Bayesian Model Comparison (arXiv:2305.11241), doi:10.48550/arXiv.2305.11241

[249] Jeffrey N., Alsing J., Lanusse F., 2021, Monthly Notices of the Royal Astronomical Society, 501, 954

[250] Jeffreys H., 1998, The Theory of Probability, third edition edn. Oxford Classic Texts in the Physical Sciences, Oxford University Press, Oxford, New York

[251] Jennings E., Wolf R., Sako M., 2016, A New Approach for Obtaining Cosmological Constraints from Type Ia Supernovae Using Approximate Bayesian Computation (arXiv:1611.03087), doi:10.48550/arXiv.1611.03087

[252] Jha S., 2002, PhD thesis, Harvard University, `https://ui.adsabs.harvard.edu/abs/2002PhDT........10J`

[253] Jha S., et al., 2006, The Astronomical Journal, 131, 527

[254] Jha S., Riess A. G., Kirshner R. P., 2007, The Astrophysical Journal, 659, 122

[255] Jia H., Seljak U., 2020, in Proceedings of The 2nd Symposium on Advances in Approximate Bayesian Inference. PMLR, pp 1–14, `https://proceedings.mlr.press/v118/jia20a.html`

[256] Jia X.-D., Hu J.-P., Wang F.-Y., Dai Z.-G., 2024, Forecast of Cosmological Constraints with Superluminous Supernovae from the Chinese Space Station Telescope (arXiv:2406.19758), `http://arxiv.org/abs/2406.19758`

[257] Johnson H. L., Morgan W. W., 1953, The Astrophysical Journal, 117, 313

[258] Jones D. O., Riess A. G., Scolnic D. M., 2015, The Astrophysical Journal, 812, 31

[259] Jones D. O., et al., 2017, The Astrophysical Journal, 843, 6

[260] Jones D. O., et al., 2018, The Astrophysical Journal, 867, 108

[261] Jones D. O., et al., 2019, The Astrophysical Journal, 881, 19

[262] Jones D. O., et al., 2022, The Astrophysical Journal, 933, 172

[263] Jordan M. I., Ghahramani Z., Jaakkola T. S., Saul L. K., 1999, Machine Learning, 37, 183

[264] Karchev K., 2023, Journal of Cosmology and Astroparticle Physics, 2023, 065

[265] Karchev K., Trotta R., 2025, Journal of Cosmology and Astroparticle Physics, 2025, 031

[266] Karchev K., Montel N. A., Coogan A., Weniger C., 2022a, Strong-Lensing Source Reconstruction with Denoising Diffusion Restoration Models (arXiv:2211.04365), doi:10.48550/arXiv.2211.04365

[267] Karchev K., Coogan A., Weniger C., 2022b, Monthly Notices of the Royal Astronomical Society, 512, 661

[268] Karchev K., Trotta R., Weniger C., 2023a, SimSIMS: Simulation-based Supernova Ia Model Selection with Thousands of Latent Variables (arXiv:2311.15650), doi:10.48550/arXiv.2311.15650

[269] Karchev K., Trotta R., Weniger C., 2023b, Monthly Notices of the Royal Astronomical Society, 520, 1056

[270] Karchev K., Grayling M., Boyd B. M., Trotta R., Mandel K. S., Weniger C., 2024, Monthly Notices of the Royal Astronomical Society, 530, 3881

[271] Karchev K., Trotta R., Jimenez R., 2025, CIGaRS I: Combined Simulation-Based Inference from SNae Ia and Host Photometry (arXiv:2508.15899), doi:10.48550/arXiv.2508.15899

[272] Kawahara H., Kawashima Y., Masuda K., Crossfield I. J. M., Pannier E., van den Bekerom D., 2022, The Astrophysical Journal Supplement Series, 258, 31

[273] Kaye S., 2024, Internet Encyclopedia of Philosophy

[274] Kelsey L., et al., 2021, Monthly Notices of the Royal Astronomical Society, 501, 4861

[275] Kelsey L., et al., 2023, Monthly Notices of the Royal Astronomical Society, 519, 3046

[276] Kenneth Mees C. E., 1931, Journal of the Optical Society of America, 21, 753

[277] Kenworthy W. D., et al., 2021, The Astrophysical Journal, 923, 265

[278] Kessler R., Scolnic D., 2017, The Astrophysical Journal, 836, 56

[279] Kessler R., et al., 2009a, Publications of the Astronomical Society of the Pacific, 121, 1028

[280] Kessler R., et al., 2009b, The Astrophysical Journal Supplement Series, 185, 32

[281] Kessler R., Conley A., Jha S., Kuhlmann S., 2010, Supernova Photometric Classification Challenge (arXiv:1001.5210), doi:10.48550/arXiv.1001.5210

[282] Kessler R., et al., 2015, The Astronomical Journal, 150, 172

[283] Kessler R., et al., 2019, Publications of the Astronomical Society of the Pacific, 131, 094501

[284] Kessler R., Vincenzi M., Armstrong P., 2023, The Astrophysical Journal, 952, L8

[285] Kim A., Goobar A., Perlmutter S., 1996, Publications of the Astronomical Society of the Pacific, 108, 190

[286] Kim A. G., et al., 2013, The Astrophysical Journal, 766, 84

[287] Kim Y.-L., Smith M., Sullivan M., Lee Y.-W., 2018, The Astrophysical Journal, 854, 24

[288] Kim Y.-L., Kang Y., Lee Y.-W., 2019, Journal of Korean Astronomical Society, 52, 181

[289] Kingma D. P., Ba J., 2017, Adam: A Method for Stochastic Optimization (arXiv:1412.6980), doi:10.48550/arXiv.1412.6980

[290] Kingma D. P., Welling M., 2014, Auto-Encoding Variational Bayes (arXiv:1312.6114), doi:10.48550/arXiv.1312.6114

[291] Kingma D. P., Salimans T., Jozefowicz R., Chen X., Sutskever I., Welling M., 2016, in Advances in Neural Information Processing Systems. Curran Associates, Inc., `https://proceedings.neurips.cc/paper/2016/hash/ddeebdeefdb7e7e7a697e1c3e3d8ef54-Abstract.html`

[292] Kirshner R. P., Kwan J., 1974, The Astrophysical Journal, 193, 27

[293] Kitaev N., Kaiser L., Levskaya A., 2019, in International Conference on Learning Representations. `https://openreview.net/forum?id=rkgNKkHtvB&utm_campaign=AI+Weekly&utm_medium=email&utm_source=Revue+newsletter`

[294] Kloek T., van Dijk H. K., 1978, Econometrica, 46, 1

[295] Korytov D., et al., 2019, The Astrophysical Journal Supplement Series, 245, 26

[296] Krause O., Tanaka M., Usuda T., Hattori T., Goto M., Birkmann S., Nomoto K., 2008, Nature, 456, 617

[297] Kriek M., Conroy C., 2013, The Astrophysical Journal, 775, L16

[298] Krisciunas K., et al., 2017, The Astronomical Journal, 154, 211

[299] Kullback S., Leibler R. A., 1951, The Annals of Mathematical Statistics, 22, 79

[300] Kunz M., Bassett B. A., Hlozek R. A., 2007, Physical Review D, 75, 103508

[301] Kvasiuk Y., Münchmeyer M., 2024, An Auto-Differentiable Likelihood Pipeline for the Cross-Correlation of CMB and Large-Scale Structure Due to the Kinetic Sunyaev-Zeldovich Effect (arXiv:2305.08903), doi:10.48550/arXiv.2305.08903

[302] LSST Dark Energy Science Collaboration (LSST DESC) et al., 2021, The Astrophysical Journal Supplement Series, 253, 31

[303] LSST Science Collaboration et al., 2009, LSST Science Book, Version 2.0 (arXiv:0912.0201), doi:10.48550/arXiv.0912.0201

[304] Lanzieri D., Zeghal J., Makinen T. L., Boucaud A., Starck J.-L., Lanusse F., 2024, Optimal Neural Summarisation for Full-Field Weak Lensing Cosmological Implicit Inference (arXiv:2407.10877), doi:10.48550/arXiv.2407.10877

[305] Le-Khac P. H., Healy G., Smeaton A. F., 2020, IEEE Access, 8, 193907

[306] Lee J., Lee Y., Kim J., Kosiorek A., Choi S., Teh Y. W., 2019, in Proceedings of the 36th International Conference on Machine Learning. PMLR, pp 3744–3753, `https://proceedings.mlr.press/v97/lee19d.html`

[307] Lee Y.-W., Chung C., Kang Y., Jee M. J., 2020, The Astrophysical Journal, 903, 22

[308] Legendre A.-M., 1825-1837, Traité Des Fonctions Elliptiques. Paris

[309] Léget P.-F., et al., 2020, Astronomy and Astrophysics, 636, A46

[310] Lehmann E. L., Casella G., 1998, Theory of Point Estimation. Springer Texts in Statistics, Springer-Verlag, New York, doi:10.1007/b98854

[311] Leibundgut B., Tammann G. A., Cadonau R., Cerrito D., 1991, Astronomy and Astrophysics Supplement Series, 89, 537

[312] Leistedt B., Mortlock D. J., Peiris H. V., 2016, Monthly Notices of the Royal Astronomical Society, 460, 4258

[313] Leja J., Johnson B. D., Conroy C., van Dokkum P. G., Byler N., 2017, The Astrophysical Journal, 837, 170

[314] Lemaître G., 1927, Annales de la Société Scientifique de Bruxelles, 47, 49

[315] Lemaître G., 1931, Monthly Notices of the Royal Astronomical Society, 91, 483

[316] Lemaître G., 1933, Annales de la Société Scientifique de Bruxelles, 53, 51

[317] Lemaître A. G., 1997, General Relativity and Gravitation, 29, 641

[318] Lemos P., Coogan A., Hezaveh Y., Perreault-Levasseur L., 2023, Sampling-Based Accuracy Testing of Posterior Estimators for General Inference (arXiv:2302.03026), doi:10.48550/arXiv.2302.03026

[319] Li Y., Modi C., Jamieson D., Zhang Y., Lu L., Feng Y., Lanusse F., Greengard L., 2022a, Differentiable Cosmological Simulation with Adjoint Method (arXiv:2211.09815), doi:10.48550/arXiv.2211.09815

[320] Li Y., et al., 2022b, Pmwd: A Differentiable Cosmological Particle-Mesh N-Body Library (arXiv:2211.09958), doi:10.48550/arXiv.2211.09958

[321] Li X., Mandelbaum R., Jarvis M., Li Y., Park A., Zhang T., 2023, A Differentiable Perturbation-based Weak Lensing Shear Estimator (arXiv:2309.06506), doi:10.48550/arXiv.2309.06506

[322] Liaudat T., Starck J.-L., Kilbinger M., Frugier P.-A., 2021, Rethinking the Modeling of the Instrumental Response of Telescopes with a Differentiable Optical Model (arXiv:2111.12541), doi:10.48550/arXiv.2111.12541

[323] Lin J., 1991, IEEE Transactions on Information Theory, 37, 145

[324] Linder E. V., 2003, Physical Review Letters, 90, 091301

[325] Linder E. V., Mitra A., 2019, Physical Review D, 100, 043542

[326] List F., Montel N. A., Weniger C., 2023, Bayesian Simulation-based Inference for Cosmological Initial Conditions (arXiv:2310.19910), doi:10.48550/arXiv.2310.19910

[327] Liu D.-Z., Ma C., Zhang T.-J., Yang Z., 2011, Monthly Notices of the Royal Astronomical Society, 412, 2685

[328] Livio M., 2011, Nature, 479, 171

[329] Lopes M., Bernui A., Franco C., Avila F., 2024, The Astrophysical Journal, 967, 47

[330] Loredo T. J., 1990, in Fougère P. F., ed., , Maximum Entropy and Bayesian Methods. Springer Netherlands, Dordrecht, pp 81–142, doi:10.1007/978-94-009-0683-9_6

[331] Lueber A., Karchev K., Fisher C., Heim M., Trotta R., Heng K., 2025, The Astrophysical Journal Letters, 984, L32

[332] Lueckmann J.-M., Bassetto G., Karaletsos T., Macke J. H., 2019, in Proceedings of The 1st Symposium on Advances in Approximate Bayesian Inference. PMLR, pp 32–53, `https://proceedings.mlr.press/v96/lueckmann19a.html`

[333] Lueckmann J.-M., Boelts J., Greenberg D., Goncalves P., Macke J., 2021, in Banerjee A., Fukumizu K., eds, Proceedings of The 24th

International Conference on Artificial Intelligence and Statistics. PMLR, pp 343–351, `https://proceedings.mlr.press/v130/lueckmann21a.html`

[334] Luger R., Bedell M., Foreman-Mackey D., Crossfield I. J. M., Zhao L. L., Hogg D. W., 2021, Mapping Stellar Surfaces III: An Efficient, Scalable, and Open-Source Doppler Imaging Model (arXiv:2110.06271), doi:10.48550/arXiv.2110.06271

[335] Lunn D., Spiegelhalter D., Thomas A., Best N., 2009, Statistics in Medicine, 28, 3049

[336] Ma C., Corasaniti P.-S., Bassett B. A., 2016, Monthly Notices of the Royal Astronomical Society, 463, 1651

[337] MacKay D. J. C., 2003, Information Theory, Inference and Learning Algorithms, illustrated edition edn. Cambridge University Press, Cambridge, UK ; New York

[338] Madau P., Dickinson M., 2014, Annual Review of Astronomy and Astrophysics, 52, 415

[339] Makinen T. L., Alsing J., Wandelt B. D., 2023, Fishnets: Information-Optimal, Scalable Aggregation for Sets and Graphs (arXiv:2310.03812), doi:10.48550/arXiv.2310.03812

[340] Malmquist K. G., 1922, Meddelanden fran Lunds Astronomiska Observatorium Serie I, 100, 1

[341] Malmquist K. G., 1925, Meddelanden fran Lunds Astronomiska Observatorium Serie I, 106, 1

[342] Mamajek E. E., et al., 2015, IAU 2015 Resolution B2 on Recommended Zero Points for the Absolute and Apparent Bolometric Magnitude Scales (arXiv:1510.06262), doi:10.48550/arXiv.1510.06262

[343] Mandel K. S., Wood-Vasey W. M., Friedman A. S., Kirshner R. P., 2009, The Astrophysical Journal, 704, 629

[344] Mandel K. S., Narayan G., Kirshner R. P., 2011, The Astrophysical Journal, 731, 120

[345] Mandel K. S., Scolnic D. M., Shariff H., Foley R. J., Kirshner R. P., 2017, The Astrophysical Journal, 842, 93

[346] Mandel K. S., Thorp S., Narayan G., Friedman A. S., Avelino A., 2022, Monthly Notices of the Royal Astronomical Society, 510, 3939

[347] March M. C., Trotta R., Berkes P., Starkman G. D., Vaudrevange P. M., 2011, Monthly Notices of the Royal Astronomical Society, 418, 2308

[348] March M. C., Wolf R. C., Sako m., D'Andrea C., Brout D., 2018, A Bayesian Approach to Truncated Data Sets: An Application to Malmquist Bias in Supernova Cosmology (arXiv:1804.02474), doi:10.48550/arXiv.1804.02474

[349] Margossian C., Blei D., 2024, in The 40th Conference on Uncertainty in Artificial Intelligence. `https://openreview.net/forum?id=mCVYIsnctr`

[350] Marjoram P., Molitor J., Plagnol V., Tavaré S., 2003, in Proceedings of the National Academy of Sciences. National Academy of Sciences, pp 15324–15328, doi:10.1073/pnas.0306899100

[351] Martin G. M., Frazier D. T., Robert C. P., 2024, Statistical Science, 39

[352] Masci F. J., et al., 2019, Publications of the Astronomical Society of the Pacific, 131, 018003

[353] Masserano L., Dorigo T., Izbicki R., Kuusela M., Lee A. B., 2022, Simulation-Based Inference with Waldo: Confidence Regions by Leveraging Prediction Algorithms or Posterior Estimators for Inverse Problems (arXiv:2205.15680), doi:10.48550/arXiv.2205.15680

[354] Mazzali P. A., Röpke F. K., Benetti S., Hillebrandt W., 2007, Science, 315, 825

[355] McEwen J. D., Wallis C. G. R., Price M. A., Mancini A. S., 2023, Machine Learning Assisted Bayesian Model Comparison: Learnt Harmonic Mean Estimator (arXiv:2111.12720), doi:10.48550/arXiv.2111.12720

[356] McKee C. F., Ostriker E. C., 2007, Annual Review of Astronomy and Astrophysics, 45, 565

[357] Mészáros A., Řípa J., 2013, Astronomy and Astrophysics, 556, A13

[358] Metropolis N., Rosenbluth A. W., Rosenbluth M. N., Teller A. H., Teller E., 1953, The Journal of Chemical Physics, 21, 1087

[359] Miles R., 2007, Journal of the British Astronomical Association, 117, 172

[360] Miller B. K., Cole A., Louppe G., Weniger C., 2020, Simulation-Efficient Marginal Posterior Estimation with Swyft: Stop Wasting Your Precious Time (arXiv:2011.13951), doi:10.48550/arXiv.2011.13951

[361] Miller B. K., Cole A., Forré P., Louppe G., Weniger C., 2021, in Ranzato M., Beygelzimer A., Dauphin Y., Liang P., Vaughan J. W., eds, Neural Information Processing Systems Vol. 34, Advances in Neural Information Processing Systems. Curran Associates, Inc., pp 129–143, `https://proceedings.neurips.cc/paper/2021/hash/01632f7b7a127233fa1188bd6c2e42e1-Abstract.html`

[362] Miller B. K., Cole A., Weniger C., Nattino F., Ku O., Grootes M. W., 2022a, Journal of Open Source Software, 7, 4205

[363] Miller B. K., Weniger C., Forré P., 2022b, Advances in Neural Information Processing Systems, 35, 3262

[364] Minkowski R., 1941, Publications of the Astronomical Society of the Pacific, 53, 224

[365] Mitra A., Kessler R., More S., Hlozek R., 2023, The Astrophysical Journal, 944, 212

[366] Mo H., van den Bosch F., White S., 2010, Galaxy Formation and Evolution. Cambridge University Press

[367] Modi C., Lanusse F., Seljak U., Spergel D. N., Perreault-Levasseur L., 2021a, CosmicRIM : Reconstructing Early Universe by Combining Differentiable Simulations with Recurrent Inference Machines (arXiv:2104.12864), doi:10.48550/arXiv.2104.12864

[368] Modi C., Lanusse F., Seljak U., 2021b, Astronomy and Computing, 37, 100505

[369] Möller A., de Boissière T., 2020, Monthly Notices of the Royal Astronomical Society, 491, 4277

[370] Moreno-Cartagena D., Cabrera-Vives G., Protopapas P., Donoso-Oliva C., Pérez-Carrasco M., Cádiz-Leyton M., 2023, Positional Encodings for Light Curve Transformers: Playing with Positions and Attention (arXiv:2308.06404), doi:10.48550/arXiv.2308.06404

[371] Moreno-Raya M. E., López-Sánchez Á. R., Mollá M., Galbany L., Vílchez J. M., Carnero A., 2016a, Monthly Notices of the Royal Astronomical Society, 462, 1281

[372] Moreno-Raya M. E., Mollá M., López-Sánchez Á. R., Galbany L., Vílchez J. M., Carnero Rosell A., Domínguez I., 2016b, The Astrophysical Journal, 818, L19

[373] Moreno-Torres J. G., Raeder T., Alaiz-Rodríguez R., Chawla N. V., Herrera F., 2012, Pattern Recognition, 45, 521

[374] Moretti C., Autenrieth M., Serra R., Trotta R., van Dyk D. A., Mesinger A., 2024, StratLearn-z: Improved Photo-z Estimation from Spectroscopic Data Subject to Selection Effects (arXiv:2409.20379), doi:10.48550/arXiv.2409.20379

[375] Morningstar W. R., et al., 2019, The Astrophysical Journal, 883, 14

[376] Narayan G., ELAsTiCC Team 2023, Bulletin of the American Astronomical Society, 55, 117.01

[377] Narayan G., et al., 2016, The Astrophysical Journal Supplement Series, 224, 3

[378] Neal R. M., 2003, The Annals of Statistics, 31, 705

[379] Neal R. M., 2011, MCMC Using Hamiltonian Dynamics. (arXiv:1206.1901), doi:10.1201/b10905

[380] Neill J. D., et al., 2009, The Astrophysical Journal, 707, 1449

[381] Newman J. A., Gruen D., 2022, Annual Review of Astronomy and Astrophysics, 60, 363

[382] Neyman J., 1935, Istituto Italiano degli Attuari, 6, 309

[383] Neyman J., Jeffreys H., 1937, Philosophical Transactions of the Royal Society of London. Series A, Mathematical and Physical Sciences, 236, 333

[384] Neyman J., Pearson E. S., Pearson K., 1933, Philosophical Transactions of the Royal Society of London. Series A, Containing Papers of a Mathematical or Physical Character, 231, 289

[385] Nicholl M., 2021, Astronomy and Geophysics, 62, 5.34

[386] Nicolas N., et al., 2021, Astronomy and Astrophysics, 649, A74

[387] Noll S., Burgarella D., Giovannoli E., Buat V., Marcillac D., Muñoz-Mateos J. C., 2009, Astronomy and Astrophysics, 507, 1793

[388] Norgaard-Nielsen H. U., Hansen L., Jorgensen H. E., Aragon Salamanca A., Ellis R. S., 1989, Nature, 339, 523

[389] Nugent P. E., et al., 2011, Nature, 480, 344

[390] Ocampo I., Alestas G., Nesseris S., Sapone D., 2024a, Enhancing Cosmological Model Selection with Interpretable Machine Learning (arXiv:2406.08351), doi:10.48550/arXiv.2406.08351

[391] Ocampo I., Cañas-Herrera G., Nesseris S., 2024b, Neural Networks for Cosmological Model Selection and Feature Importance Using Cosmic Microwave Background Data (arXiv:2410.05209), doi:10.48550/arXiv.2410.05209

[392] Oke J. B., Gunn J. E., 1983, The Astrophysical Journal, 266, 713

[393] Papamakarios G., Murray I., 2016, Advances in Neural Information Processing Systems, 29

[394] Papamakarios G., Pavlakou T., Murray I., 2017, in Advances in Neural Information Processing Systems. Curran Associates, Inc., `https://proceedings.neurips.cc/paper_files/paper/2017/hash/6c1da886822c67822bcf3679d04369fa-Abstract.html`

[395] Papamakarios G., Sterratt D., Murray I., 2019, in The 22nd International Conference on Artificial Intelligence and Statistics. PMLR, pp 837–848, `http://proceedings.mlr.press/v89/papamakarios19a.html`

[396] Papamakarios G., Nalisnick E., Jimenez Rezende D., Mohamed S., Lakshminarayanan B., 2021, Journal of Machine Learning Research, 22, 1

[397] Parisi G., 1988, Statistical Field Theory. Redwood City, CA. : Addison-Wesley Pub. Co., `http://archive.org/details/statisticalfield0000pari`

[398] Park M., Gatti M., Jain B., 2024, Dimensionality Reduction Techniques for Statistical Inference in Cosmology (arXiv:2409.02102), doi:10.48550/arXiv.2409.02102

[399] Pascale M., et al., 2024, SN H0pe: The First Measurement of H_0 from a Multiply-Imaged Type Ia Supernova, Discovered by JWST (arXiv:2403.18902), http://arxiv.org/abs/2403.18902

[400] Paszke A., et al., 2019, in Wallach H., Larochelle H., Beygelzimer A., d\textquotesingle Alché-Buc F., Fox E., Garnett R., eds, , Advances in Neural Information Processing Systems 32. Curran Associates, Inc., pp 8024–8035

[401] Peebles P. J. E., 1993, Principles of Physical Cosmology by P.J.E. Peebles. Princeton University Press, 1993. ISBN: 978-0-691-01933-8

[402] Pereyra M., 2016, Statistics and Computing, 26, 745

[403] Perlmutter S., 2012, Reviews of Modern Physics, 84, 1127

[404] Perlmutter S., et al., 1997, The Astrophysical Journal, 483, 565

[405] Perlmutter S., et al., 1999, The Astrophysical Journal, 517, 565

[406] Peterson E. R., et al., 2023, Monthly Notices of the Royal Astronomical Society, 522, 2478

[407] Petrecca V., et al., 2024, Recovered SN Ia Rate from Simulated LSST Images (arXiv:2402.17612), doi:10.48550/arXiv.2402.17612

[408] Pham K. C., Nott D. J., Chaudhuri S., 2014, Stat, 3, 218

[409] Phan D., Pradhan N., Jankowiak M., 2019, Composable Effects for Flexible and Accelerated Probabilistic Programming in NumPyro (arXiv:1912.11554), doi:10.48550/arXiv.1912.11554

[410] Phillips M. M., 1993, The Astrophysical Journal, 413, L105

[411] Phillips M. M., et al., 2019, Publications of the Astronomical Society of the Pacific, 131, 014001

[412] Planck M., Masius M., 1914, The Theory of Heat Radiation. P. Blakiston's Son & Co., Philadelphia

[413] Planck Collaboration et al., 2014, Astronomy and Astrophysics, 571, A27

[414] Planck Collaboration et al., 2020, Astronomy and Astrophysics, 644, A100

[415] Plato 1966, in , Vol. 1, Plato in Twelve Volumes. Harvard University Press, Cambridge, MA, https://www.perseus.tufts.edu/hopper/text?doc=Perseus%3Atext%3A1999.01.0170%3Atext%3DApol.

[416] Poole B., Ozair S., Oord A. V. D., Alemi A., Tucker G., 2019, in Proceedings of the 36th International Conference on Machine Learning. PMLR, pp 5171–5180, https://proceedings.mlr.press/v97/poole19a.html

[417] Pope B. J. S., Pueyo L., Xin Y., Tuthill P. G., 2021, The Astrophysical Journal, 907, 40

[418] Popovic B., Scolnic D., Kessler R., 2020, The Astrophysical Journal, 890, 172

[419] Popovic B., Brout D., Kessler R., Scolnic D., Lu L., 2021, The Astrophysical Journal, 913, 49

[420] Popovic B., Brout D., Kessler R., Scolnic D., 2023, The Astrophysical Journal, 945, 84

[421] Popovic B., et al., 2024a, ZTF SN Ia DR2: Evidence of Changing Dust Distributions With Redshift Using Type Ia Supernovae (arXiv:2406.06215), doi:10.48550/arXiv.2406.06215

[422] Popovic B., et al., 2024b, Monthly Notices of the Royal Astronomical Society, 529, 2100

[423] Porqueres N., Heavens A., Mortlock D., Lavaux G., 2022, Monthly Notices of the Royal Astronomical Society, 509, 3194

[424] Press W. H., Teukolsky S. A., Vetterling W. T., Flannery B. P., 1992, Numerical Recipes in C: The Art of Scientific Computing, 3rd edn. Cambridge University Press, USA

[425] Pritchard J. K., Seielstad M. T., Perez-Lezaun A., Feldman M. W., 1999, Molecular Biology and Evolution, 16, 1791

[426] Pritchet C. J., Howell D. A., Sullivan M., 2008, The Astrophysical Journal, 683, L25

[427] Pritchet C., Thanjavur K., Bottrell C., Gao Y., 2024, The Astronomical Journal, 167, 131

[428] Pskovskii Yu. P., 1967, Soviet Astronomy, 11, 63

[429] Pskovskii Iu. P., 1977, Soviet Astronomy, 21, 675

[430] Pskovskii Yu. P., 1984, Soviet Astronomy, 28, 658

[431] Qu H., Sako M., 2023, The Astrophysical Journal, 954, 201

[432] Qu H., Sako M., Möller A., Doux C., 2021, The Astronomical Journal, 162, 67

[433] Qu H., et al., 2023, The Dark Energy Survey Supernova Program: Cosmological Biases from Host Galaxy Mismatch of Type Ia Supernovae (arXiv:2307.13696), doi:10.48550/arXiv.2307.13696

[434] Radev S. T., D'Alessandro M., Mertens U. K., Voss A., Köthe U., Bürkner P.-C., 2021, Amortized Bayesian Model Comparison with Evidential Deep Learning (arXiv:2004.10629), doi:10.48550/arXiv.2004.10629

[435] Radev S. T., Mertens U. K., Voss A., Ardizzone L., Köthe U., 2022, IEEE Transactions on Neural Networks and Learning Systems, 33, 1452

[436] Radhakrishna Rao C., 1945, Bulletin of the Calcutta Mathematical Society, 37, 81

[437] Rahman W., Trotta R., Boruah S. S., Hudson M. J., van Dyk D. A., 2022, Monthly Notices of the Royal Astronomical Society, 514, 139

[438] Rajamuthukumar A. S., Hamers A. S., Neunteufel P., Pakmor R., de Mink S. E., 2023, The Astrophysical Journal, 950, 9

[439] Ranganath R., Gerrish S., Blei D., 2014, in Proceedings of the Seventeenth International Conference on Artificial Intelligence and Statistics. PMLR, pp 814–822, `https://proceedings.mlr.press/v33/ranganath14.html`

[440] Rasmussen C. E., Williams C. K. I., 2005, Gaussian Processes for Machine Learning. The MIT Press, doi:10.7551/mitpress/3206.001.0001

[441] Refsdal S., 1964, Monthly Notices of the Royal Astronomical Society, 128, 307

[442] Reichardt C., Jimenez R., Heavens A. F., 2001, Monthly Notices of the Royal Astronomical Society, 327, 849

[443] Rest A., et al., 2008, The Astrophysical Journal, 681, L81

[444] Revsbech E. A., Trotta R., van Dyk D. A., 2018, Monthly Notices of the Royal Astronomical Society, 473, 3969

[445] Riess A. G., Press W. H., Kirshner R. P., 1996, The Astrophysical Journal, 473, 88

[446] Riess A. G., et al., 1998, The Astronomical Journal, 116, 1009

[447] Riess A. G., et al., 1999, The Astronomical Journal, 117, 707

[448] Rigault M., et al., 2013, Astronomy and Astrophysics, 560, A66

[449] Rigault M., et al., 2015, The Astrophysical Journal, 802, 20

[450] Rigault M., et al., 2020, Astronomy and Astrophysics, 644, A176

[451] Rigault M., et al., 2024, ZTF SN Ia DR2: Overview (arXiv:2409.04346), doi:10.48550/arXiv.2409.04346

[452] Robbins H., Monro S., 1951, The Annals of Mathematical Statistics, 22, 400

[453] Roberts E., Lochner M., Fonseca J., Bassett B. A., Lablanche P.-Y., Agarwal S., 2017, Journal of Cosmology and Astroparticle Physics, 2017, 036

[454] Rodney S. A., et al., 2014, The Astronomical Journal, 148, 13

[455] Rodrigues P., Moreau T., Louppe G., Gramfort A., 2021, in Advances in Neural Information Processing Systems. Curran Associates, Inc., pp 13432–13443, `https://proceedings.neurips.cc/paper/2021/hash/6fbd841e2e4b2938351a4f9b68f12e6b-Abstract.html`

[456] Roman M., et al., 2018, Astronomy and Astrophysics, 615, A68

[457] Rosasco L., Vito E. D., Caponnetto A., Piana M., Verri A., 2004, Neural Computation, 16, 1063

[458] Rose B. M., Garnavich P. M., Berg M. A., 2019, The Astrophysical Journal, 874, 32

[459] Roy R., Olver F. W. J., Askey R. A., Wong R., 2022, NIST digital library of mathematical functions

[460] Rubin D., 2020, The Astrophysical Journal, 897, 40

[461] Rubin D., et al., 2015, The Astrophysical Journal, 813, 137

[462] Rubin D., et al., 2025, The Astrophysical Journal, 986, 231

[463] Ruhlmann-Kleider V., Lidman C., Möller A., 2022, Journal of Cosmology and Astroparticle Physics, 2022, 065

[464] Sako M., et al., 2018, Publications of the Astronomical Society of the Pacific, 130, 064002

[465] Sánchez B. O., et al., 2022, The Astrophysical Journal, 934, 96

[466] Sánchez B. O., et al., 2024, The Astrophysical Journal, 975, 5

[467] Sanguinetti G., 2021, TSDS Entrance Exam

[468] Saul L. K., Jaakkola T., Jordan M. I., 1996, Journal of Artificial Intelligence Research, 4, 61

[469] Saunders C., et al., 2018, The Astrophysical Journal, 869, 167

[470] Schlafly E. F., Finkbeiner D. P., 2011, The Astrophysical Journal, 737, 103

[471] Schosser B., Röspel T., Schaefer B. M., 2024, Markov Walk Exploration of Model Spaces: Bayesian Selection of Dark Energy Models with Supernovae (arXiv:2407.06259), doi:10.48550/arXiv.2407.06259

[472] Schwarz G., 1978, The Annals of Statistics, 6, 461

[473] Scolnic D., et al., 2015, The Astrophysical Journal, 815, 117

[474] Scolnic D. M., et al., 2018, The Astrophysical Journal, 859, 101

[475] Scolnic D., et al., 2022, The Astrophysical Journal, 938, 113

[476] Shannon C. E., 1948, The Bell System Technical Journal, 27, 623

[477] Shapley H., Curtis H. D., 1921, Bulletin of the National Research Council, 2, 171

[478] Shariff H., Dhawan S., Jiao X., Leibundgut B., Trotta R., van Dyk D. A., 2016a, Monthly Notices of the Royal Astronomical Society, 463, 4311

[479] Shariff H., Jiao X., Trotta R., van Dyk D. A., 2016b, The Astrophysical Journal, 827, 1

[480] Sharrock L., Simons J., Liu S., Beaumont M., 2024, in Forty-First International Conference on Machine Learning. https://openreview.net/forum?id=8viuf9PdzU

[481] Sisson S. A., Fan Y., Tanaka M. M., 2007, Proceedings of the National Academy of Sciences, 104, 1760

[482] Sisson S., Fan Y., Beaumont M., 2018, Handbook of Approximate Bayesian Computation, 1st edn. Handbooks of Modern Statistical Methods, Chapman and Hall/CRC

[483] Skilling J., 2004, AIP Conference Proceedings, 735, 395

[484] Skilling J., 2006, Bayesian Analysis, 1, 833

[485] Smith L. N., Topin N., 2018, Super-Convergence: Very Fast Training of Neural Networks Using Large Learning Rates (arXiv:1708.07120), doi:10.48550/arXiv.1708.07120

[486] Sohl-Dickstein J., Weiss E., Maheswaranathan N., Ganguli S., 2015, in Proceedings of the 32nd International Conference on Machine Learning. PMLR, pp 2256–2265, https://proceedings.mlr.press/v37/sohl-dickstein15.html

[487] Song Y., Ermon S., 2019, in Advances in Neural Information Processing Systems. https://proceedings.neurips.cc/paper/2019/hash/3001ef257407d5a371a96dcd947c7d93-Abstract.html?ref=https://githubhelp.com

[488] Song Y., Sohl-Dickstein J., Kingma D. P., Kumar A., Ermon S., Poole B., 2020, in International Conference on Learning Representations. https://openreview.net/forum?id=PxTIG12RRHS&utm_campaign=NLP%20News&utm_medium=email&utm_source=Revue%20newsletter

[489] Song J., Meng C., Ermon S., 2022, in International Conference on Learning Representations. https://openreview.net/forum?id=St1giarCHLP

[490] Sorrenti F., Durrer R., Kunz M., 2024, The Low Multipoles in the Pantheon+SH0ES Data (arXiv:2403.17741), doi:10.48550/arXiv.2403.17741

[491] Speagle J. S., 2020, Monthly Notices of the Royal Astronomical Society, 493, 3132

[492] Spergel D., et al., 2015, Wide-Field InfrarRed Survey Telescope-Astrophysics Focused Telescope Assets WFIRST-AFTA 2015 Report (arXiv:1503.03757), doi:10.48550/arXiv.1503.03757

[493] Srinivasan R., Crisostomi M., Trotta R., Barausse E., Breschi M., 2024, $floZ$: Improved Bayesian Evidence Estimation from Posterior Samples with Normalizing Flows (arXiv:2404.12294), doi:10.48550/arXiv.2404.12294

[494] Stace W. T., 1939, The Philosophical Review, 48, 296

[495] Stahl B. E., et al., 2019, Monthly Notices of the Royal Astronomical Society, 490, 3882

[496] Stevanovich D., Hearin A. P., Nagai D., 2023, A Differentiable Model of the Evolution of Dark Matter Halo Concentration (arXiv:2309.07854), doi:10.48550/arXiv.2309.07854

[497] Stone C., Courteau S., Cuillandre J.-C., Hezaveh Y., Perreault-Levasseur L., Arora N., 2023, AutoPhot: Fitting Everything Everywhere All at Once in Astronomical Images (arXiv:2308.01957), doi:10.48550/arXiv.2308.01957

[498] Stothers R., 1977, Isis. Journal of the History of Science Society, 68, 443

[499] Stoye M., Brehmer J., Louppe G., Pavez J., Cranmer K., 2018, Likelihood-Free Inference with an Improved Cross-Entropy Estimator (arXiv:1808.00973), doi:10.48550/arXiv.1808.00973

[500] Sui C., Zhao X., Jing T., Mao Y., 2023, Evaluating Summary Statistics with Mutual Information for Cosmological Inference (arXiv:2307.04994), doi:10.48550/arXiv.2307.04994

[501] Tak H., Chen Y., Kashyap V. L., Mandel K. S., Meng X.-L., Siemiginowska A., van Dyk D. A., 2024, Six Maxims of Statistical Acumen for Astronomical Data Analysis (arXiv:2408.16179), doi:10.48550/arXiv.2408.16179

[502] Talts S., Betancourt M., Simpson D., Vehtari A., Gelman A., 2020, Validating Bayesian Inference Algorithms with Simulation-Based Calibration (arXiv:1804.06788), doi:10.48550/arXiv.1804.06788

[503] Tegmark M., 1997, Physical Review D, 55, 5895

[504] Tegmark M., Taylor A. N., Heavens A. F., 1997, The Astrophysical Journal, 480, 22

[505] Tejero-Cantero A., Boelts J., Deistler M., Lueckmann J.-M., Durkan C., Gonçalves P. J., Greenberg D. S., Macke J. H., 2020, Journal of Open Source Software, 5, 2505

[506] The PLAsTiCC team et al., 2018, The Photometric LSST Astronomical Time-series Classification Challenge (PLAsTiCC): Data Set (arXiv:1810.00001), doi:10.48550/arXiv.1810.00001

[507] Thielemann F. K., Brachwitz F., Höflich P., Martinez-Pinedo G., Nomoto K., 2004, New Astronomy Reviews, 48, 605

[508] Thomas R. C., Kantowski R., 2000, Physical Review D, 62, 103507

[509] Thomas O., Dutta R., Corander J., Kaski S., Gutmann M. U., 2022, Bayesian Analysis, 17, 1

[510] Thorp S., Mandel K. S., 2022, Monthly Notices of the Royal Astronomical Society, 517, 2360

[511] Thorp S., Mandel K. S., Jones D. O., Ward S. M., Narayan G., 2021, Monthly Notices of the Royal Astronomical Society, 508, 4310

[512] Thorp S., Mandel K. S., Jones D. O., Kirshner R. P., Challis P. M., 2024, Monthly Notices of the Royal Astronomical Society, 530, 4016

[513] Tolman R. C., 1934, Proceedings of the National Academy of Sciences, 20, 169

[514] Tran J. H., Kleppe T. S., 2024, Tuning Diagonal Scale Matrices for HMC (arXiv:2403.07495), doi:10.48550/arXiv.2403.07495

[515] Tripp R., 1997, Astronomy and Astrophysics, 325, 871

[516] Tripp R., 1998, Astronomy and Astrophysics, 331, 815

[517] Uddin S. A., et al., 2020, The Astrophysical Journal, 901, 143

[518] Uzsoy A. S., 2022, PhD thesis, University of Cambridge, Cambridge, UK, `https://www.mlmi.eng.cam.ac.uk/files/2021-2022_dissertations/scalable_bayesian_inference_for_probabilistic_spectrotemporal_models.pdf`

[519] Uzsoy A. S. M., Thorp S., Grayling M., Mandel K. S., 2024, Variational Inference for Acceleration of SN Ia Photometric Distance Estimation with BayeSN (arXiv:2405.06013), `http://arxiv.org/abs/2405.06013`

[520] Valkenburg W., 2012, General Relativity and Gravitation, 44, 2449

[521] Verde L., Bernal J. L., Heavens A. F., Jimenez R., 2017, Monthly Notices of the Royal Astronomical Society, 467, 731

[522] Villar V. A., 2022, Amortized Bayesian Inference for Supernovae in the Era of the Vera Rubin Observatory Using Normalizing Flows (arXiv:2211.04480), doi:10.48550/arXiv.2211.04480

[523] Villar V. A., et al., 2020, The Astrophysical Journal, 905, 94

[524] Vincent P., 2011, Neural Computation, 23, 1661

[525] Vincenzi M., et al., 2024, The Astrophysical Journal, 975, 86

[526] Vink J., Bleeker J., van der Heyden K., Bykov A., Bamba A., Yamazaki R., 2006, The Astrophysical Journal, 648, L33

[527] Virtanen P., et al., 2020, Nature Methods, 17, 261

[528] Wagner-Carena S., Aalbers J., Birrer S., Nadler E. O., Darragh-Ford E., Marshall P. J., Wechsler R. H., 2022, arXiv:2203.00690 [astro-ph]

[529] Wagoner R. V., 1977, The Astrophysical Journal, 214, L5

[530] Wagstaff E., Fuchs F. B., Engelcke M., Osborne M. A., Posner I., 2022, Journal of Machine Learning Research, 23, 1

[531] Wang T., Melchior P., 2022, Machine Learning: Science and Technology, 3, 015023

[532] Wang Z. R., Qu Q.-Y., Chen Y., 1997, Astronomy and Astrophysics, 318, L59

[533] Wang G.-J., Cheng C., Ma Y.-Z., Xia J.-Q., 2022, The Astrophysical Journal Supplement Series, 262, 24

[534] Wang G.-J., Cheng C., Ma Y.-Z., Xia J.-Q., Abebe A., Beesham A., 2023a, CoLFI: Cosmological Likelihood-free Inference with Neural Density Estimators (arXiv:2306.11102), doi:10.48550/arXiv.2306.11102

[535] Wang J., Huang Z., Huang L., 2023b, Science China Physics, Mechanics, and Astronomy, 66, 129511

[536] Wang B., et al., 2023c, The Astrophysical Journal, 944, L58

[537] Ward S. M., et al., 2022, SN 2021hpr and Its Two Siblings in the Cepheid Calibrator Galaxy NGC 3147: A Hierarchical BayeSN Analysis of a Type Ia Supernova Trio, and a Hubble Constant Constraint (arXiv:2209.10558), doi:10.48550/arXiv.2209.10558

[538] Ward S. M., Dhawan S., Mandel K. S., Grayling M., Thorp S., 2023, Monthly Notices of the Royal Astronomical Society, 526, 5715

[539] Warner B., 2003, Cataclysmic Variable Stars. Cambridge University Press

[540] Webb S., Goliński A., Zinkov R., Siddharth N., Rainforth T., Teh Y. W., Wood F., 2018, in Proceedings of the 32nd International Conference on Neural Information Processing Systems. NIPS'18. Curran Associates Inc., Red Hook, NY, USA, pp 3074–3084

[541] Weinberg S., 2008, Cosmology, illustrated edition edn. Oxford University Press, Oxford ; New York

[542] Weyant A., Schafer C., Wood-Vasey W. M., 2013, The Astrophysical Journal, 764, 116

[543] Weyant A., et al., 2018, The Astronomical Journal, 155, 201

[544] Williams B. J., et al., 2011, The Astrophysical Journal, 741, 96

[545] Winkler C., Worrall D., Hoogeboom E., Welling M., 2023, Learning Likelihoods with Conditional Normalizing Flows (arXiv:1912.00042), doi:10.48550/arXiv.1912.00042

[546] Wirtz C., 1918, Astronomische Nachrichten, 206, 109

[547] Wojtak R., Davis T. M., Wiis J., 2015, Journal of Cosmology and Astroparticle Physics, 2015, 025

[548] Wolfram Research 2021, CarlsonRJ, https://reference.wolfram.com/language/ref/CarlsonRJ.html

[549] Wolpert D. H., 1995, in Petsche T., Hanson S. J., Shavlik J., eds, , Vol. 3, Computational Learning Theory and Natural Learning Systems: Selecting Good Models. The MIT Press, p. 0, doi:10.7551/mitpress/2007.003.0018

[550] Wong A., Pope B., Desdoigts L., Tuthill P., Norris B., Betters C., 2021, Journal of the Optical Society of America B Optical Physics, 38, 2465

[551] Wong K. W. K., Isi M., Edwards T. D. P., 2023, Fast Gravitational Wave Parameter Estimation without Compromises (arXiv:2302.05333), doi:10.48550/arXiv.2302.05333

[552] Yoachim P., et al., 2016, in Observatory Operations: Strategies, Processes, and Systems VI. SPIE, pp 406–420, doi:10.1117/12.2232947

[553] Yoachim P., et al., 2023, Lsst/Rubin_sim, Zenodo, doi:10.5281/zenodo.7087822

[554] Zaheer M., Kottur S., Ravanbakhsh S., Poczos B., Salakhutdinov R. R., Smola A. J., 2017, in Guyon I., Von Luxburg U., Bengio S., Wallach H., Fergus R., Vishwanathan S., Garnett R., eds, Neural Information Processing Systems Vol. 30, Advances in Neural Information Processing Systems. Curran Associates, Inc., https://proceedings.neurips.cc/paper/2017/hash/f22e4747da1aa27e363d86d40ff442fe-Abstract.html

[555] Zaninetti L., 2016, Journal of High Energy Physics, Gravitation and Cosmology, 02, 581

[556] Zaninetti L., 2019, International Journal of Astronomy and Astrophysics, 9, 51

[557] Zhang K., 2022, Analytic Simplifications to Planetary Microlensing under the Generalized Perturbative Picture (arXiv:2207.12412), doi:10.48550/arXiv.2207.12412

[558] Zhang C., Bütepage J., Kjellström H., Mandt S., 2019, IEEE Transactions on Pattern Analysis and Machine Intelligence, 41, 2008

[559] Zhang R., Yuan H., Chen B., 2023, The Astrophysical Journal Supplement Series, 269, 6

[560] Zhang G., Helfer T., Gagliano A. T., Mishra-Sharma S., Villar V. A., 2024, Maven: A Multimodal Foundation Model for Supernova Science (arXiv:2408.16829), doi:10.48550/arXiv.2408.16829

[561] Zwicky F., 1940, Reviews of Modern Physics, 12, 66

[562] da Silva L. L. A., 1993, Astrophysics and Space Science, 202, 215

[563] de Finetti B., 1937, Annales de l'institut Henri Poincaré, 7, 1

[564] van Haasteren R., 2024, Use Model Averaging Instead of Model Selection in Pulsar Timing (arXiv:2409.06050), doi:10.48550/arXiv.2409.06050

The manufacturer's authorised representative in the EU is Springer Nature Customer Service Centre GmbH, Europaplatz 3, 69115 Heidelberg, Germany. If you have any concerns regarding our products, please contact ProductSafety@springernature.com

Printed and bound by CPI Group (UK) Ltd, Croydon, CR0 4YY
01/10/2026
02226267-0001